NORTHERN AUSTRALIA

This publication has been supported by a grant from the Northern Australia Development Council. The contents of the volume are entirely the reponsibility of the Editor and the Contributors.

NORTHERN AUSTRALIA

The Arenas of Life and
Ecosystems on Half a Continent

Edited by
DON PARKES

DEPARTMENT OF GEOGRAPHY
UNIVERSITY OF NEWCASTLE, AUSTRALIA

1984

ACADEMIC PRESS

A Subsidiary of Harcourt Brace Jovanovich, Publishers
Sydney   Orlando   San Diego   San Francisco   New York
London   Toronto   Montreal   Tokyo   São Paulo

ACADEMIC PRESS AUSTRALIA
Centrecourt, 25–27 Paul Street North
North Ryde, N.S.W. 2113

United States Edition published by
ACADEMIC PRESS INC.
Orlando, Florida 32887

United Kingdom Edition published by
ACADEMIC PRESS, INC. (LONDON) LTD.
24/28 Oval Road, London NW1 7DX

Printed in Australia

National Library of Australia Cataloguing-in-Publication Data

Northern Australia.

Includes index.
ISBN 0 12 545080 X.

1. Human ecology — Australia, Northern.
I. Parkes, Don.

304.2'0994

Library of Congress Catalog Card Number: 84-70638

Academic Press Rapid Manuscript Reproduction

CONTENTS

FOREWORD

The tropical savannas of the world and their more arid extensions continue to challenge development, in the sense of the application of modern science and technology for increased and sustained production and improved living standards. The environmental constraints of these 'difficult' latitudes are widely-shared, but each region differs in its social, economic and historical circumstances, and in its geographical context.

There has long been a need for a geographical study of the interplay of environmental challenge and human endeavour in the vast arena of Northern Australia, and issues such as lands rights, mining impacts, national parks and political developments among the Aboriginal and European communities have recently emphasised that need. Professor Parkes deserves our gratitude for his initiative and persistence, and his success in bringing together effectively so many experienced collaborators from a range of disciplines to focus on the region of Northern Australia, its broad natural environments, its extensive pastoral and maritime land use, and its scatter of isolated mining and pastoral settlements. The authors are to be congratulated on the results, and the Northern Australia Development Council thanked for making the book possible.

J. A. Mabbutt
School of Geography,
University of New South Wales

CONTRIBUTORS

Numbers in parentheses indicate the pages on which the authors' contributions begin.

Andris Auliciems (17), Department of Geography, University of Queensland, St Lucia, Queensland 4067, Australia.

F. H. Bauer (1), Department of Economic History, Australian National University, P.O. Box 4, Canberra City, Australian Capital Territory 2601, Australia.

T. B. Brealey (363), CSIRO Division of Building Research, P.O. Box 56, Highett, Victoria 3190, Australia.

J. S. Bunt (125), Australian Institute of Marine Science, P.M.B. No. 3, Townsville, Queensland 4810, Australia.

I. H. Burnley (115), School of Geography, University of New South Wales, P.O. Box 1, Kensington, New South Wales 2033, Australia.

E. K. Christie (157), School of Australian Environmental Studies, Griffith University, Nathan, Queensland 4111, Australia.

Richard Dedear (295), Department of Geography, University of Queensland, St Lucia, Queensland 4067, Australia.

C. D. Ellyett (75), c.o. Faculty of Economics and Commerce, University of Newcastle, Shortland, New South Wales 2308, Australia.

Dean Graetz (173), CSIRO Division of Wildlife and Rangelands Research, Deniliquin, New South Wales 2710, Australia.

J. H. Holmes (209), Department of Geography, University of Queensland, St Lucia, Queensland 4067, Australia.

F. R. Honey (21), CSIRO Institute of Energy and Earth Resources, Division of Groundwater Research, Private Bag, P.O. Wembley, Western Australia 6014, Australia.

J. A. Jones(363), CSIRO Division of Building Research, P.O. Box 56, Highett, Victoria 3190, Australia.

D. M. Lee (29), Bureau of Meteorology, Department of Science and Technology, P.O. Box 1289K, Melbourne, Victoria 3001, Australia.

J. J. Mott (147), CSIRO Division of Tropical Pastures, Cunningham Laboratories, St Lucia, Queensland 4067, Australia.

A. B. Neal (29), Bureau of Meteorology, Department of Science and Technology, P.O. Box 1289K, Melbourne, Victoria 3001, Australia.

C. C. Neil (363), CSIRO Division of Building Research, P.O. Box 56, Highett, Victoria 3190, Australia.

P. W. Newton (363), CSIRO Division of Building Research, P.O. Box 56, Highett, Victoria 3190, Australia.

C. D. Ollier (233), Department of Geography, University of New England, Armidale, New South Wales 2350, Australia.

J. Oliver (57), Department of Geography, James Cook University of North Queensland, Townsville, Queensland 4811, Australia.

Don Parkes (89, 333), Department of Geography, University of Newcastle, Shortland, New South Wales 2308, Australia.

Roger Pech (173), CSIRO Division of Wildlife and Rangelands Research, Deniliquin, New South Wales 2710, Australia.

R. A. Perry (147), CSIRO Division of Groundwater Research, Private Bag, P.O. Wembley, Western Australia 6014, Australia.

Balwant Singh Saini (253), Department of Architecture, University of Queensland, St Lucia, Queensland 4067, Australia.

Neville Stanley (395), Department of Microbiology, Queen Elizabeth II Medical Centre, University of Western Australia, Nedlands, Western Australia 6009, Australia.

S. V. Szokolay (263), Architectural Science Unit, University of Queensland, St Lucia, Queensland 4067, Australia.

Michael D. Young (191), CSIRO Division of Wildlife and Rangelands Research, Deniliquin, New South Wales 2710, Australia

PREFACE

Northern Australia, as defined in this volume, extends north from -26° latitude south to about -9°30' south. It extends east and west along the southern limits of Australia's Northern Territory. The membrane of the waters of the Indian and Pacific Oceans and the Timor, Arafura and Coral Seas are also included. They provide a diverse marine environment for human activity and climatic influence. The area so encompassed occupies about half the area of the world's largest island.

The *arena* of northern Australia has been settled by people for about 50 000 years. But the past 200 years have witnessed the most profound changes in its use; changes which have altered the appearance of hundreds of thousands of square kilometres of territory, including land and water; and changes which have altered the ecological mechanisms that operate to bring living and non-living elements into relation with each other.

This book is organized into three parts. Part A, *Arena* presents the contextual setting for Parts B and C. It has seven chapters. They include a historical geographer's perspective on the ecological impact of 200 years of European settlement, a description of the use of satellite imagery, a description of climate, discussion of some of the interactions among natural subsystems as they impinge on human activities (especially in the extensive rangelands), a summary of the principal renewable and non-renewable energy resources located in and available to northern Australia, a description of population characteristics, distribution and ecological structure, and, finally, a short chapter which concentrates on the Aboriginal population in more detail than is found in the chapter on population as a whole.

Part B, *Extensive Ecosystems* discusses some of the human ecosystems which extend over a very large geographical territory. In these ecosystems the human population is small in absolute number and small relative to the population of other living things (other biota). Energy for these ecosystems is derived essentially from endogenous sources, but increasingly its use is controlled by humans. These are *human* ecosystems because of intended intervention in their mechanisms. They include the tropical marine ecosystems and their growing utilization for mariculture; and rangeland ecosytems dominated by cattle and the overlapping semi-arid grasslands, their productivity and ecological stability. Two chapters consider aspects of land management and administration. One of these chapters demonstrates the utility of Landsat imagery for monitoring and managing rangeland ecosystems and the other discusses land administration, leases and tenure systems from a social and economic viewpoint.

In the *Intensive Ecosystems* discussed in Part C, the human population is dominant in number. Unlike the extensive ecosystems, energy is imported into the territory or habitat. It is exogenous to the immediate area of habitation and movement. However, in the first chapter, where the nucleated rural settement is considered in relation to the large cattle station, this numerical domination is less well defined, except in the precinct of the homestead and its associated structures. In the strictest sense the nucleated settlement on the large cattle station occupies a transitional position between the extensive and the intensive ecosystems. Attention then turns from the environmental impact of mining to aspects of human habitat in materially structured places, human settlement including settlement design, building efficiency and human behaviour in stressed climates, and to other ecological and attitudinal aspects of life in remote towns. The final chapter addresses the incidence of infectious disease in relation to human ecosystems and health in northern Australia, with most of the discussion centred on the Aboriginal population because, among other reasons, this demonstrates the increasing complexity of human interactions and changes to the environment wrought by intensifying urban processes and international movement.

The volume opens with the statement, "What man hath wrought", (Bauer Chapter 1, title). It concludes with an implicit question, "What hath man wrought?" (Stanley, Chapter 20).

As editor, and on behalf of the contributors, I hope that this volume will contribute in some small way at least to knowledge and interest in northern Australia. Less than a million people live on half a continent. Possibly less than a million of Australia's residents have visited northern Australia, but the gates to the arena are open and population is steadily increasing.

To all the contributors I extend my sincerest thanks. As a human geographer with a particular interest in the human ecology of settlement, I have learned much from their chapters.

For help in preparing this volume I wish to thank: Ken Jones and Rod Davies (Hamersley Iron Pty Ltd); the many residents of Paraburdoo who accepted me; Father Patrick Speed, Anglican Rector of Tennant Creek; the people of Tennant Creek; Chris and Netta Knott, of Warabri, now Ali Curuing; Geoff Taylor; Mrs Sharon Parry, who typed all but one of the 20 chapters; Laurie Henderson, cartographer to the Department of Geography at Newcastle University; and finally the Northern Australia Development Council which provided essential support towards the publication costs of the book. As editor I was free to include material as I saw fit. I hope that this volume matches up to the support which they so generously provided.

1

WHAT MAN HATH WROUGHT:
GEOGRAPHY AND CHANGE IN NORTHERN AUSTRALIA

F. H. BAUER

Department of Economic History
Research School of Social Sciences
Australian National University
Canberra, Australian Capital Territory

"...hard on horses and men, hell on women and dogs."
Anonymous.

I. INTRODUCTION

The permanent non-aboriginal settlement of northern
Australia, as designated in this volume, began in the first
years of the 1840s. The present cultural landscape,
therefore, is the result of less than 150 years of change
wrought largely by Europeans.[1] Considering the nature of the
land to which they came and the cultural heritage they brought
to it, the extent of those changes is remarkable.

When European man came to northern Australia he entered
a sparsely populated but by no means empty or pristine land.
For millenia this vast and diverse part of the continent had
been sporadically occupied by a people who managed to procure
a living from the land as it stood, a feat no others have been
able to emulate. In so doing they altered the ecosystems in
varying ways and to an extent which is as yet not fully
understood. It remained for the newcomers to ring the changes
which have created both comfort and concern for today's
inhabitants of the region.

This chapter aims at sketching in some of the major
elements in the process of development, a frequently used
term which most often seems to mean *changing the character of*

1

*a region so that it takes its place as a producing factor in
an economy.*

II. PROLOGUE

The first verifiable contacts were Dutch. Between 1606
and 1644 their navigators, through a combination of accident
and design, sketched in most of the coast west of Cape York,
but finding neither solace nor potential profit, they wisely
left it alone. Their meetings with Aborigines were fleeting
and unfriendly. Much more extensive were those of Macassans
who, from the late 17th century to early in this, visited
various parts of the north coast to collect beche-de-mer
(McKnight, 1976). They left a legacy of tamarind trees at
many of their beach campsites and, to paraphrase McKnight,
their influence in some aspects of aboriginal life was quite
extensive, but it did not transform the fundamental basis of
their society (McKnight, 1971).

Between 1824 and 1838 the British Colonial Office
established three settlements on Melville Island and the
Cobourg Peninsula. Basically strategic with hopeful commercial undertones, no attempts were made to explore inland. All
fell victim to distance, sickness, boredom and a bad press;
the last, Port Essington, was abandoned in November 1849.
Their principal bequests were a pronounced taint of failure
and the water buffalo, which had been imported and released,
thus setting the stage for a future major ecological problem.
Their relations with the Aborigines were mixed and their
lasting influence small.

Late in November 1837, Lieutenant George Grey and party
were landed near Port George IV, on the far northwestern
coast of Western Australia with the mission to travel south,
overland, to the Swan River settlement (Perth), keeping a
sharp eye out for good pastoral land. They floundered about
on ravine-riven plateau country for most of the wet season,
never got more than 100 km. from their starting point, and
were very happy indeed to be picked up the following April
(Grey, 1841). It was not an auspicious start for northern
land exploration. Nevertheless, the curtain was rising on
actual settlement in northern Australia, a play which had
no script and a cast which had not even seen the stage.

And a vast stage it was, about 4,500,000 km² vast,
fifteen times the size of England, Scotland and Wales. It
was a land almost inconceivable to northwest Europeans.
Topographically, plains predominated, and although truly high

ranges were absent, there were very large areas of rough,
rugged terrain. By and large the soils were poorly structured
for European crops, and there were few large, contiguous
tracts of really good alluvial soils. Grey-green sclerophytic
open woodland, desert shrub and/or grassland covered most of
it, while fringes of mangrove and tropical rainforest clung
to the more exposed and wetter eastern and northern margins.
Herbivores were the dominant animal species, most of them
scarcely believable marsupials; the dingo was the only
carnivore to have an economic impact.

These biota and soils developed under climatic conditions
which in themselves posed frustrating problems for the
newcomers. Rainfall, its amount and seasonal and areal
distributions, was from the first a major concern. Over 80%
of the region receives insufficient rainfall to permit the
growth of crops without irrigation; fully 50% is climatic
desert. Over most of the 'North' summer rainfall is strongly
predominant, although this trait decreases inland. On the
northern continental margins rainfall is higher and its
seasonality most pronounced, combining with high temperatures
and humidity to cause discomfort, while the wetter conditions
seriously hamper many activities during the summer months.
Although good falls are assured, there is considerable
variation in when the wet season begins and ends, making
planning difficult, especially for cropping. The trite
comment, "There is no such thing as a normal season" is a
truism. Much of the east coast is backed by a comparatively
narrow lowland behind which slopes rise abruptly,
circumstances which, combined with proximity to a warm ocean
and onshore winds, are responsible for a more even seasonal
distribution and much higher annual totals.

(The climate of northern Australia is treated in more
detail by Lee and Neal in chapter 3, and Auliciems and Dedear
discuss aspects of thermal 'comfort', in the Darwin region,
in chapter 17).

A corollary of climate is water, and water, its presence
and permanence, was the major localizing factor for living
sites and activities of all kinds. Lacking the materials and
technology to store even quite modest quantities of potable
water, (see Ollier chapter 14), those who ventured into the
new lands were dependant on natural surface supplies and
what could be obtained from wells dug by hand. The most
common supply was the waterhole. In response to the seasonal
nature of the rainfall, the flow of northern Australian
streams and rivers varies greatly. Many cease to flow during
the dry season and those in the interior may not flow for
several years. Deeper parts of the beds retain water, and

while some of these waterholes were virtually permanent, most had a limited 'life' between replenishing flows.

From the European standpoint the human inhabitants proved to be a far less serious problem than did the physical environment. A hunting and gathering folk organized into closely knit kinship groupings, their movements over the landscape were mostly confined to unmarked but well recognized ranges with which they had a spiritual bond incomprehensible to, and largely unrecognized by, 19th century Europeans. Suspicious of outsiders, even their own kind, divided by language differences and lacking weapons which enabled them to compete fairly against the invaders, they were unable to mount concerted resistance to invasion. They relied on stealth and surprise, tactics which Europeans most often interpreted as treachery.

It was into this situation that European man, largely British, with his Victorian mores, morals and technology, enthusiastically projected himself. Culturally ill-prepared to deal with the environment and the huge distances over which he was forced to operate, and wishing to do things to and in this land which had never been done here before, he has spent a great deal of time, energy and money trying to bring his cultural heritage into some sort of accommodation with that environment. In the process he has changed both it and himself.

And so the changes – the development of northern Australia – began. It is doubtfull if their full extent can yet be assessed; certainly their ultimate results were unforseen by those who began them. From our pinnacle of over a century of hindsight it is easy to condemn those pioneers. Yet they were but following the technology and beliefs of their day to make a success of their undertakings, and in most cases this meant making a none-too-secure living from an unfamiliar, unfriendly and recalcitrant land. To them a thousand acres of ringbarked, dead but still standing timber gave a profound sense of accomplishment, and therein lies one of Australia's major dilemmas: *how to develop without destroying.*

III. THE FIRST ACT

Three pursuits – pastoralism, mining and agriculture[2] – attracted Europeans to these new lands. In most cases pastoralism came first, but not infrequently all three were prosecuted, or at least attempted, contemporaneously.

Certainly the desire for new land upon which to depasture
sheep was responsible for the first permanent settlement north
of 26°S, although the impetus came from farther south.
     In spite of a New South Wales regulation forbidding
settlement more than 50 miles from the outpost for incor-
rigibles at Moreton Bay (Brisbane), in mid-1840 the Leslie
brothers formed the first stations on the Darling Downs.
In February 1842 Governor Gipps, recognizing the inevitable,
opened what is now southeastern Queensland to pastoral
occupation, and for the next decade or two exploration and
settlement became thoroughly mixed. Small, almost secretive,
private parties went out looking for land, and it is quite
possible that the first ones north of 26°S have not been
recorded. An attempt to form a station near Tiaro (25°44'S,
152°35'E) in 1842 failed because of hostile Aborigines
(Cilento & Lack, 1959), but not before wool had been shipped
from a river landing in 1843 (Meston, 1895).
     More traditional explorers were also in the field, and
in 1845-46 Mitchell reported good pastoral country along
the upper Warrego and Belyando Rivers, but his thunder was
stolen by Leichhardt, whose private party completed
northern Australia's first major overland exploration by
going from the Darling Downs to Port Essington. Given up
for dead, Leichhardt's sudden reappearance focussed attention
on the 'North' as never before. Meanwhile, far to the west,
poor Sturt struggled to the edge of the Simpson Desert in
his search for an inland sea, but found only gibbers, sand
and blazing sun.
     An attempt (1847) to set up the new penal colony of
Northern Australia at Port Curtis (Gladstone) failed miserably
but shortly thereafter pastoralists, notably the Archer
brothers, moved up the Burnett to take up stations in the
Eidsvold district. They soon found it much easier to send
their wool south by vessel from a landing on the river than
to haul it by dray, and thus, in 1847, Maryborough came into
being, the first permanent town north of 26°S. They also
found their Burnett country unsuitable for sheep, and in
1850, acting belatedly on letters Leichhardt had written
them in 1846 telling of good country along the Dawson, they
were off land hunting again; first to the Callide valley
and then, still unsatisfied, on north until, in 1853 they
came upon the plains along the Fitzroy River and established
(1855) the first station (Gracemere) north of the Tropic.
Rockhampton followed the next year (McDonald, 1981).
     The decade of the 1860s was when significant northern
movement got under way. In 1859 Queensland became a colony
in her own right and a burst of pastoral expansion, fueled

in part by Victorian gold, took occupation to Cape York, the Gulf and the Channel Country by the end of the decade. Towns to support the inland movement grew up, mushroom-like, along the coast: Bowen (1861), Mackay (1862), Townsville (1864), Burketown (1865) and Normanton (1868). Queensland was on the move.

On the other side of the continent there were no northward gestures until after Gregory's 1861 expedition to Nickol Bay reported good, although patchy, pastoral country, and in 1863 James Padbury's party landed sheep, bullocks and horses; the first town, Roebourne, soon followed. Later in the same year land was taken up in the Gascoyne district, to the south of the Pilbara (Battye, 1924).

In the Centre, Stuart finally crossed the continent, a feat which was used to consummate the biggest land grab in Australian history: in 1863 South Australia gained control of the Northern Territory and promptly embarked on a program of settlement from the north, based on expectations of tropical agriculture and horse raising. The first settlement, in 1864 and called Palmerston, was a fiasco, following which several years were spent in obtaining explorers' reports which were ignored. Finally, in desperation, Goyder was sent to found the first permanent town, another Palmerston (1869), now Darwin (Bauer, 1964).

Thus European man and his animals moved into the North to begin an onslaught on the natural resources which still continues. And now it is "time to talk of many things", of the effects on the landscape, the flora and fauna and, not least, of man himself.

IV.   EFFECTS OF PASTORALISM

The depasturing of European livestock on lands and plants which had never before known a hoof brought changes, subtle at first but all too often cumulative. Initially they were small because the natural carrying capacity of the country was substantial, stocking densities were low and in the absence of fences the stock, particularly sheep which had to be shepherded, were moved frequently. Sheep were the preferred livestock, partly from tradition and partly because wool brought a good price per unit weight and could be stored without serious deterioration. There was also a widely held theory that both pasture and water supply would improve after a few seasons' grazing, especially with sheep; the opposite was, of course, nearer the mark, for the nibbling eating

habit of sheep meant very close cropping and pasture plants
suffered accordingly. Some plants could not withstand
trampling by hooved animals, while others were particularly
palatable and were grazed so heavily that they disappeared
from many areas. Annuals, which formed an important part of
the pastures, were especially vulnerable because grazing
prevented seed set.

Fire deserves special mention. In pre-European times
fires, both natural and man-caused, were an integral part of
the Australian ecology; many plant species developed special
traits which ensured their survival. There can be no doubt
that for millenia the Aborigines had been changing the flora
by burning; the question of whether they actually 'managed'
country by using fire is still arguable. So, too, is that of
whether European man was more or less pyromaniacal, but in
any event, the burning continued. The practice of burning
pasture lands to eliminate dry, unpalatable growth and to
induce a green shoot came from Europe and was greatly reinforced by colonial experience, especially in open woodland
with a grass understorey. Frequent, uncontrolled and extensive burning encouraged some species, most of them of limited
grazing value.

Open woodland was widespread and since most of it carried
an understorey of herbage, stock were run under the trees. At
first the trees were relatively undisturbed except as bush
timber was needed for yards, fences and buildings. Removal
of the shade cast by trees encouraged herbaceous plants, and
because there was no market for the wood and labour was
scarce, trees were simply ringbarked and left standing. Since
care was rarely taken to leave trees on the steeper slopes
and in drainage paths, soil erosion and gullying resulted.
As stock numbers increased overgrazing was common. Indeed,
given the long, dry (winter) season and droughts it was
inevitable, with the result that the incidence of sheetwash,
soil erosion and gullying increased. River frontages where
permanent water was available were the most seriously
affected. For example, by the mid-1950s hundreds of square
kilometres along the Ord River frontages in northwestern
Western Australia were completely denuded of all herbage,
shrubs and most trees. The vast open grassland downs of
central Queensland and the Barkly Tableland of the Northern
Territory were less seriously affected.

New plants, most of them exotics adapted to disturbed
conditions, established themselves, particularly on overgrazed
areas or after drought. The classic example was prickly pear
(*Opuntia* spp.), but rubber vine (*Cryptostegia grandiflora*)
and noogoora burr (*Xanthium pungens*) have infested thousands

of kilometres of stream frontages, especially in Queensland, crowding out and replacing nutritious pasture plants. Similarly in the Top End of the Northern Territory hyptis (*Hyptis suaveolens*) has become a serious plant pest.

Some native species behaved in much the same way. As early as the 1870s sheep could no longer be run successfully on large parts of north Queensland's eastern watershed because overgrazing and burning had favoured the spread of spear grass (*Heteropogon contortus*), the sharp seeds of which carry a devilish corkscrew tail or barb which literally twists the seeds through wool, hide and flesh. Native tussock grasses, for example some of the *Stipa* spp., generally of low grazing value, have followed overgrazing, burning and drought.

Not all of the plant introductions have been disadvantageous. Prominent among those which have proved to be useful is Townsville lucerne or stylo (*Stylothanses humilis*). This tropical legume from South America was an accidental introduction which proved to have considerable value as a pasture plant, and although it failed to live up to the high expectations its advocates entertained for it, that very failure led to a wide and partly successful search for other species which might be used to improve northern Australian pastures.

The *surface water supply* was also seriously affected by changes in the vegetative cover which increased both the quantity and intensity of runoff. The sediment load carried by streams and rivers increased to the point of overloading and deposition took place, particularly in the waterholes, greatly reducing their useful life. Livestock coming to waterholes aided and abetted the silting process by breaking down and loosening the material of the banks, and at the next rain the loose material washed into the waterholes.

Animal life also saw some changes. While the rabbit never became a serious problem in northern Australia, the dingo did. Early accounts rarely mention the dingo as more than a nuisance, but with the advent of livestock, especially sheep, and the additional water supplies which gradually came with it, the dingo increased in numbers and became a factor of some economic importance. There is similar, although not as strong, evidence for the deleterious impact of birds, notably hawks.

V.   EFFECTS OF MINING

The search for and extraction of minerals has been a
major pursuit since the earliest days of settlement in
northern Australia;   indeed, in many instances it was the
first.   The industry has been responsible for bringing people
and amenities to remote areas which otherwise would have
remained under pastoral tenure or would not have been
developed, and in the process the natural landscape has been
altered.

It is necessary to distinguish between the fossicker and
the miner, although the one often merged with the other.
Traditionally the *fossicker* was a man who preferred to work
alone or in small groups.   Fossickers made a significant
contribution to the exploration of northern Australia for
which they rarely got, or wanted, recognition.   The effects
they had on the country were minor, although if successful
they attracted the *miner*, who operated on quite a different
level and with quite different results.   Even so, I feel the
miner, particularly those of earlier days, has received rather
more than his share of censure for what he has done to alter
the northern landscape and ecology.

The principle reason for this is that his activities are
highly localized, intensive and obvious.   Another reason, and
one for which he can sometimes be fairly censured, is that
when finished in an area he often leaves an unsightly and
sometimes dangerous mess which persists for a long time.   Over
the past decade or two, however, increased awareness of
environmental matters has led to stricter legislation aimed at
controlling the worst excesses of the past.   Most mining
operations in northern Australia have, in themselves, affected
very small areas, usually less than half a square kilometre.
The area disturbed and the degree of that disturbance
depended, then as now, on the mineral exploited, the
techniques employed and the period over which it was worked.
(Ollier discusses the impact of mining, at the present time,
in Chapter 14.)

Without exception *gold* was the first metal sought and
worked, and almost always the first methods used were the
various forms of alluvial mining, which disturbed only small
areas.   While entire stream beds and adjacent areas might be
overturned and the streams themselves interfered with, few
alluvial fields lasted more than a year and the effects of the
actual mining operations were not great.   Far greater changes
were wrought in the places where the miners lived, but while
population densities were sometimes high, again the areas
involved were small and the period of occupation short.

As soon as *reef mining* began, however, changes
multiplied. Wood and timber were needed for power and mine
supports, and some districts were denuded of trees for miles
around. Piles of rock tailings accumulated outside shafts,
while the finer tailings from stamp mills clogged streams
or formed sterile piles. Reef mining lasted longer, involved
more labour and put heavy demands on the natural resources.
Furthermore, almost every mining town had its mob of goats,
often outnumbering the human inhabitants, and they were vastly
more damaging to vegetation than sheep. Following the
introduction (in the early 1890s) of the cyanide process for
extracting gold from crushed ore, contamination of streams,
already apparent, increased. A great deal of very obvious
environmental damage was certainly done in the immediate
vicinity of the mines and towns, but aside from stream pollution the effects did not extend far. However, when ores of
the *base metals*, such as lead, zinc and copper, began to be
worked the demands on some natural resources became greater;
more space was required, as was more water and, with the
coming of the smelters which such ores required, air pollution
was added to that of the streams. Rather paradoxically,
demands on the local timber supply may have diminished, since
such operations used fossil fuels, and bush timber was
unsuitable for either underground or surface requirements.
        There were other, more positive, aspects to the mining
industry. Roads, railways and other communication lines were
built to mining strikes through country which could not have
supported them otherwise; services such as schools, mail and
medical facilities followed, and these benefitted all those
living within their range. On the whole it seems likely that
the positive contributions made by mining to the quality of
life in the North at least equalled its effects on the natural
environment.

VI.   EFFECTS OF AGRICULTURE

        Perhaps it is as well for the physical environment that
European style agriculture has not become truly widespread
in northern Australia, because its potential for major
ecological change is far greater than that of either pastoralism or mining. It demands the removal of the native
vegetation and turning of the topsoil over large areas.
Detailed analysis of such wholesale change is beyond the scope
of this writing and only brief mention of the historical
course of events and major changes will be made.

Ever since European settlement was first attempted
tropical agriculture has been one of the major desiderata of
those hoping to develop northern Australia. Expectably the
first efforts at cultivation were at the lower end of the
horticultural scale, *gardening*. In an attempt to be somewhat
self-sufficient, a variety of plants was tried at the early
settlements on Melville Island and the Cobourg Peninsula, but
the physical constraints already noted, plus ignorance of
tropical agriculture generally, brought failure (Bauer, 1980).
The early graziers and miners had little interest in, or
time for, gardening; furthermore, theirs was a predominantly
male population, and I do not believe the house or station
garden, so prominent later on, became common until women came
north to live permanently. In the gold mining centres it
was men of another culture, the Chinese, who undertook the
task. While of considerable importance to the populations
concerned, the areas involved in production and the effects
on the natural environment were small. It was not until the
introduction of *commercial field cropping* that substantial
changes occurred.
Even then progress was slow; by the turn of the century
there were less than 40,000 ha. of crops in the whole of
northern Australia. The 1981 figure was slightly over
992,000 ha.^{3}, of which 98% was in Queensland (Figure 1). Most
of the cropland lies in a roughly triangular zone extending
inland from Maryborough to Roma (420 km.) and narrowing
sharply to the coastal lowlands north of Rockhampton. Even
within this zone agriculture is distinctly discontinuous.
Greatest development has taken place in several discrete
districts where a combination of suitable soils, topography
and climate have offered opportunities found nowhere else in
the North. Cropping began in the late 1860s with sugar cane
and cotton, but sugar cane has always predominated; in 1981
cane was grown on 34% of all of the land under crop in
northern Australia. An industry requiring substantial capital
investment, it brought people, stability and wealth to the
districts in which it could be grown.
The Atherton Tableland of north Queensland and the
surrounding district is an anomoly to the general pattern of
agriculture in coastal Queensland. Standing between 750 and
1,000 metres above sea level and exposed to moisture-laden
winds from the ocean, this moist, sub-tropical volcanic
upland was originally covered by very dense rainforest which,
during the 1880s and 1890s, supported an active timber
industry. As the forest was cut farmers moved in to complete
the clearing. Maize was the principle crop for many years,

to be joined first by dairying and then, with the completion
of the Tineroo Falls Dam in 1958, tobacco was grown under
irrigation (Birtles, 1982;   Courtenay, 1982).

As local population grew and the farmers gained
experience, other crops were grown in the sub-tropical coastal
districts, a development encouraged by better transport to
the southern states.   Gradually a complex cropping pattern has
evolved in which specialty crops, such as out-of-season vegetables, tropical and sub-tropical fruits and rice figure
strongly.

The other major agricultural region of northern Australia
centres on the Clermont - Rolleston district of central
Queensland.   From the 1870s onward sporadic but persistent
attempts to grow cereal grains met with little success
because of normally low rainfall and recurrent drought.   In
spite of this record, in 1947 a major project was funded by
the Queensland and British governments to grow grain sorghum
and to raise cattle and pigs on nearly 300,000 ha. of the
broadly undulating, lightly timbered plains (Skerman, 1978).
Although the scheme failed it did encourage research and
renewed efforts, with the result that in the last two decades
this has become the most diversified agricultural district of
northern Australia, accounting, in 1981, for one third of all
the North's cropped land.   Clearing of extensive areas of
brigalow (*Acacia harpophylla*) for both pasture and grain has
added to the available land.   Crops include grain sorghum,
wheat, sunflower, safflower, maize and a variety of beans.
Cotton, grown to some extent since the 1920s, has expanded
considerably since the completion, in 1971, of the Fairbairn
Dam on the Nogoa River (Courtenay, 1982).

Attempts to grow crops in the Northern Territory have
been more notable for their failure than otherwise.   Sugar
cane and a number of other sub-tropical crops were tried near
Darwin during the 1880s with marked lack of success, while an
ambitious promotion of agriculture during the second decade
of this century fared no better.   Peanuts showed some promise
in the Katherine district after the First World War, but
succumbed to a combination of variable quality, inadequate
labour and marketing problems (Bauer, 1964).   During the
Second World War the Australian Army established a number of
farms, totalling several hundred hectares, at various places
north of Barrow Creek to provide fresh fruits and vegetables
for the servicemen stationed in that part of the Territory.
Relieved of the necessity to turn a profit, and with ample
labour and fertilizers, the farms proved to be the most
successful agricultural venture the Territory had known, but
they did not persist into peacetime (Bauer, 1982).   Much more

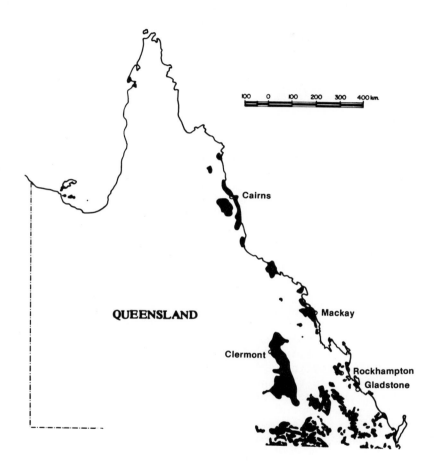

*FIGURE 1.   Principal areas of crop land in Queensland,
north of 26°south.* [a]

[a] *Adapted from a map by the Land Resources Branch, Queensland
Department of Primary Industry.*

recently, in the 1960s and 1970s, the names Humpty Doo (rice)
and Tipperary and Willeroo (grain sorghum) have become
synonyms for large-scale agricultural failure. Market
gardening near Darwin, Katherine, Tennant Creek and Alice
Springs has met with some success recently, but new initiatives involving sorghum, beans, maize and mangoes have not
been under way long enough to be judged.

Aside from efforts by local gardeners around towns, only
two districts in Western Australia have been in any way successful. Along the lower Gascoyne River good soils, a satisfactory water supply and the Perth market 1,000 km to the
south have combined to permit growing of vegetables and
sub-tropical fruits. Perhaps the most controversial of all
Australian irrigation projects, the Ord River Scheme, has made
water available to approximately 70,000 ha. of potentially
irrigable alluvial soils, but the largest area cropped in any
one year since the diversion dam was completed in 1965 has
been 5,785 ha. in 1971-72 (Young, 1979). Thus far no crop
or assemblage of crops has been found which have proved to be
economic, but the search continues. At Camballin, on the
lower Fitzroy River 260 km. east of Broome, large-scale rice
growing in the 1950s failed because of lack of suitable rice
varieties and inability to control both irrigation water and
floods. In 1965 the project was taken over by overseas
interests which began an even larger and more ambitious grain
sorghum development, only to pass into receivership in 1982.
The project is now on a care and maintenance basis pending
the outcome of attempts to find further finance.

While the data noted above make it quite apparent that
very substantial areas in some parts of northern Australia
have been cleared for agriculture, they do not tell the whole
story. There are considerable areas which have been cleared
and either abandoned (for agriculture) or are not cropped in
any given year. An accurate estimate of this area has not
been possible, but it must raise the total to over 1,000,000
ha. It means, of course, that effect of agriculture on the
original landscape has been much greater than is at once
apparent.

VII.   MAN'S METAMORPHOSIS

From the discussion above it will be clear that changes
in what may be fairly called the *human ecology* of northern
Australia have been scarcely less profound than those affecting the physical world, but because they are less easily

observed and documented they have received less attention.
Most obvious have been the effects on the aboriginal
population.

For more than a century after initial contact the predominant view held by the European newcomers was that the
northern Aborigines were simply another aspect of the natural
environment which had to be dealt with in some fashion.
If the Aborigines could be adapted to some useful role without
undue effort and expense, this was the best solution possible.
If this failed, as it often did, they were (at best) ignored
or (at worst) eliminated. The results were predictable:
their cultural organization was disrupted at all levels, they
were often forced off their traditional lands and malnutrition, unfamiliar diseases and direct attack drastically
reduced populations, sometimes to the vanishing point (See
Stanley chapter 20). It was not one of European man's finer
moments; while the argument that such behaviour was merely
the ethos of the time may offer explanation, it cannot also
provide justification. Whatever one's opinions on the morals
of the matter, the outcome was that the Aborigines became
mendicants in their own land and significant parts of their
cultural heritage were lost. They become a different people.

The actual processes responsible for these changes and
how they operated in different situations have, until
recently, been largely a matter for conjecture, for little
came from the Aborigines themselves and European accounts
were tainted by self-interest. Only within the last two
decades has the sorry story been given widespread airing[4].
Only in the pastoral industry did the Aborigines find a place
in the society which replaced theirs, and even there they
were relegated to inferior positions as stockmen and household servants.

Europeans were affected very differently. Many of those
taking up and stocking new land had had experience in colonial
Australia before pushing north, but even these were sorely
pressed by the unfamiliar environmental conditions encountered
and the economic and social realities of distance. As men of
substance they, and the sugar planters who soon joined them in
Queensland, formed the most responsible and stable segment of
the developing society, and their confidence and influence
grew accordingly. Those not so fortunate - the 'new chums',
and they were in the majority - had to adapt to even more
unfamiliar conditions. Those associated with climate and
physical comfort were certainly stringent, but they paled
beside the social and cultural changes which were necessary.

Frontiers everywhere have attracted men (and some women)
who prefer to move frequently, as though afraid of staying in

one place too long, and northern Australia acquired this kind of shifting population, a characteristic it still retains to a considerable degree (See Neil *et al*, chapter 19, and Parkes, chapter 18). Northern Australia was, perhaps, the last frontier to which a man could go safely with the clothes in which he stood and be reasonably sure of making a living of some sort.

The demand for labour in all industries, particularly mining and agriculture, brought men less likely to contribute to stability, while the very remoteness and lack of social and legal restraints attracted still others even less so inclined. Men accustomed to gregarious urban life were thrust on their own resources in a heavily male society which had few social pretensions. Some sought solitude, many others had it forced on them. Such personal relationships as were possible had special significance and the concept of 'mateship', already strong in the southern colonies, was reinforced. Few personal questions were asked and fewer answers given.

Women came into the north at a marked remove from their men, a circumstance which reflected masculine opinions on female stamina and the general 'fitness of things' rather than those of the women themselves. Theirs was the real hardship. Amenities, even of the simplest sort, were few, companionship of their own sex rare. Often alone for long periods on sprawling, isolated stations or in mining camps, they bore and raised their children and made family life possible. Whereas many men welcomed the challenge and adventure of northern development, the hardest parts of bush life - heat, dust, flies, boredom, lack of water, medical attention, education facilities and familial backup - bore most heavily on the women. Small wonder that corrugated iron, the Flying Doctor radio and the kerosene refrigerator were in turn considered the most important inventions of their day. It was no accident that the families to which women came early and stayed, became the most stable.

People coming to the towns also had adaptations to make, for the towns were small, dirty, gossipy, rough and introspective; few had an assured water supply. Their designations as one, two or three pub towns accurately reflected their main attractions. A substantial part of the population was mobile. As governmental and commercial services became necessary a new factor was added, a managerial class. These were officials of various sorts, bank managers, stock and station agents and school teachers, most of whom came north because promotion was more rapid in the back blocks. Some of them brought a semblance of sophistication, but most were intent only in serving out their two or three year term and getting away. They were, in a sense, participating bystanders. Population transience is still a significant component in the human ecology of the 'North', particularly those back from the east coast.

Into this developing society there were deliberately introduced two groups of quite different cultures, the *Chinese* and *Pacific Islanders*. The Chinese were brought as labour for gold mines or were attracted to the successive 'rushes' from previous ones. Despite suspicion and discrimination, they came to form one of the most stable, conservative and successful elements of the northern population. Many took to gardening or shopkeeping, or performed the multitudinous menial tasks Europeans liked to pass over to others.

The Islanders were largely Melanesians, first from the New Hebrides and later from the Solomon Islands. The first group was brought by Robert Towns in 1863 to work on cotton plantations south of Brisbane. Sugar planters, eager for dependable and tractable labour, were soon importing them by the shipload under a form of indentured labour which, while fairly well controlled, was nevertheless open to abuse and excesses. The southern-backed White Australia policy put an end to such immigration, and while many members of both groups returned to their homelands, many who had been in Australia for long periods were permitted to stay, to the undoubted benefit of the society generally.

They were by no means the only folk of non-Anglo origins who now form the northern Australian population. Indeed, if there were an 'index of ethnicity' the northern part of the continent would rank high, especially in such places as Broome, Darwin and Mt. Isa. While Mediterraneans, especially Greeks, Italians and Spanish, came in substantial numbers before the Second World War, it was, as elsewhere in Australia, the post-war immigration policy which brought the greatest variety of peoples and cultures; the latest have been the Vietnamese. Most of these people, who were urban dwellers in their homelands, have gravitated to the larger towns, where they tend to live and associate with their own kind. They have, however, made significant contributions to the society, especially gastronomically!

It would be gratifying to report that this remarkable mix of cultures and people has evolved into a cohesive, stable society; such is not the case. Two broadly based and paradoxical attitudes pervade northern thinking. Northern Australians are suspicious and critical of most of the people and policies which come from anywhere south of where they

happen to live. Well aware of the lower level of amenities
they possess, of the higher cost of living in a remote part
of the country, of being ignored by the nation at large and of
their meagre political clout, they feel conspired against by
the well populated southern states[5] At the same time, and
despite their wish to control their own destinies and
development, they characteristically look to governments to
provide the solutions to the many problems which that
development entails. 'Aggressive dependence' may sum it up.
Such community feeling as exists rarely transcends local
boundaries and concerns. Those parts of the population most
obviously willing to help themselves are the comparative
newcomers.

And so the human and natural resources of northern
Australia, indigenous and extra-Australian, have been altered
by a complex and changing set of interactions; it would be
surprising only had this not occurred. There is a tendency
to assume that all the changes have been made, but we have
arrived only at a point along a continuum. The country
and the people are still dynamic, perhaps more so than ever
before, and changes will continue. I believe that we now
know enough about the physical nature of northern Australia to
avoid the worst excesses of modern development. I am even
more convinced that the real challenge for the future lies in
how well the social and cultural elements are understood and
manipulated. Basic to this is research into people problems
and the development of appropriate settlement. A number of
the chapters which follow address these issues as they apply
to the contemporary arena of northern Australia.

NOTES

[1]The term 'European' will be used to refer collectively to
all non-aboriginal folk who have contributed to what is now
the dominant culture of Northern Australia. It is
recognised that people from other parts of the world have
played a significant role.

[2]The term 'agriculture' will be used in a narrow sense to
include only those activities which involve cultivation of
the soil.

[3]This figure may be put into perspective by considering that
the area of the seventh largest pastoral property in the
Northern Territory, Waterloo Station, has an area of
1,001,300 ha. or 10,103 sq. kilometres. Unless otherwise
stated, crop areas include sown pasture harvested for seed
or cut for hay, green fodder or silage.

[4]See, for example:
Rowley, C.D. 1970. *The destruction of aboriginal society*,
    A.N.U. Press, Canberra;
Rowley, C.D. 1971. *Outcasts in Australia*, A.N.U. Press,
    Canberra;
Loos, N. 1982. *Invasion and resistance: aboriginal-
    European relations on the North Queensland frontier,
    1861-1897*, A.N.U. Press, Canberra.

[5]These generalizations apply less to the better populated,
more affluent eastern parts of Queensland than to the rest
of the North. Still, even in those towns and agricultural
districts sectionalism is a very real force.

REFERENCES

Battye, J.S. 1924. *Western Australia: a history from its
    discovery to the inauguration of the Commonwealth*,
    Clarendon Press, Oxford, pp. 262-4.
Bauer, F.H. 1964. *Historical geography of white settlement
    in part of northern Australia, Part 2, The Katherine -
    Darwin region.* CSIRO Division of Land Research and
    Regional Survey, Div. Rpt. No. 64/1, Canberra,
    pp. 42-66, 95-103, 233-35.
Bauer, F.H. 1980. Trial and mostly error. In *Of Time and
    Place Essays in Honour of OHK Spate,* (eds.)
    J.N. Jennings and G.J.R. Linge, A.N.U. Press, Canberra,
    pp. 9-30.
Bauer, June B. 1982. *Army farms in northern Australia,
    1942 - 45,* unpublished ms. in possession of author.
Birtles, T.G. 1982. Trees to burn: settlement in the
    Atherton - Evelyn rainforest. 1880-1900. *North
    Australia Res. Bull.* No. 8, Sep 1982, pp. 31-85.
Cilento, R. and Lack, C. 1959. *Triumph in the Tropics,*
    Smith & Peterson, Brisbane, p. 116.
Courtenay, P.P. 1982. *Northern Australia,* Longman Chesire,
    Melbourne, pp. 141-55.

Grey, Sir George,   1841.   *Journal of two expeditions of
    discovery in north-western and western Australia during
    the years 1837, 38 and 39,* etc. 2 vols. T. & W. Boone,
    London, 1:76-237.

McDonald, Lorna,   1981.   *Rockhampton, a history of city and
    district,* University of Queensland Press, Brisbane,
    pp. 17-20.

McKnight, C.C.   1971.   Macassans and Aborigines, *Oceania,*
    42 pp. 283-321.

McKnight, C.C.   1976.   *The voyage to Marege',* Melbourne
    University Press, Melbourne, pp.93-126.

Meston, A.   1895.   *Geographic history of Queensland,*
    Government Printer, Brisbane, p.54.

Skerman, P.J.   1978.   Cultivation in western Queensland,
    North Australia Res. Bull. No. 2, October, 1978,
    pp. 23-58.

Young, Sir Norman,   1979.   *Ord River irrigation area review,*
    Australian Government Printing Service, pp. 12, 42.

2

NORTHERN AUSTRALIA: A VIEW FROM SPACE

F. R. HONEY

CSIRO Division of Groundwater Research
Wembley, Western Australia

I. INTRODUCTION

Satellite imagery has provided a synoptic, repetitive
view of large portions of the surface of the Earth since the
early flights of the manned and unmanned spacecraft.
Satellites are an ideal platform for observation of either
rapidly varying phenomena such as floods and fires, or of
slowly changing patterns such as degradation of the landscape
due to soil erosion, overgrazing of rangelands, deforestation,
or invasion of unpalatable shrub and grass types in an
otherwise productive area. With the computer techniques which
have been developed it is possible to monitor broadly the
type of vegetation, and the density and vigour. The principal
source of data for these developments has been the Landsat
series of satellites, although since 1980 there has been an
increasing application of data and imagery from satellites
with significantly less spatial resolution than the Landsat
system, but with more frequent coverage, and different spectral information. The resolution required for mapping and
monitoring is problem specific. For example, to map the
rapidly diminishing areas of tropical rainforest and vine
thickets along the western two thirds of the northern
coastline, it is necessary to use the highest spatial resolution data which is available regularly, which at present in
Australia is the Landsat multispectral scanner data. To
monitor the extensive rangeland areas of Australia, the
Landsat system provides suitable data, although satisfactory
results can be achieved over very large areas, and at greatly
reduced cost using the data from the NOAA (U.S. National

NORTHERN AUSTRALIA
ISBN 0 12 545080 X

Oceanic and Atmospheric Administration) environmental
satellites. Monitoring the extensive and frequent fires which
occur in the northern half of the Australian continent may
be done at either large or small scale, although the frequency
of overpass of the higher resolution Landsat satellites
precludes daily monitoring of fire area. The lower spatial
resolution of the orbiting environmental satellites allows
only large area fires to be monitored, although smoke plumes
may be observed extending many tens of kilometres. To
determine the most appropriate satellite system, and the
time for gathering data, it is important to understand the
spectral properties of the target materials under investigation, the contribution of each component in a sampled area,
and the extent and detail required in the survey. An
understanding of the spectral properties of vegetation, rocks
and soils is applied when sensor systems are designed for
satellites.

II.   APPLICABLE SATELLITE SYSTEMS

The principal factors determining the suitability of
different satellites for regular observation of areas of
interest on the surface of the Earth are:

(1)   the operational status of the satellite,
(2)   the area of the surface imaged,
(3)   the spatial resolution of the sensors imaging the
area of interest,
(4)   the spectral bandpasses of the satellite sensor,
(5)   the time of image acquisition, and
(6)   frequency of coverage.

The satellites which provide imagery of Australia
range from the operational Japanese geosynchronous meteorological satellites (GMS-1 and GMS-2), through the operational
polar orbiting environmental satellites in the NOAA series,
operated by the United States National Oceanic and Atmospheric Administration, to the quasi-operational Landsat
satellites operated by the National Aeronautics and Space
Administration (NASA) of the United States. The most
familiar of these satellites is the GMS, which provides the
images seen on television for the weather segments of the
news. The Landsat satellites, which have received the most
intensive research, and have been applied to a wide range of
mapping and monitoring problems, are research satellites, with

no guarantee of continued operation. However, the Landsat
system, more than any other satellite or aircraft system,
have brought the attention of the community at large to
the tremendous potential of synoptic, temporal coverage of
the continent.

Radar imagery from satellite sensors, although only
available till 1986 on a limited, research basis, has
significant capabilities which differ from the imagery
recorded in the visible and infrared. Microwave radiation
is not scattered or absorbed by clouds, and the active nature
of radar systems allows them to image the Earth night and
day, independent of the weather. Although the interaction
of microwave radiation with target materials is considerably
different to that of visible and infrared radiation, interpretation procedures are well established. Planned
operational satellites which include radar systems will
image areas in northern Australia commencing 1986.

A.   *Geostationary Meteorological Satellites.*

The GMS satellites operated by Japan, located over
Papua New Guinea, provide synoptic coverage of approximately
one half of the surface of the Earth. They provide complete
coverage of Australia as frequently as once every fifteen
minutes, although the image data recorded by the Bureau
of Meteorology, which is acquired by the satellite, transmitted to Japan, processed, then retransmitted to Australia,
is received once every three hours, principally for
meteorological purposes. The imaging sensors on the GMS
satellites are Visible and Infrared Spin Scan Radiometres
(VISSR), with two channels, 0.5-0.9 micrometres (visible -
near infrared), and 10.5 and 12.5 micrometres (thermal
infrared). The spatial resolutions for each of these
channels on the surface are 2.5 and 5 kilometres respectively.
The data from the GMS has not been applied to mapping or
environmental monitoring in northern Australia, although it
has a particularly important role in monitoring severe
weather conditions such as cyclones.

B. *NOAA Satellite Imagery*

The NOAA series of polar orbiting environmental
satellites are filling an increasing role in broad scale
mapping and monitoring in Australia. The two most significant
advantages of these satellites are that they are operational
satellites, with guaranteed availibility till at least the
mid-1990's, and the diurnal coverage which each satellite
provides of the entire Australian continent, at a practical
resolution, in spectral regions which allow computer

enhancement to highlight a range of important resource and
environmental parameters.

The NOAA series of satellites are polar orbiting,
approximately sun synchronous, satellites operating at
nominal altitudes between 833 and 870 kilometres.   The
satellite's pass can image any location on the surface of
the Earth twice per day, with equator crossing times of
approximately 0730 and 1930 hours local time, or 0330 and
1530 hours local time, depending on the particular satellite.
Two satellites are generally operational at the same time,
producing a combined coverage four times per day.

The instrument providing image data of most value on
the NOAA satellites is the Advanced Very High Resolution
Radiometer (AVHRR), which is either a four or five spectral
channel imaging spectroradiometer, the number of channels
being determined by the particular satellite.   The spectral
characteristics of the AVHRR sensors are:

Channel	Wavelength (micrometres)
1	0.55 - 0.68
2	0.725 - 1.10
3	3.55 - 3.95
4	10.3 - 11.3
5 (NOAA-7 only)	11.5 - 12.5

Data from these sensors are digitised and transmitted as
10 bit data (0-1023).   This range is important as it allows
the sensors to record data with high precision over a range
of targets with varying albedos and temperatures.   The
scanner records data for a continuous swath 2,900 kms wide,
with 2,048 picture elements per scan line.   The spatial
resolution, or pixel size of the scanner is 1.1 kms at nadir.

Image data from the satellites is received in Perth,
at the Western Australian Institute of Technology, and in
Melbourne Victoria, at the CSIRO Division of Atmospheric
Research.   The Australian Bureau of Meteorology plans
purchase of receiving facilities to be located around
Australia.

Applications have been investigated and for which the
AVHRR data has considerable promise in northern Australia
include regional geologic and soils mapping, delineation of
paleodrainage networks, observation of active and past fire
patterns, and broad scale vegetation assessment.   Most of
these applications are apparent from an examination of
computer enhanced images from the four spectral bands of a
daytime image, from a night-time thermal image, and a
"vegetation index" image.   These images are presented in

Plates 1 - 6.[*] Each image covers an area extending E-W from
approximately Port Hedland Western Australia to Darwin,
Northern Territory, and from the northern coastline to Lake
Amadeus in central Australia.

The potential of the NOAA satellites for delineation
of areas affected by flooding after heavy rains associated
with tropical cyclones, and the ability to determine regional
soil moisture patterns using diurnal and annual surface
temperature cycles are under investigation, and while the
results indicate some promise, the most serious limitation,
particularly for monitoring the progress of floods, is cloud
cover.

C.   *The Landsat Satellites*

The data and imagery which have been provided by the
Landsat series of satellites since 1972 have become familiar
to an extremely diverse range of researchers, users and
managers, with applications ranging from geologic mapping,
mineral exploration, water resources through to range assessment, soil erosion studies, and detailed vegetation mapping.
In Australia there has been a concentrated research effort
into techniques for enhancement and interpretation of the
imagery, and extension of the knowledge gained through to
the day to day inventory, monitoring and management of
resources and the environment (see chapter 11 following).
Although the Landsat satellites can not be considered as
operational satellites, and there have been problems in
acquiring imagery from the system, the research into an
application of the data from these satellites has highlighted
the extremely important role satellite remote sensins has
in management of Australian resources.

The early Landsat satellites, Landsats 1, 2 and 3,
orbited in near polar, sun synchronous orbits at altitudes
of 917 km, with an equator crossing time in the descending
mode of 9:38 am local time, repeating coverage once every
eighteen days. The sensors on the multispectral scanner
(MSS) on Landsat, which measured reflected solar radiation
were:

Channel	Wavelength (micrometres)
4	0.5 - 0.6
5	0.6 - 0.7
6	0.7 - 0.8
7	0.8 - 1.1

The scanner images a continuous swath 185 km wide, the

* Plates are contained in Appendix

data stream being segmented into image frames 185 km by 185
km at the ground processing facility. The data from the
sensors is transmitted as six bit information (0-63),
limiting partially the sensitivity and dynamic range which
can be achieved. The spatial resolution of the MSS is 79
metres.

The most recent of the Landsat satellites, Landsat 4,
was launched late 1982. The scanner on this, the first of
a new series, has considerably enhanced spectral and spatial
characteristics, with seven selected spectral channels, and
a spatial resolution of 30 metres. Data from this satellite
will become available with the proposed upgrade of the
Australian Landsat Station in Alice Springs. The spectral
channels of the advanced sensor, the Thematic Mapper are:

Channel	Wavelength (micrometres)
1	0.45 - 0.52
2	0.52 - 0.62
3	0.63 - 0.69
4	0.76 - 0.90
5	1.55 - 1.75
6	10.4 - 12.5
7	2.08 - 2.35

These refined channels, combined with the improved
spatial resolution of the satellite will enable more refined
mapping and monitoring.

Examples of enhanced Landsat imagery for an area in the
Kimberley Basin, northern Western Australia, are presented
in plates 7, 8 and 9.[*] These images, for bands 5 and 7, and
for a vegetation index (7-5)/(7+5), illustrate the significant
differences in spectral reflectance of healthy, dense vegetation, soils, and water. Band 4, which correlates closely
with band 5, and generally provides poorer contrast than the
other bands do to atmospheric backscatter contributions, is
not presented. Band 6 correlates closely with band 7, and is
not displayed. Although the four bands generally correlate
in two pairs, 4 and 5, and 6 and 7, the gain in information
content with all four bands justifies the recording and
processing of all four, and the use of all four, either
separately or as linear combinations, to discriminate subtle
variations in vegetation type, density and vigour; to
determine some changes in soil colour, and to estimate
turbidity in broad streams. Graetz and Pech discuss applications of satellite imagery to rangeland management in *The
Channel Country* of Queensland, in chapter 11 following.

* Plates are contained in Appendix

D.   Shuttle Imaging Radar - A

During November 1981, the synthetic aperture imaging
radar on Shuttle recorded three swaths of imagery across
Australia. The instrument, Shuttle Imaging Radar - A
(SIR-A), recorded 23 cm wavelength (L-band) data for
continuous swaths 50 km wide, at a spatial resolution of
approximately 40 metres. The data, recorded in photographic
form on-board the spacecraft, was processed at the Jet
Propulsion Laboratory, Pasadena California, before analysis
by investigators in Australia and other countries. Radar
imagery is particularly valuable for geologic and geomorpho-
logic mapping, discriminating water areas, and delineating
vegetation. The example of the SIR-A imagery presented in
Plate 10 corresponds approximately to the area imaged by
Landsat (Plate 7).[*]

III.   COMPUTER PROCESSING

Almost all satellite imagery which is recorded and
applied to mapping and monitoring of northern Australia is
computer processed and enhanced to highlight particular
features, and to increase capability for discrimination.
The computer techniques which have been developed enable a
potential user of the data to work interactively with the
satellite data, enhancing the imagery specifically for a
particular problem. The systems providing these processing
facilities are located in most capital cities in Australia,
and have similar capabilities.

The basic processing procedures include:

(1)   Histogram modification, to either improve the overall presentation of the image, or to enhance particular
features,

(2)   interband ratios, to enhance spectral properties
and differences of materials,

(3)   generation of linear and non-linear combinations
of the spectral channels, based on a statistical procedure
such as principal components analysis, canonical analysis
or linear and non-linear regression,

(4)   classification procedures,

(5)   spatial filtering procedures to enhance spatial
and textural patterns,

(6)   merging of different images and image types, and
combination of image information with ancillary data sets

[*]   Plates are contained in Appendix

such as digital elevation data, soil maps, or land use
data, and
        (7)   temporal analysis of imagery to highlight vegetation
growth patterns.

IV.   SUMMARY

        The satellite imagery which is available for northern
Australia, combined with the powerful computing techniques
which have been developed to maximise the utility of the
data, are extremely important tools for the inventory
monitoring and management of the renewable and non-renewable
resources, and for effective planning of future development.
The techniques now being developed will become a routine tool
for the resource managers in these vast areas, and will
complement the presently accepted procedures.   Continued
application of the satellite data seems assured with the
increasing interest being shown, and with the development of
facilities for the reception, processing and archiving of the
data.

3

THE CLIMATE OF NORTHERN AUSTRALIA

D. M. LEE
A. B. NEAL

Bureau of Meteorology
Melbourne, Victoria

I. INTRODUCTION

In preparing a climatological summary for northern
Australia the first problem to be considered is what southern
latitude limit should be applied. In this instance a southern
limit of 26° has been chosen; consequently the area being
considered is about half the land area of Australia.

Climate can be described in two ways – firstly as the
aggregation of statistics relating to meteorological elements
presented as a concise set of tables and maps from which
information can be derived; and secondly as a physical entity
which can have a profound effect on the nourishment, welfare,
safety and quality of life of human society. By presenting a
series of tables and maps describing the climatological
elements, this chapter attempts to provide sufficient
information so that personal or management decisions can be
made regarding the effects of the physical entity 'climate' on
the life and actions of people in northern Australia.

There are a number of major works on the climate of
Australia as a whole (e.g. Hunt *et al*. 1913, Taylor 1932,
Gentilli 1971, Linacre and Hobbs 1977, Bureau of Meteorology
1978) but none deal exclusively with northern Australia.
These studies present the reader with a variety of climate
classification systems based on air mass frequency, thermal
efficiency etc., but not one of these systems is entirely
satisfactory by itself. Here we do not attempt an overall
classification of the climate of northern Australia,

NORTHERN AUSTRALIA
ISBN 0 12 545080 X

preferring instead to describe the main climatic elements and briefly discuss atmospheric controls. There are two general points worth noting at the outset, however: one concerning the climate of the arid region of northern Australia, and the other relating to the definition of monsoons. Trewartha (1962) draws attention to the climatic anomaly (in global terms) of the tropical Australian dry zone. Although it is one of the world's largest deserts its rainfall is not as meagre as the other low latitude deserts. Furthermore in northern Australia the minimum rainfall is in the interior whereas in other low latitude continents the very driest area is along the west coast. The main reasons for this are the shape of the Western Australian coastline and the absence of a cold ocean current and associated upwelling along the northwest tropical coast. This enables tropical cyclones and tropical rain depressions to exist in this region and they contribute greatly to the relatively high seasonal rainfall there.

The term 'monsoon' means a seasonal persistence of a given wind direction in a climate where there is a pronounced change in prevailing wind direction from one season to another. In the tropics in summer the seasonal wind blows from the ocean, (summer monsoon) while in winter it is of continental origin (winter monsoon). In popular use, however, the term 'monsoon' is often used to denote seasonal rains, without reference to the winds. In the Australian context Ramage (1971) points out that only a small portion of northern Australia (viz: the area north of a line from about Port Hedland to Cairns) is really monsoonal. The remainder of northern Australia is dominated by the southeast trade winds.

Climate in northern Australia differs greatly from the climate in the south as most of the area we are considering lies within the tropics. Here there are basically only two seasons, known locally as 'The Wet' and 'The Dry'. The wet season generally starts late in the year and ends sometime about late March or April, but its time of onset and cessation varies from place to place and from year to year. During 'The Wet' the most significant weather, apart from thunderstorms, is produced by tropical cyclones and tropical rain depressions. In the dry season (usually May to October) bushfires represent the main hazard. Daytime temperatures are generally warm and relative humidity remains at comfortable levels. In southern inland parts, frosts are not uncommon and days are often quite cool.

II. CLIMATE CONTROLS

In the dry season high pressure systems pass from west to east across southern Australia but often remain stationary over the interior for several days. Northern Australia is then influenced by dry southeast trade winds while southern Australia experiences moister westerlies. During the other half of the year (November to April) the highs track much further south and areas of relatively low pressure develop over northern Australia. This allows an inflow of warm air from the surrounding tropical oceans and produces a hot rainy season. Periodically the northwesterly monsoonal winds of lower latitudes extend well south into the continent and help maintain tropical rain disturbances. During the transitional months and extending into the wet season thunderstorms are common. Frequently they are limited in area but can produce very heavy rainfall in quite a short time. Tropical depressions and tropical cyclones develop over the seas around northern Australia during the wet season. On the average three cyclones per season affect Queensland, two to three affect Western Australia, and one affects the Northern Territory. Cyclones usually produce very heavy rain and gales in coastal areas. Some move inland, losing strength but still producing widespread heavy rainfall. Some cyclones control the weather over northern Australia for a week or more, others only for a day or so.

TABLE 1 : Mean monthly rainfall (mm)

STATION	Jan	Feb	Mar	Apr	May	Jun	Jul	Aug	Sep	Oct	Nov	Dec	Year
Wyndham	189	172	127	27	9	4	5	1	3	12	47	108	701
Halls Creek	152	146	79	20	10	6	6	3	5	18	48	94	580
Derby	185	155	111	31	22	11	6	2	1	3	17	86	633
Broome	165	153	97	28	25	23	5	2	2	1	12	65	573
Port Hedland	57	72	69	25	30	27	10	7	1	1	2	15	316
Carnarvon	11	20	17	14	39	56	42	18	5	4	2	3	231
Mundiwindi	44	45	46	22	20	21	10	9	3	7	11	26	261
Darwin	399	337	279	96	16	3	1	3	14	59	130	241	1575
Yirrkala	214	246	263	259	86	30	12	5	3	6	48	142	1308
Katherine	234	214	166	33	5	2	1	1	6	30	85	142	970
Newcastle Waters	122	116	79	20	8	5	3	2	5	18	35	70	483
Tennant Creek	90	90	48	14	12	7	6	3	7	15	26	57	377
Alice Springs	43	41	30	16	17	14	11	11	9	20	25	35	272
Weipa	442	402	332	113	15	4	2	2	6	28	103	255	1687
Normanton	268	257	158	32	7	9	3	2	3	10	44	143	933
Richmond	121	105	65	20	15	14	9	3	7	16	29	72	475
Cairns	428	430	461	247	105	63	35	36	38	46	91	200	2167
Townsville	288	297	199	74	32	30	16	13	16	30	47	131	1169
Longreach	69	89	64	26	24	19	20	9	13	24	26	51	435
Mt Isa	96	90	59	17	19	14	6	4	7	20	22	57	414
Birdsville	27	30	18	9	11	11	10	7	6	11	11	15	162
Rockhampton	179	185	110	54	42	54	40	22	29	47	65	113	941

TABLE 2 : *Median monthly rainfall (mm)*

STATION	Jan	Feb	Mar	Apr	May	Jun	Jul	Aug	Sep	Oct	Nov	Dec
Wyndham	146	161	116	10	0	0	0	0	0	5	41	105
Halls Creek	132	126	66	6	0	0	0	0	0	8	38	77
Derby	154	135	90	7	*	1	0	0	0	0	5	51
Broome	115	118	71	5	4	4	*	0	0	0	1	33
Port Hedland	23	43	24	1	8	8	1	1	0	0	0	1
Carnarvon	1	2	2	4	24	41	31	12	3	2	0	0
Mundiwindi	27	28	24	7	9	8	3	2	0	1	5	17
Darwin	401	340	252	71	1	0	0	0	7	46	120	214
Yirrkala	191	216	230	178	63	19	6	2	1	0	11	100
Katherine	209	205	140	10	0	0	0	0	0	18	70	176
Newcastle Waters	108	94	45	1	0	0	0	0	0	8	26	56
Tennant Creek	74	67	19	2	0	0	0	0	0	10	20	39
Alice Springs	18	17	12	5	3	6	1	1	2	16	19	19
Weipa	417	375	321	95	6	2	*	0	1	16	98	256
Normanton	246	247	148	17	0	0	0	0	0	1	36	120
Richmond	91	80	40	4	3	4	0	0	0	6	17	55
Cairns	399	400	400	177	90	47	26	27	23	28	61	126
Townsville	238	242	179	36	15	13	3	3	4	13	26	82
Longreach	43	45	38	11	6	6	6	3	4	12	17	44
Mt Isa	66	76	33	1	1	*	0	0	0	9	19	49
Birdsville	8	8	3	*	1	3	1	1	2	3	4	5
Rockhampton	124	114	80	35	28	32	21	14	17	34	59	92

*= less than 0.4 mm

TABLE 3 : *Median rainfall for the wet season and water year (mm)*

STATION	6 Month Period	Wet Season Median	Water Year Median
Wyndham	Nov - Apr	634	675
Halls Creek	Nov - Apr	548	599
Derby	Dec - May	567	605
Broome	Dec - May	472	514
Port Hedland	Jan - Jun	249	287
Carnarvon	Mar - Aug	178	208
Mundiwindi	Dec - May	185	232
Darwin	Nov - Apr	1494	1586
Yirrkala	Dec - May	1094	1153
Katherine	Oct - Mar	916	966
Newcastle Waters	Nov - Apr	407	447
Tennant Creek	Nov - Apr	278	334
Alice Springs	Nov - Apr	175	248
Weipa	Nov - Apr	1638	1712
Normanton	Nov - Apr	872	894
Richmond	Nov - Apr	385	433
Cairns	Dec - May	1849	2142
Townsville	Nov - Apr	987	1110
Longreach	Nov - Apr	291	391
Mt Isa	Dec - May	317	368
Birdsville	Oct - Mar	86	134
Rockhampton	Nov - Apr	665	856

Rainfall

Tables 1 and 2 show the mean and median monthly rainfalls
for a selection of stations throughout northern Australia.
The median is that value not exceeded on half the total
occasions while the mean is the simple arithmetic average.
Median figures cannot be aggregated to obtain a seasonal
total.   Instead separate seasonal values for each water year
(1 September to 31 August) are calculated and then ranked in
increasing order of magnitude.   Table 3 shows the median
rainfall for both the wet season and the water year for
selected stations.

Figure 1 shows the distribution of median rainfall for the
calendar year.   Features to note are the minor maximum in the
Pilbara in Western Australia, the general south to north
rainfall gradient over the remainder of Western Australia and
the Northern Territory, and the southwest to northeast
gradient across Queensland.   The acknowledged wettest place in
Australia is on the northeast Queensland coast at Tully, which
has an annual median rainfall of 4204 millimetres, but close
by Bellinden Ker Top Station probably has an even wetter
regime.   Recordings have only been made at Bellinden Ker Top
Station since 1973, but already it has set new records for the
highest daily (1140 mm 4 January 1979), monthly (5387 mm
January 1979) and yearly (11251 mm 1979) rainfall totals in
Australia.

The adequate presentation of rainfall variability over an
extensive area such as northern Australia is difficult.

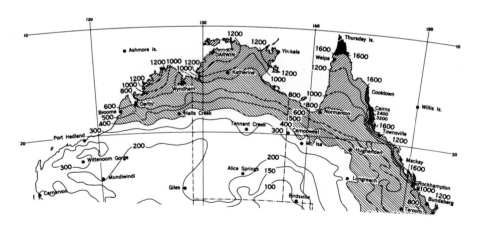

*FIGURE 1   Median annual (mm) for all years of record
to 1973 inclusive*

Numerous index values have been developed, but none seem to be able to clearly describe all the vagaries of rainfall. One useful presentation (Gibbs and Maher 1967) consists of tables of the decile values of rainfall in monthly, seasonal or yearly time spans. The deciles indicate the frequency of occurrence of rainfall below specified values. For example the decile 4 value indicates that 40 per cent of rainfalls are below the given value. For rainfall to be in decile range 3 the total for the period being considered would be between the value shown as Decile 2 and the value shown as Decile 3. Table 4 shows decile values for Darwin and Alice Springs. Rainfall in various decile ranges can be classified as:

Decile range 1   —   very much below 'average'
Decile range 2   —   much below 'average'
Decile range 3   —   below 'average'
Decile range 4   —   slightly 'below' average
Decile ranges 5 and 6 — 'average'
Decile range 7   —   slightly above 'average'
Decile range 8   —   above 'average'
Decile range 9   —   much above 'average'
Decile range 10   —   very much above 'average'

TABLE 4 : *Decile values of monthly, seasonal and yearly rainfall (mm)*

DARWIN

	Jan	Feb	Mar	Apr	May	Jun	Jul	Aug	Sep	Oct	Nov	Dec	Nov to Apr	Sep to Aug
Lowest	68	14	21	1	0	0	0	0	0	0	10	25	613	699
Decile 1	190	159	106	11	0	0	0	0	0	5	48	111	1109	1151
Decile 2	237	207	157	26	0	0	0	0	0	17	63	150	1218	1325
Decile 3	287	246	201	37	0	*	0	0	0	1	25	77	170	1325	1415
Decile 4	337	298	230	50	*	0	0	0	3	38	103	198	1403	1498
Decile 5	401	340	252	71	1	0	0	0	7	46	120	214	1494	1586
Decile 6	444	369	307	88	4	0	0	0	11	58	133	241	1568	1639
Decile 7	499	398	346	104	9	1	0	*	16	76	155	269	620	1725
Decile 8	537	446	383	164	18	2	*	1	24	93	186	322	1723	1837
Decile 9	598	546	471	220	46	9	3	4	41	126	229	411	1825	1975
Highest	906	815	1014	603	299	41	65	84	108	339	399	665	2197	2252
Mean	399	377	279	96	16	3	1	3	14	59	130	241	1478	1574
No of OBS	112	111	111	112	112	112	112	112	112	112	112	112	111	111

ALICE SPRINGS

	Jan	Feb	Mar	Apr	May	Jun	Jul	Aug	Sep	Oct	Nov	Dec	Nov to Apr	Sep to Aug
Lowest	0	0	0	0	0	0	0	0	0	0	0	0	27	31
Decile 1	1	0	0	0	0	0	0	0	0	0	1	0	78	131
Decile 2	3	*	1	0	0	0	0	0	0	2	4	4	100	166
Decile 3	7	3	3	0	0	1	0	0	0	5	9	8	121	187
Decile 4	11	9	7	*	1	2	0	*	*	8	13	13	140	218
Decile 5	18	17	12	5	3	6	1	1	2	16	19	19	175	248
Decile 6	32	23	23	13	10	11	4	3	3	20	25	24	195	285
Decile 7	51	50	39	17	22	17	10	7	7	24	30	40	213	305
Decile 8	76	87	55	28	34	24	19	17	17	35	40	48	262	327
Decile 9	122	137	70	53	54	44	35	33	32	49	57	100	298	440
Highest	281	236	227	117	109	101	108	158	90	116	139	288	752	923
Mean	43	41	30	16	17	14	11	11	9	20	25	35	189	272
No of OBS	108	107	107	107	107	107	108	108	108	108	108	108	107	107

*= less than 0.4 mm

Examples of decile values for Mt Isa and Port Hedland will be found in Lee and Neal (1981).

In meteorology a rainday is defined as a day when 0.1 or more millimetres of rain were recorded in the 24-hour period ending at 9 a.m. The distribution of the mean number of raindays per year for northern Australia is shown in Figure 2, while details for each month for selected stations are provided in Table 5.

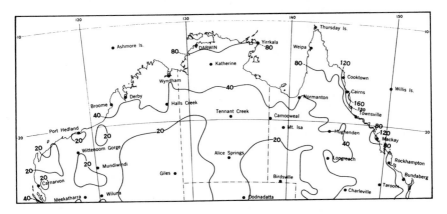

FIGURE 2   Average annual rainday frequency

TABLE 5 : Mean monthly raindays

	Jan	Feb	Mar	Apr	May	Jun	Jul	Aug	Sep	Oct	Nov	Dec	Year
Wyndham	13	11	9	3	1	1	0	0	0	2	6	10	56
Halls Creek	13	11	8	3	2	1	1	1	1	3	6	10	60
Derby	11	10	7	1	1	1	0	0	0	0	1	4	36
Broome	11	10	7	3	3	2	2	2	1	1	1	5	47
Port Hedland	5	7	4	2	3	3	2	1	1	1	1	1	31
Carnarvon	2	3	2	3	6	8	8	5	3	3	1	1	45
Mundiwindi	6	5	5	3	3	3	2	2	1	1	3	4	38
Darwin	20	20	19	9	2	0	0	1	2	6	12	17	108
Yirrkala	14	15	15	13	8	5	4	2	1	1	3	8	89
Katherine	14	13	10	2	1	0	0	0	1	3	7	12	63
Newcastle Waters	10	10	6	2	1	0	1	0	1	3	5	8	47
Tennant Creek	6	6	4	1	1	1	1	0	1	2	4	6	33
Alice Springs	4	4	3	2	2	2	2	1	2	3	4	4	33
Weipa	20	19	18	8	3	1	1	1	1	2	8	14	96
Normanton	14	14	10	2	1	1	1	0	0	1	4	9	57
Richmond	8	8	5	2	1	1	1	1	1	2	4	6	40
Cairns	17	17	18	15	12	10	9	8	7	7	9	12	141
Townsville	16	17	15	8	6	4	2	3	2	4	7	10	94
Longreach	7	6	5	3	3	2	3	2	2	4	5	5	47
Mt Isa	6	7	5	3	3	3	2	1	2	4	4	5	45
Birdsville	3	4	3	2	3	3	2	2	3	3	3	3	34
Rockhampton	12	11	11	6	5	6	5	4	4	6	7	9	86

Temperature

Figure 3 shows the distribution of average annual temperature while Figure 4 shows the month of occurrence of the highest mean maximum temperature. The relationship with the north-south movement of the sun is obvious.

In January the average maximum temperature exceeds 30°C over virtually all of northern Australia and exceeds 36°C over a large proportion of the area. Mean January maximum temperatures above 40°C are common over inland Western Australia. Marked gradients exist along coastal areas due mainly to the penetration of cooling sea breezes. In July (which is the coldest month throughout) there is a more general south to north gradient in maximum temperature, and

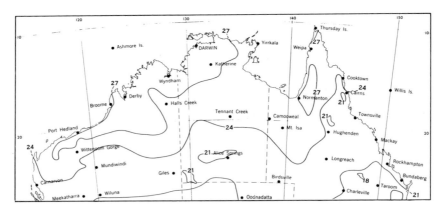

FIGURE 3   *Average annual temperature (°C)*

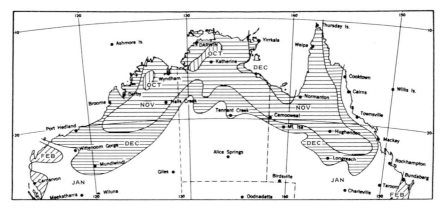

FIGURE 4   *Month of highest average maximum temperature*

overnight radiative cooling resulting from clear skies over inland parts can produce some very low temperatures.

Table 6 shows the 86 percentile, mean and 14 percentile values of maximum and minimum temperature for selected stations. The percentile values are used to show the variation or spread of temperature throughout the month, and we can say for example that on the average, one day per week the maximum temperature for a particular place remains lower than the 14 percentile value, and one day per week the maximum temperature exceeds the 86 percentile value. The same interpretation applies to the minimum temperature percentiles. A more detailed selection of percentile values appears in Lee and Neal (1981).

TABLE 6   :   *Maximum and minimum temperature data for selected stations*

	Jan	Feb	Mar	Apr	May	Jun	Jul	Aug	Sep	Oct	Nov	Dec
BROOME												
Maximum temperature												
86 percentile	35.4	34.6	36.1	36.7	34.4	31.8	31.3	32.8	35.2	38.0	36.9	36.1
Mean	33.3	32.9	33.9	34.4	31.3	29.2	28.5	30.0	31.8	32.9	33.6	33.9
14 percentile	31.6	31.1	31.9	32.2	28.1	26.6	25.8	26.9	28.4	29.1	31.0	32.1
Minimum temperature												
85 percentile	28.3	28.3	28.1	26.3	22.2	19.5	17.6	18.5	21.7	24.6	26.7	28.2
Mean	26.2	26.0	25.5	22.8	18.5	15.5	13.6	14.8	18.3	22.1	25.0	26.6
14 percentile	23.9	·23.4	23.2	19.4	14.9	11.8	9.4	11.4	15.0	19.3	23.3	24.6
DARWIN												
Maximum temperature												
86 percentile	33.4	33.2	33.4	33.9	33.7	32.3	32.0	32.8	34.3	34.6	34.6	34.2
Mean	31.8	31.5	31.9	32.7	31.9	30.6	30.3	31.2	32.5	33.3	33.4	32.8
14 percentile	29.7	29.6	30.2	31.5	30.0	28.9	28.7	29.5	31.1	32.2	32.3	31.2
Minimum temperature												
86 percentile	27.1	26.7	25.7	25.1	23.9	22.6	21.6	22.4	24.6	26.3	27.2	27.3
mean	24.8	24.6	24.3	23.9	21.8	19.9	18.9	20.3	22.9	24.9	25.3	25.1
14 percentile	22.8	22.9	22.8	22.4	19.1	17.3	16.3	17.9	21.2	23.4	23.3	23.0
ALICE SPRINGS												
Maximum temperature												
86 percentile	40.6	39.4	36.6	32.9	27.4	24.9	24.1	27.6	31.9	36.6	38.8	39.5
Mean	36.6	35.5	32.8	28.5	22.7	20.2	19.3	21.8	26.4	31.2	33.8	35.1
14 percentile	32.6	31.2	28.6	23.9	18.1	15.9	15.3	16.6	21.0	25.6	28.7	30.6
Minimum temperature												
86 percentile	26.7	25.7	22.2	18.4	13.9	11.6	9.3	12.0	15.5	20.6	23.3	24.7
Mean	22.2	21.2	18.2	13.6	8.9	6.3	4.5	6.6	10.4	15.4	18.2	20.2
14 percentile	17.3	16.6	13.9	8.8	4.4	1.3	0.2	1.8	5.5	10.0	13.4	15.5
TOWNSVILLE												
Maximum temperature												
85 percentile	32.9	32.2	31.6	30.5	28.9	27.0	26.6	27.7	29.1	30.7	32.0	32.6
Mean	31.3	30.8	30.1	29.4	27.3	25.5	24.9	26.0	27.5	29.3	30.7	31.2
14 percentile	29.5	29.3	28.7	28.3	25.7	23.9	23.1	24.3	26.1	27.9	29.3	29.9
Minimum temperature												
86 percentile	25.7	25.1	24.1	22.1	20.5	18.4	16.8	18.8	20.4	22.9	24.9	25.6
Mean	23.8	23.7	32.4	20.0	17.3	14.7	13.1	15.0	17.2	20.5	22.7	23.6
14 percentile	21.8	22.4	20.6	17.9	13.9	11.1	8.8	11.3	14.3	17.9	20.6	21.7

In Figures 5 and 6 the variation of temperatures during
the day by months is presented for Darwin and Alice Springs.
As would be expected from its proximity to the sea, Darwin
exhibits much less variability and smaller range than does
Alice Springs with its extremely continental desert type
environment.

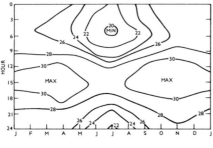

*FIGURE 5   Darwin:
Mean temperature (°C)*

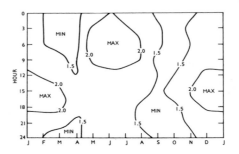

*Darwin:
Standard deviation of
temperature (°C)*

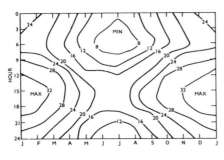

*FIGURE 6   Alice Springs:
Mean temperature (°C)*

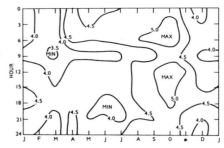

*Alice Springs:
Standard deviation of
temperature (°C)*

Relative Humidity
     Mean relative humidities for 9 a.m. and 3 p.m. are
presented in Table 7 for a selection of stations. These data
must be considered in conjunction with corresponding
temperature and pressure values before the significance of
relative humidity to human comfort can be properly assessed.

*TABLE 7   :   Mean relative humidity (per cent)
at 9 a.m. and 3 p.m.*

9 AM

STATION	Jan	Feb	Mar	Apr	May	Jun	Jul	Aug	Sep	Oct	Nov	Dec	Year
Wyndham	64	66	58	44	36	36	33	34	36	41	45	54	46
Halls Creek	49	51	47	36	38	38	34	29	24	27	32	41	37
Derby	70	70	60	48	32	36	41	43	38	36	51	52	48
Broome	68	73	67	54	47	48	46	43	46	51	56	62	55
Port Hedland	52	54	50	39	39	47	42	37	31	29	33	40	41
Carnarvon	62	59	58	59	59	73	69	63	55	53	56	58	60
Mundiwindi	29	35	31	35	41	50	48	38	28	20	21	23	33
Darwin	79	81	81	75	66	62	60	65	68	68	70	74	71
Yirrkala	78	81	80	72	70	70	58	70	68	68	68	72	72
Katherine	76	80	76	65	56	54	50	49	50	56	60	70	62
Newcastle Waters	57	60	57	51	45	48	41	34	36	35	37	46	46
Tennant Creek	46	47	42	36	41	44	39	34	27	31	30	38	38
Alice Springs	31	32	36	41	55	63	60	46	33	24	24	24	39
Weipa	83	86	85	79	79	79	77	72	67	65	68	75	76
Normanton	71	74	68	57	49	51	44	40	43	45	48	60	54
Richmond	55	62	57	51	55	57	50	41	36	33	37	43	48
Cairns	71	75	78	76	75	74	72	70	65	64	64	66	71
Townsville	68	74	71	66	65	65	63	62	57	58	59	62	64
Longreach	43	49	48	43	50	54	50	44	36	35	29	39	43
Mt Isa	48	51	48	41	46	51	44	38	31	27	28	37	41
Birdsville	28	31	33	40	54	62	59	47	32	26	21	24	38
Rockhampton	69	73	71	69	70	72	70	68	63	58	60	62	67

3 PM

STATION	Jan	Feb	Mar	Apr	May	Jun	Jul	Aug	Sep	Oct	Nov	Dec	Year
Wyndham	49	50	44	34	32	29	27	28	32	37	39	43	37
Halls Creek	31	32	30	32	27	24	22	19	16	17	20	24	24
Derby	53	45	42	32	25	25	25	26	24	25	35	34	33
Broome	62	65	57	42	37	35	32	31	40	51	55	59	47
Port Hedland	50	49	44	35	35	36	31	30	29	31	37	41	37
Carnarvon	61	59	59	57	53	55	54	53	53	54	57	61	56
Mundiwindi	18	20	19	22	27	31	28	23	16	12	12	14	20
Darwin	68	69	65	51	42	39	35	40	45	50	55	63	52
Yirrkala	72	77	74	68	68	66	65	64	62	63	63	68	68
Katherine	52	54	49	37	34	31	27	24	23	27	32	43	36
Newcastle Waters	33	36	33	28	38	27	24	20	19	20	21	26	26
Tennant Creek	25	27	24	23	26	27	24	21	17	29	18	23	23
Alice Springs	19	19	21	23	31	34	30	25	19	15	15	16	22
Weipa	73	77	73	60	56	53	48	45	42	43	49	61	57
Normanton	53	55	48	37	32	32	27	25	27	31	34	42	37
Richmond	31	37	33	29	32	31	27	22	20	19	22	26	27
Cairns	62	65	65	63	62	59	56	54	52	53	57	59	59
Townsville	62	66	63	57	54	52	47	51	51	52	56	58	56
Longreach	24	29	29	25	31	32	28	25	20	18	16	21	25
Mt Isa	30	30	31	26	29	29	25	22	19	18	18	24	25
Birdsville	17	18	19	23	29	32	28	23	16	14	12	15	21
Rockhampton	52	54	51	47	44	43	39	37	36	38	43	47	44

Pressure

With a few exceptions, (e.g. tropical cyclones) weather
systems in northern Australia are not well described using the
customary surface 'weather chart' with its lines of equal
pressure (isobars). Wind flow analyses (called 'streamlines')
give a better definition of most systems. However for some
special applications (mainly in aviation) the rather large

diurnal variation of pressure is important. Both inter-month
and diurnal pressure variation at Darwin and Alice Springs are
shown in Figure 7. Other locations throughout the north have
similar variations.

Major deviations from normal pressures are most common
when locations come under the influence of tropical cyclones
when the pressure at mean sea level can fall to well below 950
millibars (normal MSL pressure is 1013.25 mb) for short
periods in the vicinity of the centre. At the other end of
the scale high MSL pressures (about 1040 mb) can occur in the
southern parts of northern Australia when an intense anticyclone becomes established over the Australian continent in
winter.

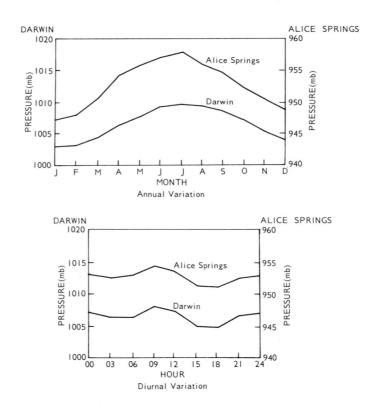

FIGURE 7   Inter-month and diurnal pressure variation
            (station level pressure)

General Wind Circulation Patterns

Mean low level wind circulation patterns for January and June are given in Figures 8 and 9.

In January the dominant feature is the monsoonal shear line (shown as a dot-dash line in Figure 8) which lies roughly along latitude 12°S. The northwest monsoon prevails to the north of this shear line bringing moist air and rainy conditions to the far north of Australia. The heaviest rains frequently occur in the vicinity of the monsoon shear line where the air streams coverage. Monsoonal rain disturbances and tropical depressions develop in the line, although tropical cyclones form only over the ocean. South of the shear line the prevailing winds have an easterly trajectory and are relatively moist as they strike the east coast. Occasional disturbances in the easterlies (or upper air disturbances resulting from interactions with higher latitude troughs) can bring heavy rain to the east coast, particularly north from Townsville where the mountains are quite close to the sea. The easterly flow may dry out considerably west of the Dividing Range, but periodically the low level moisture penetrates into the interior from the northeast. Aided by fierce surface heating, sporadic shower and thunderstorm activity results, which is enhanced from time to time by the passage of upper air troughs associated with higher latitude weather systems. The cyclonic circulation south of the Kimberlies is a shallow 'heat low', not usually an active monsoonal weather disturbance.

In June, (Figure 9) the pattern is quite different. The dry southeasterly trade winds, driven by the anticyclones over the south of the continent, now prevail over all of northern Australia. By and large the air is of dry southern origin but occasionally in the far north, trajectories are from the east and a few showers develop over the far north and northeast coasts. Winter rains can occur from upper air disturbances and cold frontal passages but usually these are confined to areas south from about latitude 20°S.

III. PHENOMENA

Thunderstorms and Hail

Figure 10 shows the average frequency of thunderstorms per year. Nearly all thunderstorms occur during the period from October through to April. The appearance of storms in October usually heralds the build-up to the wet season, while in April the occurrence of storms without prolonged rainy spells indicates that the dry season is approaching.

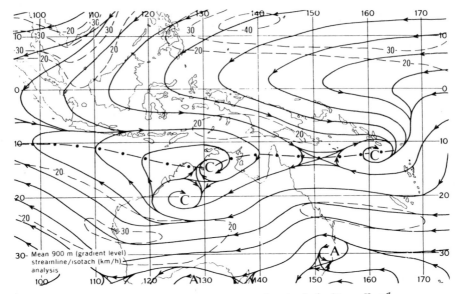

FIGURE 8   *Mean January low level wind (after Neal,*
             Holland et al. 1978)

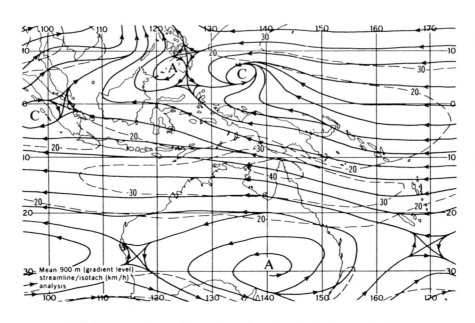

FIGURE 9   *Mean June low level wind (op. cit.)*

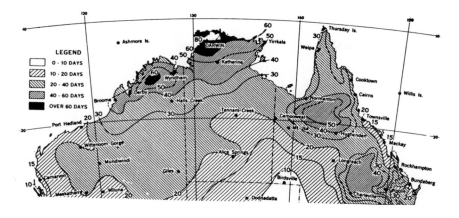

FIGURE 10   Average annual thunder days

The thunder and lightning in northern Australian
thunderstorms can be spectacular at times. Rainfall is often
intense although in the tropical interior the cloud base can
be very high with considerable thunder and lightning but
little or no rain. These 'dry storms' are a frequent cause of
bushfires. The onset of most storms is associated with gusty
winds although typically these winds are not strong enough to
do significant damage. Sometimes, however, very severe wind
squalls occur. The anemograph recording of a gust during a
severe thunderstorm squall at Tennant Creek is given in Figure
11.

The incidence of hail in association with thunderstorms
decreases with latitude. Hail has been reported as far north
as Katherine and Thursday Island (just off Cape York) but in
general hail occurrence is rare in the north. In the Darling
Downs and Central Highlands districts of Queensland hail is
often reported, especially in spring and early summer but
further north the frequency is assessed at about once in ten
years.

Frosts

In northern Australia the most frost prone areas are in
the south of the Northern Territory and over the highlands of
Queensland as far north as the Atherton Tableland. In the
southern parts of northern Australia frost can occur as early
as April and has been recorded as late as October. The length
of the frost period (i.e. the number of days between the first
and last recording of an air temperature of 2°C or less) is
shown in Figure 12.

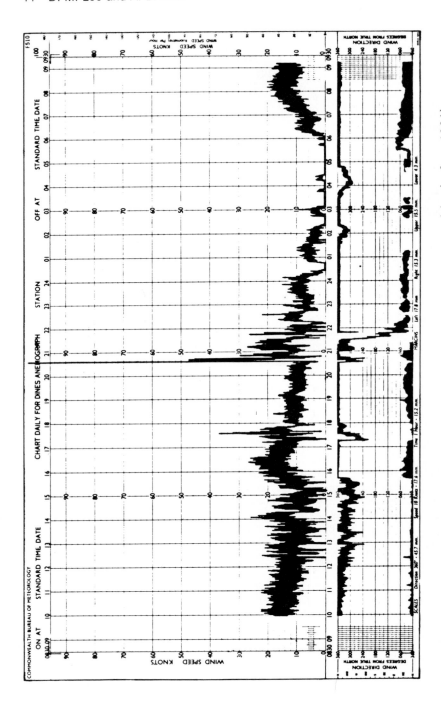

FIGURE 11   Tennant Creek anemogram of 12 December 1961 (after Whittingham 1964)

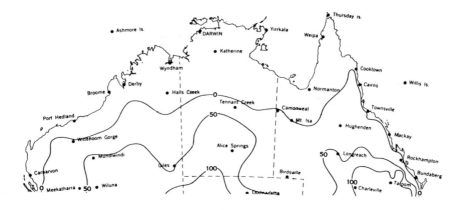

FIGURE 12   *Frost period-mean annual length (days per year)*

Sunshine and Radiation

Northern Australia receives relatively large amounts of sunshine although seasonal cloud formations have a notable effect on its spatial and temperal distribution. Cloud cover reduces both incoming and outgoing radiation and thus affects sunshine, air temperature and other climate elements at the earth's surface.

Average daily sunshine and global radiation values are shown in Figures 13 and 14. A high correlation exists between daily global radiation and daily hours of sunshine, but sunshine is more dependent on variations in cloud cover than

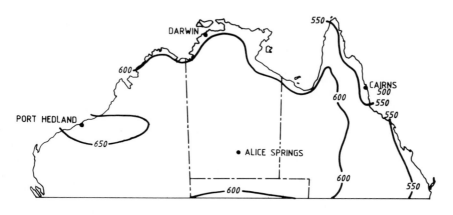

FIGURE 13   *Annual average daily hours of sunshine*

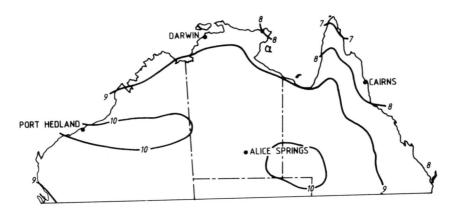

FIGURE 14   *Average daily global radiation* $(mWhs/cm^2)$

is global radiation, since the latter includes diffuse
radiaton from the sky as well as direct radiation from the
sun. In Darwin, hours of sunshine during the dry month of
July approach twice that of the cloudy month of January, but
global radiation figures for the two months are comparable.

Cloud
      Particularly strong seasonal variability of cloud cover
exists over northern Australia where skies are clouded during
the wet season and often cloudless in the dry season. Cloud
coverage is greater near the coasts and on the windward slopes
of the eastern uplands, and is far less over the dry interior.
The average number of clear and cloudy days per month are
presented in Tables 8 and 9.

Evaporation
      Potential evaporation is determined by measuring the
amount of water evaporated from a free water surface exposed
in a pan; actual evaporation is of course highly dependent on
the availability of moisture. Evaporation depends on a number
of climatic elements, mainly temperature, humidity and wind.
      The average annual Class A pan evaporation is shown in
Figure 15. Maximum evaporation occurs in the dry interior of
Western Australia and Northern Territory.

TABLE 8 : *Mean number of clear days*

STATION	Jan	Feb	Mar	Apr	May	Jun	Jul	Aug	Sep	Oct	Nov	Dec
Wyndham	3	2	4	13	17	23	26	25	18	15	9	4
Halls Creek	2	3	6	11	13	16	19	19	17	12	7	3
Derby	0	2	1	8	9	13	15	12	13	13	7	3
Broome	3	3	7	14	15	20	21	22	21	18	14	8
Port Hedland	8	6	11	14	15	18	21	23	23	23	20	15
Carnarvon	16	13	17	12	14	14	16	19	20	19	18	19
Mundiwindi	8	8	11	12	14	16	20	22	22	20	15	11
Darwin	0	0	1	4	8	13	15	15	11	6	2	1
Yirrkala	0	1	0	2	2	3	5	7	7	7	5	2
Katherine	1	1	2	9	13	19	21	23	16	9	5	2
Newcastle Waters	3	3	6	13	17	22	24	24	19	14	8	5
Tennant Creek	6	3	7	15	15	20	22	20	17	16	10	8
Alice Springs	13	12	16	16	15	17	21	22	21	17	14	13
Weipa	1	1	1	2	5	8	7	7	8	8	5	2
Normanton	2	1	4	10	14	20	22	23	19	16	7	4
Richmond	5	4	9	15	15	20	22	23	20	17	12	8
Cairns	2	2	3	4	5	7	8	9	8	7	5	3
Townsville	3	2	5	7	8	12	14	15	12	8	6	4
Longreach	8	5	11	14	13	16	21	30	19	17	14	11
Mt Isa	6	2	7	14	13	17	22	21	18	16	11	8
Birdsville	16	15	19	19	17	18	22	22	20	20	16	16
Rockhampton	2	1	3	6	9	12	15	15	11	8	5	4

TABLE 9 : *Mean number of cloudy days*

STATION	Jan	Feb	Mar	Apr	May	Jun	Jul	Aug	Sep	Oct	Nov	Dec
Wyndham	14	13	12	4	3	1	1	1	1	2	4	9
Halls Creek	12	11	8	4	5	2	2	2	1	2	4	8
Derby	15	11	14	3	6	2	2	2	1	1	3	7
Broome	13	13	8	4	6	3	2	1	1	1	2	7
Port Hedland	5	5	4	2	4	3	2	1	1	1	1	2
Carnarvon	3	4	3	6	6	5	5	3	2	2	2	2
Mundiwindi	5	5	3	4	5	5	3	2	2	2	3	3
Darwin	21	19	17	10	7	3	3	2	3	4	9	17
Yirrkala	20	20	20	12	9	7	6	5	4	4	8	15
Katherine	18	17	15	7	5	3	2	1	3	5	9	15
Newcastle Waters	11	12	8	3	3	1	1	0	2	2	4	7
Tennant Creek	11	12	10	3	5	3	1	1	2	3	5	6
Alice Springs	5	5	4	4	7	5	4	3	2	4	4	5
Weipa	20	17	17	12	5	5	5	4	5	4	6	15
Normanton	14	14	11	4	3	2	1	0	1	1	4	9
Richmond	8	7	6	2	4	2	1	1	1	2	2	5
Cairns	14	15	15	11	11	9	7	6	5	4	6	10
Townsville	13	14	12	8	8	6	5	4	3	4	4	7
Longreach	8	9	5	2	5	4	3	3	2	3	3	6
Mt Isa	9	12	7	3	5	2	2	2	2	3	4	6
Birdsville	4	3	3	2	4	3	3	3	2	3	4	4
Rockhampton	12	13	11	6	7	5	6	4	3	5	6	8

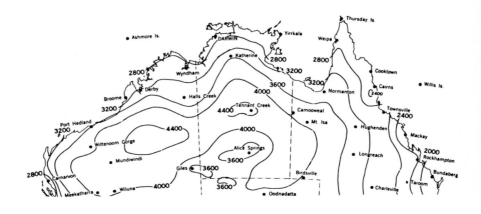

FIGURE 15   Average annual potential evaporation (mm)

Over about 75 per cent of the continent, comprising most inland areas, rainfall does not exceed evaporation loss from a free water surface in any month of the year. However in central and northwest parts of Australia the annual evaporation exceeds ten times the rainfall. Data for some selected stations showing mean daily evaporation by month are shown in Table 10.

TABLE 10 : Mean daily evaporation (mm) from Class A pan

STATION	Jan	Feb	Mar	Apr	May	Jun	Jul	Aug	Sep	Oct	Nov	Dec
Halls Creek	9.6	8.3	7.9	8.2	6.4	5.8	6.2	7.9	10.2	-12.9	11.9	10.9
Derby	8.9	8.1	7.4	8.3	8.1	7.6	7.9	8.6	10.7	11.7	12.5	11.4
Broome	8.5	7.5	7.1	7.6	6.5	6.0	6.5	7.4	8.8	9.4	9.8	9.5
Port Hedland	10.8	10.1	9.2	8.8	7.2	6.4	6.8	7.3	9.3	11.1	11.8	11.7
Carnarvon	9.9	9.7	8.4	6.2	4.9	3.6	3.5	4.7	6.5	7.9	8.9	9.8
Darwin	6.2	5.9	5.6	6.6	6.7	6.6	6.8	7.2	7.9	8.3	7.8	6.8
Katherine	6.1	6.7	5.5	6.5	5.7	5.2	5.7	6.3	8.4	8.3	8.9	5.9
Newcastle Waters	8.4	7.4	6.5	6.5	5.0	4.6	4.9	6.6	8.4	9.4	9.7	9.2
Tennant Creek	13.3	11.9	10.9	9.7	9.0	7.7	7.4	9.6	12.1	13.9	14.1	14.5
Alice Springs	12.8	12.1	9.7	7.7	4.7	3.7	3.9	5.1	7.5	9.5	11.2	12.4
Weipa	4.3	4.1	4.5	5.3	4.8	4.8	5.0	5.8	6.9	7.5	7.0	6.1
Normanton	7.7	6.6	7.0	7.7	8.2	7.5	8.0	9.0	10.4	11.1	10.4	9.0
Richmond	6.9	6.2	6.6	5.6	5.0	4.4	5.0	5.6	7.3	8.6	9.0	8.4
Cairns	7.2	6.0	5.7	5.4	4.9	4.8	5.4	5.9	7.1	8.1	8.3	7.9
Townsville	8.5	7.1	6.6	6.6	5.9	5.3	5.7	6.5	8.3	9.4	9.8	9.4
Longreach	11.4	9.4	8.6	8.0	5.5	4.5	4.9	6.6	8.6	11.4	12.9	12.4
Mt Isa	8.7	6.9	8.3	7.8	5.8	5.3	5.4	6.5	8.3	9.9	11.4	10.7
Birdsville	16.6	13.2	11.6	8.7	5.5	4.2	4.4	5.9	8.3	11.4	13.9	16.5
Rockhampton	6.9	6.1	5.8	5.2	4.1	3.6	3.5	4.3	5.6	6.6	7.4	7.4

IV.   MAJOR NATURAL HAZARDS

A comprehensive review of natural hazards (including earthquakes, biological and health hazards, and man-made hazards) is given by Oliver (1980), while a discussion of atmospheric factors in natural disasters is given by Gentilli (1976).

Tropical Cyclones
Tropical cyclones are one of the few meteorological hazards in Australia that are exclusive to the northern half of the continent. Cyclones can be both malevolent and benevolent: near the coast they bring destructive winds, torrential rain, flooding, and sometimes storm surges (inundation by the sea) but their structure is modified by passage over land, so that further inland thay can become very efficient rain depressions that provide essential rainfall for primary industry. According to Irish and Devin (1978) nearly 500 000 dwellings are exposed to the 'malevolent' effects of cyclones while the total mean annual damage bill is about 17 million dollars. (These figures incorporate areas south of 26°S, such as Brisbane and therefore are not strictly representative of 'Northern Australia').

The Australian tropical cyclone season is nominally from November to April but there is considerable year to year variation in the starting and finishing dates. Cyclones have been recorded as early as October and as late as June, but more than 90 per cent occur between 1 December and 30 April with a little over 70 per cent occurring in the months of January, February and March. A map on the average number of tropical cyclone occurrences in northern Australia is given in Figure 16.

Cyclones vary widely in size, intensity, speed and direction of movement, and impact. All cyclones are dangerous but few are as small and intense as Tracy which devastated Darwin in December 1974 (Wilkie and Neal 1976, Mottram 1977). On the other hand, relatively weak cyclones (so far as the peak wind speeds are concerned) can have a big impact. In 1974 the weak cyclone Wanda contributed to the worst flooding in the Brisbane area this century (Shields and Neal 1974).

Floods
Floods occur in every northern Australian river but their spatial and temporal frequency depends on the type of rain producing event and on the nature of the river drainage basin. Irish and Devin (1978) estimated that in Queensland alone about 75 000 dwellings were prone to flooding, with a total

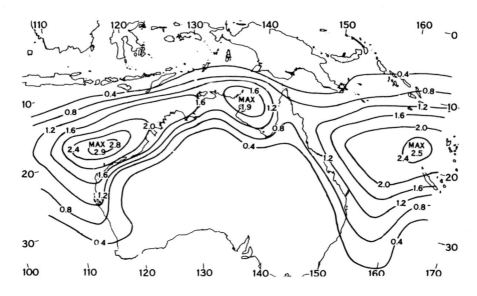

FIGURE 16   Average number of tropical cyclones per year
            (after Neal, Holland et al. 1978)

mean annual flood damage bill of just over seven million
dollars. (This value does not appear to include rural
losses.) No estimates are available for other parts of
northern Australia.

The highest incidence of flooding occurs across the far
north and in the flood plains of the east coast streams.
Cyclones and the monsoon depressions of the wet season cause
flooding in these areas almost every year, 90 per cent of
occurrences being in the period January to March. In the
northwest most flooding also occurs in the normal wet season
but according to Volprecht (1970) heavy rain and flooding can
occur in the Fortesque and De Grey areas of Western Australia
at almost any time of the year.

According to Gordon (1971) coastal streams from Cooktown
to about Townsville flood about six or seven years in every
ten. On the Atherton Tablelands and adjacent areas to the
west of the Dividing Range flooding is less frequent - three
to five years in every ten. Gibbs (1975) describes the
extensive flooding in 1973-74 when major floods (some to
record levels) occurred in almost every river system in
northern and eastern Australia and Lake Eyre was filled for
the first time in many years.

Drought

Drought is difficult to define. The most accepted definition at this time is that drought exists when demand for water exceeds supply, or there is a lack of sufficient water to meet requirements. Although many factors are involved in considering water availability, the single most useful index is rainfall and most Australian droughts tend to be defined by a failure of expected rainfall.

It is necessary to distinguish between aridity and drought. Stated simply, aridity implies a high probability of rainfall below an arbitrary but low threshold; drought, on the other hand, implies a low probability of rainfall below a somewhat more arbitrary threshold (Coughlan and Lee 1978). Thus in arid or semi-arid areas, in the absence of an alternative water supply, the rainfall is insufficient to sustain any, or at best, very limited levels of agriculture and pastoral activity.

The probability of drought in northern Australia is shown in Figure 17. This can be interpreted as the return period expectation of minor, moderate and severe rainfall deficiencies during the wet season.

Most places in northern Australia are susceptible to drought, but the growing factor is the success or failure of the wet season. Care must be taken all the time to distinguish between the normal aridity and real drought resulting from the failure of 'normal' or expected seasonal rainfall.

Bushfires

According to Luke and McArthur (1978) bushfires are rare in arid areas of northern Australia because the fuel is usually too sparse to sustain combustion. On the other hand in the tropical rain forests, the vegetation is usually too moist to burn. Nevertheless a serious seasonal fire problem can exist in the savannah forests and grasslands of northern Australia. The pattern of seasonal fire occurrence for northern Australia is shown in Figure 18.

The 1974-75 fire season in northern Australia resulted in fires on a scale not experienced for at least fifty years (Luke and McArthur 1978). Approximately seventy million hectares of northern Australia was burnt out, including thirty three per cent of the total land area of the Northern Territory. One fire on the Barkly Tableland eventually covered 2.4 million hectares. Some fires were caused by man

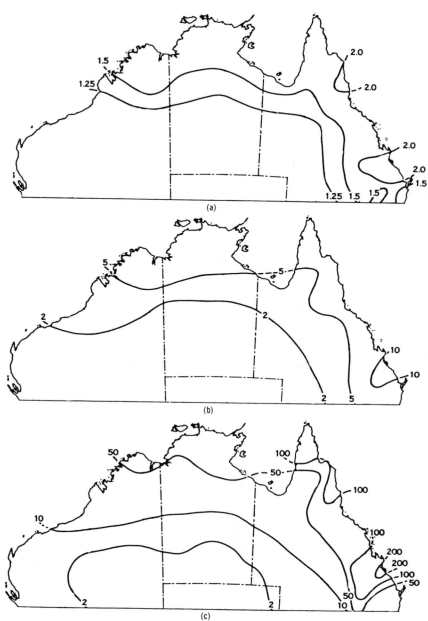

FIGURE 17   Return period expectation for drought
              (a) 1,2 or 3 failures of 2 month periods
              of wettest 6 months; (b) 2 or 3 failures;
              (c) 3 failures.

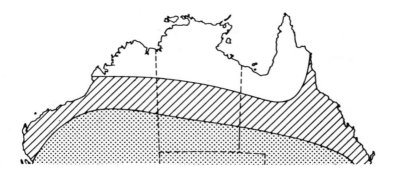

LEGEND

MONTHS OF THE FIRE SEASON

FIRE SEASON ZONE	Jul	Aug	Sep	Oct	Nov	Dec	Jan	Feb	Mar	Apr	May	Jun
Winter and Spring ☐	N	N	N	N	O	O	O	–	–	–	O	O
Spring ▨	–	O	N	N	N	N	O	O	–	–	–	–
Spring and Summer ▒	–	–	O	N	N	N	N	O	O	–	–	–
Summer ▧	–	–	–	–	O	N	N	N	O	O	–	–

N Months when serious fires are normally likely to occur
O Months when occasional serious fires are likely to occur
– Months when fire occurrence is unlikely

FIGURE 18   Pattern of seasonal fire occurrences in
             Australia (after Luke and McArthur 1978)

but the majority were ignited by lightning strikes. A map of
the area burnt during 1974-75 is given in Figure 19. A review
of other severe fire seasons in Australia is given by Cheney
(1976).

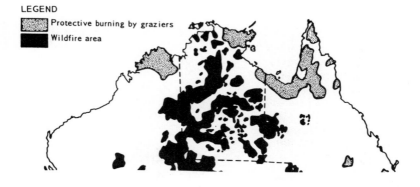

LEGEND
▒ Protective burning by graziers
■ Wildfire area

FIGURE 19   Areas burnt during the 1974-75 fire season
             in Australia (op.cit.)

ACKNOWLEDGEMENTS

The authors and editor wish to thank The Trustees and Executive of the Sir Robert Menzies Foundation for allowing us to reproduce this chapter from Transactions of the Menzies Foundation, Volume 2 (1981). Some minor changes have been made by the editor in order to accommodate the camera-ready format.

REFERENCES

Bureau of Meteorology. 1975. *Climatic Averages Australia,* Metric Edition, AGPS Canberra.

Bureau of Meteorology. 1977. *Rainfall Statistics Australia,* Metric Edition, AGPS Canberra.

Bureau of Meteorology. 1978. *Climate of Australia,* 1977-78 edition, AGPS Canberra, p.62.

Cheney, N.P. 1976. Bushfire Disasters in Australia 1945-76, in *Natural Hazards in Australia,* (ed.) R.L. Heathcote and B.G. Thom, Aust. Acad. of Sci. Canberra, pp.72-93.

Coughlan, M.J. and Lee, D.M. 1978. The Assessment of Drought Risk in Northern Australia, *Proc. Second N.A.R.U. Seminar on Natural Hazards Management in North Australia,* (ed.) G. Pickup, Australian National University, Canberra.

Douglas, I. 1976. Flooding in Australia: a Review, in *Natural Hazards in Australia,* (eds.) R.L. Heathcote and B.G. Thom, Aust. Acad. of Sci. Canberra, pp.143-163.

Gentilli, J. 1971. Climate of Australia and New Zealand, *World Survey of Climatology, Vol. 13,* Elsevier Pub. Co, pp.35-21.

Gentilli, J. 1976. Atmospheric Factors in Disaster: An Appraisal of their role, in *Natural Hazards in Australia,* (eds.) R.L. Heathcote and B.G. Thom, Aust. Acad. of Sci. Canberra, pp.34-50.

Gibbs, W.J. and Maher, J.V. 1976. Rainfall Deciles as Drought Indicators, *Bulletin 48,* Bur. Met., Australia.

Gibbs, W.J. 1975. Weather patterns in 1973-74 - were they unusual in regard to floods? *Symp. on Floods,* Melbourne, August, 1975.

Gordon, B. 1971. *Climatic Survey, Northern, Region 16 Queensland,* Bur. Met., Australia, pp.53-56.

Hunt, H.A., Taylor, G. and Quale, E.T. 1913. *The Climate and Weather of Australia,* Bur. Met., Australia, p.93.

Irish, J.L. and Devin, L.   1978.   Annual Flood Damage to
     Dwellings in Australia in Relation to other Natural
     Hazard Risks, *Proc. Second N.A.R.U. Seminar on Natural
     Hazards Management in North Australia*, (ed.) G. Pickup,
     Australian National University, Canberra.
Lee, D.M. and Neal, A.B.   1981.   The Climate of Northern
     Australia, *Transactions of the Menzies Foundation, Vol.
     2*, pp.3-29.
Linacre, E. and Hobbs, J.   1977.   *The Australian Climatic
     Environment*, John Wiley & Sons, Brisbane, p.354.
Luke, R.H. and McArthur, A.G.   1978.   *Bushfires in Australia*,
     AGPS Canberra, p.359.
Mottram, J.P.   1977.   *Report on Cyclone Tracy - December 1974*,
     AGPS Canberra, P.82.
Neil, A.B., Holland, G.J. *et al*.   1978.   *Australian Tropical
     Cyclone Forecasting Manual*, Bur. Met., Australia, p.274.
Oliver, J.   1980.   The Disaster Potential, in *Response to
     Disaster*, (ed.) J. Oliver, Centre for Disaster Studies,
     James Cook University of North Queensland, pp.3-27.
Ramage, C.S.   1971.   Monsoon Meteorology, *International
     Geophysics Series Vol. 15*, Academic Press, London, p.296.
Shields, A.J. and Neil, A.B.   1974.   *Brisbane Floods January
     1974*, AGPS Canberra, p.63.
Taylor, G.   1932.   Climatology of Australia, in *Handbuch der
     Klimatologie Band IV Feil S., Australien und Neuseeland*,
     p.107.
Treawartha, G.T.   1962.   *The Earth's Problem Climates*,
     University of Wisconsin Press, Methuen & Co., London,
     p.334.
Volprecht, R.   1972.   *Climatic Survey, Northwest, Region 6
     Western Australia*, Bur. Met., Australia, p.82.
Wittingham, H.E.   1964.   Extreme Wind Gusts in Australia,
     Bulletin 46, Bur. Met., Australia, p.133.
Wilkie, W.R. and Neal, A.B.   1976.   Meteorological Features of
     Cyclone Tracy, in *Natural Hazards in Australia*, (eds.)
     R.L. Heathcote and B.G. Thom, Aust. Acad. of Sci.
     Canberra, pp.373-487.

4

INTERACTIVE NATURAL SUB-SYSTEMS IN NORTHERN AUSTRALIA

J. OLIVER

Department of Geography
James Cook University of North Queensland
Townsville, Queensland

I.   INTRODUCTION

Whether the concern is with the natural environment or the socio-economic circumstances, or their combined totality, the principles of interrelationship and of spatial diversity are implicit in geographical study. Northern Australia must initially be identified in arbitrary terms but, once designated as the area for study, it demonstrates, like any other selected part of the earth's surface, a set of characteristics and qualities which permit it to be distinguished in varying degrees from adjacent areas. Furthermore, within the accepted confines of the specified area, landscapes, environments and human activity patterns can be differentiated on a variety of scales.

Interaction between the features and processes occurring in different areas, and between the different scales of subdivision, is of fundamental significance. This interaction is both complex and dynamic. Whilst it may be helpful and simplifying to dissect initially the components of a system under specific headings in order to identify their particular characteristics, this process must not be allowed to obscure the fact that any system involves the integration of a number of individual elements. The elements interact with each other so that the whole system can only be fully comprehended in the context of the operation and mutual association of them all. In a major system it is possible that, at different levels of interaction, a number of sub-systems will exist.

NORTHERN AUSTRALIA
ISBN 0 12 545080 X

These may have, to a varying extent, independent existences
of their own as well as forming a necessary and integral part
of the larger system.

In this chapter the focus is upon the natural sub-systems
of the major area, northern Australia. The subject matter
encompasses what is conventionally described as physical
geography. From the viewpoint of human activity and resources
the emphasis is directed to the understanding of the nature
and processes of the different ecosystems that are involved.
In these interactive systems man is a functional element,
as well being influenced by and responding to character and
behaviour of the ecological environment in which he lives.

Some implications for human society and economic activity
of the conditions comprising the natural environment are
discussed in greater detail elsewhere in this book (chapters
9, 16 and 17 for instance). Nevertheless it is considered
desirable that, in discussing the subject matter of this
chapter, a close and a two-way association be shown to exist
between the physical geography of the arena and man, the
animals and the plants living in it. In practical terms
there are a number of relationships in which the natural
circumstances are deterministic. The extent of this deter-
minism varies from factor to factor. Any deterministic
situation is subject to change depending upon the dynamic
intra-environmental processes that are forever operating.
Human understanding, decision-making and response are also
powerful forces for change, and in the interpretation of the
value and relevance of the systems within which man operates.
The broad framework of the major system is usually more fixed
and persistent than are the sub-systems that make it up.

II.   THE GENERAL PATTERNS OF MAN'S RESPONSE TO THE NATURAL POSSIBILITIES OF AUSTRALIA

In demographic terms northern Australia is an empty
land (chapter 6). Over perhaps 50,000 years the aboriginal
populations have developed ways of life that could maximise
the relatively limited resources available to them (chapters
1 and 7). Although these early inhabitants were able to
establish a viable occupation of areas which have discouraged
later European settlement, or have been utilised only under
the special circumstances of mineral exploitation, nevertheless a general pattern of a fringe of human occupation around
a difficult centre and western area was established. The
wetter north and north-east offered the greatest prospects
(chapter 3). Aridity and sparce populations coincide.

The present concentration of settlement and activity is
dominantly coastal. The sea provided the access from overseas or from south-east or south-west Australia. Based upon
limited resources on the coast, or in restricted hinterlands
inland from the coast, small towns struggled to grow. From
Carnarvon, via the small ports or the mining towns of the
north-west coast to Darwin people have struggled over the
last century or so to meet the challenges facing them. An
even sparser utilisation of the continental margins
characterised the coastal margins of Arnham Land, the lowlands
fringing the Gulf of Carpentaria and the Cape York Peninsula.
From Mossman south larger settlements have competed with each
other and grown as the coastal lowlands have offered more
favourable prospects for economic expansion. Cairns, Townsville, Bowen, Mackay, Rockhampton and Bundaberg as well as
a number of smaller settlements have established the basis
for the most populated and cultivated area in northern
Australia. Away from the coast generally the opportunities
and attractions for European settlement have been few.
Special circumstances may have provided the basis for somewhat
anomalous urban growth in the inland, such as Alice Springs
or Mt. Isa, but elsewhere a widely dispersed pattern of
pastoral properties and small towns servicing them prevails
where the climate is not too arid and distances are not too
great.

Here, then, is a strongly developed geographical pattern
that requires explanation. A number of analyses have
emphasized the peripheral nature of northern Australia with
respect to the main core of south-eastern Australia where
the population, the activity and the wealth are concentrated
(Courtenay, 1982). The north is remote, access to it is
difficult, transport and communications within the area are
deficient, the general infrastructure for economic activity
is poor, cultural and social amenities are often lacking
and the material quality of life is below that of the south-
east of Australia. It has been argued that the sum total
of these drawbacks holds back the development of the north.
This view is undoubtedly correct for large parts of the area,
but it is suggested that the failure to make the desirable
progress and to translate the beliefs of the great resources
of the north into reality is due to more than the historical
accident of a later and slow exploration and settlement or
to the sole impact of distance and isolation from the social
and economic core of Australia.

Early Portuguese and Dutch contacts were made with the
north and west coasts of Australia (chapter 1). The
experiences gained did not encourage enthusiasm to seek

further or to settle.   Later nine arduous and often ill-fated
penetrations were made about the middle of the nineteenth
century from the south or from the east into the inland.
Some brought back exaggerated reports of the prospects of
the interior;   others saw little to encourage a follow-up
of their expeditions.   Much depended upon how good the rains
had been in the period immediately preceding the exploration.
In general, however, the problems of arid Australia soon
became all too apparent to many an early pastoralist when
poor rain seasons occurred.   Different though they were from
their European homeland of the late eighteenth or nineteenth
centuries, immigrants found more familiar opportunities in
the south-east of the continent.   Once this pattern was
firmly established the northern periphery was at a disadvantage in competing for people, capital and in general for
strong support for such activities that were not markedly
different from those of the south or were not dependent upon
localised resources of significant value.

From the viewpoint of evaluating the opportunities and
difficulties of the natural environment of northern Australia
this discussion will seek to identify the characteristics
and problems of the physical geography.   Where these became
responsible for limitations or slowness in the socio-economic
developments they contributed, according to the ways they
were perceived and assessed by earlier settlers, is the
distinction between core and periphery.   Once this became
established it has tended to perpetuate itself.   It was
more difficult than in the south-east to find the resources
and opportunities needed to develop ways of counteracting
environmental drawbacks and of initiating life-styles and
economic activities more appropriate to the conditions found
in northern Australia.

The observed characteristics of the natural environment
laid down the broad patterns of the early economy.   Apart
from the better watered eastern coastal lowlands pastoralism,
based first on sheep and later increasingly on cattle,
offered the most suitable prospects.   The lack of local
demand and the distance from other markets, added to the
environmental difficulties, held back development until the
more or less contemporary gold mining expansion brought
people, capital and infrastructural improvements.   The early
mining activities were often short-lived and have disappeared
from the scene.   Other new, large scale but different mining
ventures have replaced them.   However, the early mining
enabled the pastoral industry to become the major economic
activity outside the crop producing lands, primarily located
in the wetter eastern areas of the north.   Even today cropped

land has a very localized distribution and its total area is
a minute proportion of the total land area.   For northern
Australia, as defined by Courtenay, the 1977 cropped area
amounted to 0.18% of the total area and of this 94% was in
Queensland (Courtenay, 1982, p.122).

III.   THE ENVIRONMENTAL INFLUENCE

     The part of Australia that is being considered is
primarily tropical in climatic terms.   This is so even for
those areas just south of the Tropic.   The dominating influence is the availability of moisture.   The relief of the
land is most significant when its effects on rainfall or
run-off are concerned.   Soils are significant in the way
they affect the effectiveness of the water supply and in
their influence upon vegetation patterns.   Water and plant
cover contribute importantly to the development of soil
characteristics in conjunction with parent material.   Given
sufficient time and freedom from sudden changes or disturbance
the different physical factors will interact with each other
to develop a number of sub-systems or different ecosystems.
These are likely to demonstrate progressive changes which
theoretically might be expected to attain a steady state
equilibrium were it not for the fact that there are many
potential causes of disturbance;   geological, climatic or
human.   The following sections dissect the natural sub-
systems that make-up the northern Australian area.   They
should be read keeping continually in mind the two-way
relationships that integrate them into individual ecosystems.

IV.   ASPECTS OF THE GEOLOGICAL STRUCTURE AND
   PHYSIOGRAPHY

     Northern Australia is predominantly a land of plateaux
and plains.   The relief of the locally occurring mountains
is relatively subdued and their altitudes low.   The area has
escaped the tectonic disturbances on a large scale of recent
orogenies, and many areas have experienced a long period
of exposure to sub-aerial denudation.   On a more detailed
basis elevated plateaux or ancient mountain ranges have
experienced dissection by running water even in areas now
classed as arid.   Rugged terrain also presents difficulties,
even in plateau areas.
     Only a simplified picture can be presented here.   A

major three-fold structural and physiographic pattern can
be delineated:

A.   *The Stable Pre-Cambrian Shield*

     The basement structure of granites, schists and gneisses
dominates western Queensland, all the Northern Territory and
most of tropical western Australia except where it is
bordered by basins of younger rocks;   the Barkly (Cambrian)
Tableland, the Canning Basin (Palaeozoic and Mesozoic).   Few
shallow transgressions have affected the Shield.   Desert
landscapes extend across basins and basement alike.   Over
half of the Shield has a varying depth of sands (15 cm to
30 m) as sand plains or long narrow dune systems.   The
drainage patterns are poorly organised and disconnected
and in some areas, for example over the Great Sandy Desert
of the north-west extensive areas of low relief and highly
permeable sands are devoid of an identifiable hydrographic
system.   Vast sandplains, active dunefields and wide alluvial
clay plains are characteristic of the dry inland, for instance
the Gibson, Tanami and Simpson Deserts.   Playas and residual
salt lakes (salines) add further diversity in detail to the
landscape.
     The prevailing trend in the Shield is from north-east
to south-west.   Contrasting conditions extend to the north-
west coast.   In some areas (the Great Sandy Desert) dune
sands reach the coast.   Where rivers reach the coast mangrove
flats line their estuaries.   The Arnhem Land Plateau presents
low cliffs to the sea.   The coastal characteristics are not
conducive to good natural harbours.   Two main upstanding
regions are affected by the deeply dissected, but generally
flat lying sandstones and quartzites of the Kimberley Plateau
(summits generally between 650 and 850 m) and the rather less
dissected Hamersley Plateau (up to 1,225 m).   Lower plateaux
are represented by the Pilbara Plateau, the Arnhem Land
Plateau and the Barkly Tableland (Heathcote, 1975, p.13,
fig.2.2).   Locally, often the outcome of more humid part
climates, deep valleys dissect the plateaux but most slopes
are gentle except where one level steps down to another.
Extensive plains are the characteristic landform.   In the
centre of the Shield an effective southern limit to our
area of interest is provided by the west to east folded and
tilted Macdonnell and Musgrave Ranges.   Both are deeply
dissected by narrow valleys between steep sided narrow ranges
up to over 1,000 m.   These ranges and valleys provide somewhat
less arid micro-climates and ecological niches between the
eastern and western desert areas.

B.   The Lowland Basin

This gently warped lowland formed by rocks of Palaeozoic to Tertiary age, but particularly sedimentary Mesozoic claystones and sandstones, extends from the Gulf of Carpentaria southwards. Wide river valleys separated by gentle slopes chain north to the flat marshy lowlands fringing the Gulf or south towards the interior drainage of Lake Eyre. The Great Artesian Basin underlies most of this area.

C.   The Eastern Uplands

Locally these uplands exceed 1,000 m and include northern Australia's highest point of Mt. Bartle Frere (1,611 m). The rocks are mainly metamorphic or igeneous in the south and sedimentary in the north, mostly of younger Palaeozoic age but the final shaping crustal movements occurred in late Tertiary. The eastern face is steep with many cliffs, gorges and waterfalls often associated with marked faulting. This eastern scarp has presented a continuous and difficult barrier to movement inland from the narrow coastal plain. The Great Dividing Range comprises ranges, hills plateaux and some lowlands. The actual watershed between short incissed rivers flowing east to the Coral Sea and those that flow either north-west to the Gulf or Carpentaria or south-west to the braided river systems of Queensland's Channel Country (chapter 11) is often further west than the Dividing Range. The plateaux slope gently from east to west. Within the Eastern Upland region there are a number of areas of past volcanic activity (some very recent) notably on the Atherton Tableland, inland from Cairns. The coastal plain, of which the width varies from a few kilometres to thirty or more, is well watered, often has good alluvial or basaltic soils and has a much greater agricultural potential than most of northern Australia.

V.   NORTH AUSTRALIAN SOILS

The soil variations owe a considerable amount to parent material and geomorphic processes. The latter in turn relate to both present and palaeoclimates, notably water availability and movements and their effects on weathering, erosion and deposition. Vegetation reflects the soil characteristics rather than playing a major role in soil formation, except in so far that it provides a protective cover. If drought, fire

or overgrazing disrupts this cover, water or wind transport
can cause serious soil erosion. The strongly developed
seasonality of the rainfall and the long dry season in most
parts of the north influences both the physical and chemical
processes of soil development. See also chapters 9, 10 and
11 in relation to soils and vegetation.

Some soils have evolved over long periods on old land
surfaces subjected to marked weathering. In more limited
areas, especially in the wetter east where the soil patterns
tend to be more complex, much younger landscapes occur. Late
Tertiary or recent basalts and Quarternary alluvium have soils
which are much less developed and also can be much more
fertile. In western and central areas extensive areas of
more uniforms soil conditions reflect areas sharing similar
relief, parent material and climate. In the most featureless
of areas however, quite significant but local and small
scale soil contrasts occur.

Five major soil groups have been distinguished for the
north (Isbell, 1981a, see maps p.14 fig.2.4):

a. *Shallow, coarse,* stoney soils which are widely
distributed especially in mine elevated areas. Old lateritic
surfaces and rugged dissected plateaux often have such soils.
Though developed on a wide range of rocks, parent material
has a much greater influence than the present climate.
Sandy and loamy textures prevail. In some of the desert
regions very stony, gibber, surfaces occur.

b. *Deep sands* include the extensive areas of siliceous
red sands of the sandy desert and also yellow sands of dunes
and sheets fringing river lines of the Channel Country. Also
coming into this category are earthy sands with a small
increase in clay content with depth. These latter form
extensive plains particularly in the arid interior parts of
the west. Except in the wet season these soils are drought-
prone.

c. *Cracking clays* predominate east of 135°E. The black
clays derive especially from basic rocks such as basalt or
from alluvium from such rocks. They occur particularly in
sub-humid eastern Queensland, with smaller areas in the north
of the Northern Territory (but see chapter 9, Perry and
Mott). The more widespread grey, brown and red cracking clays
of western Queensland and the Barkly region of the Northern
Territory derive from argillaceous sedimentary rocks and
alluvium. Some of these soils can be several metres in depth,
they are usually humus rich and provide fertile conditions.

Gilgai, hummocking micro-relief is often associated with the deeper cracking clays.

   d. *Massive sesquioxide red and yellow earths* are wide-spread over northern Australia in association with siliceous parent materials. The upper part of the profile is sandy or loamy with the clay content gradually increasing with depth. At least in the upper parts they are permeable. Some are relict soils. Others have formed on deeply weathered and truneated lateritic surfaces, or on material from such areas. They often include ironstone nodules or are cemented into an indurated or durierust layer.

   e. *Sodic soils* with a marked vertical texture-contrast show an abrupt change from sandy or loamy upper horizons to deeper horizons with an increased clay ratio, characterised by a high proportion of exchangeable sodium. They are often increasingly saline with depth especially in drier areas. Depth is very variable. Such soils, though developing from various parent materials, are particularly associated with intermediate to acid rocks and derived alluvium. They occur mainly in Queensland and include the desert loam of the south-west and the solodic and solodized-solonetz soils of the eastern coastal and sub-coastal areas (see chapter 9).

   The small scale of soil maps (often 1:1,000,000 to 1:2,000,000) gives the impression of a simple and uniform soil pattern. Isbell (1981a) draws attention to smaller areas of different soils which include podzolic soils in more humid often higher parts of eastern Queensland and in the Northern Territory; non-calcic brown soils and euchrozems in eastern Queensland, the Kimberley and parts of Northern Territory; kaolin-sesquioxide soils (krasnozems and xanothosems) in wet areas of eastern Queensland; alluvial, often deltaic, soils in widely dispersed but small areas such as the Burdekin (Queensland, Ord (Western Australia) and the Gascoyne (Western Australia).

VI.   SOIL MANAGEMENT

   Land-use potential reflects the joint influence of soils (Manageability and nutrients) and climate (water availability and reliability). Soils characterised by limiting conditions of a physical or chemical nature become more difficult if the water supply is inadequate. Given sufficient economic incentive some or most of these soil constraints can be

reduced or overcome. There is a cost, however, and soil
characteristics have a strong influence in a pastoral situation upon the potential of an area and the most desirable
form of land management. (Chapters 9, 11 and 12 discuss
further aspects of land management.)
    Some of the problems are:

    (1) Those associated with the entry, transmission and
storage of water in the soil especially where this is light-
textured. Soil moisture status is an important consideration
in northern Australian areas.
    (2) Poor water permeability is linked to soils which
have a marked clay content, especially with sodic properties,
in the subsoil. Waterlogging or flooding is a wet season
occurrence and this in turn restricts movement over the soil,
cultivation and root growth. Even arid areas, if a protective
vegetation cover is lacking can experience these difficulties
after intensive rain-storms.
    (3) Foundation of soil crusts or sealing impedes
drainage through the soil thus increasing surface run-off
and consequential erosion. Cracking clays on slopes and
where the plant cover is lacking (e.g. summer fallow) are
especially vulnerable. Surface seals also inhibit the
germination and establishment of small seed species.
    (4) Saline conditions can develop especially in the
sub-soil in drier areas.
    (5) Trace element deficiencies have been noted but still
much needs to be observed in the field and most of the
specific information applies to Queensland. Deficiencies of
cobalt (northernmost Cape York), molybdenum (very wet coastal
north Queensland) and copper zinc in some light textured
soils, potassium in wet areas, sulphur in moderate but not
very wet areas have been identified. Where leaching and
weathering are very active the base status may be low.
Strongly acid soils with a high level of toxic aluminium have
been associated with wetter tropical areas but major land-use
problems have not been linked with such conditions.

    Soils need careful management in tropical or near trop-
ical climates. On the whole good soils are the exception
rather than the rule. Organic matter in the soil is usually
low and soil nitrogen deficient. Though in the wetter areas
in Queensland rainforests or the leguminous brigalow and
gidgee forest cover on black and grey clays have valuable
amounts of organic matter and nitrogen, clearance followed
by intensive cropping or grazing rapidly depletes these
reserves (see also chapter 9). Phosphorous is often deficient

in the soils of much of northern Australia, though wetter
areas in Queensland on black earths, grey earths and
euchrozems are usually better endowed. Some soils, such as
krasnozems in the high rainfall areas, may fix phosphorus so
that it becomes unavailable to plants.

Soil erosion from water or wind is an ever present
threat. There is a great need to learn more concerning soil
stability in northern Australia, particularly where unwise
management occurs such as overstocking or overgrazing or
major changes in land use occur (Isbell, 1981b, p.9). Solodic
texture-contrast coils with highly dispersive sodic subsoils,
as for example in the Upper Burdekin and Upper Nogoa (central
Australia) rivers, have shown themselves especially prone to
erosion. Black earths in the Central Highlands of Queensland
when being fallowed, or soils in the wet sugar cane areas,
have demonstrated the risks of severe erosion under heavy
rain in cultivated lands. Technically greater use of arable
lands or of adequately watered pastoral areas is possible,
but it requires a keen awareness of the interactive aspects
of the soil elements in the ecosystem and a cautious increase
in the demands that are placed upon the soil.

VII.   CLIMATE AND WATER MANAGEMENT

A considerable emphasis has been placed upon soils in the
belief that their role in the natural ecosystems of northern
Australia or in the utilization of resources is a critical
one. The climate, more particularly the effectiveness of
rainfall, should be afforded a similar heavy emphasis.

Conventionally the characteristic associated with low
latitude areas is heat. Heat is not the main control of the
growing season, which is primarily dependent upon moisture
availability, but it strongly influences rates of growth of
plants. Pests and diseases may well thrive in a warm
environment though here too moisture is usually a necessary
accompaniment.

Distance from maritime influences increases the thermal
range of the inland areas. High diurnal maxima in summer
(>38°C) are more frequent inland and westwards (Marble Bar
in Western Australia's Pilbara has gained the notorious
reputation of having recorded 160 consecutive daily maxima
over 37.8°C). High surface temperatures affect both the
establishment and yield of crops (McCown *et al*, 1980) and
influence rates of organic decomposition. In winter, night
minima inland drop to much lower values when the sky is

cloud-free and atmospheric humidities are low.   Frost risks
extend further north than most would expect.   The northern
limit parallels and does not recede far from the north-
west coast from the Tropic eastwards to Wyndham, whence it
dips south to about 20°S but with two northern projections
south of the Gulf of Carpentaria and over eastern Cape York
to about the latitude of Laura (see previous chapter, 3).

Solar control of the daylight thermal and radiation
conditions is strong outside the wet season, and is accentuated inland with fewer hours when rain falls and reduced
cloud coverage.   These climatic contrasts emphasize the
differences between the north and the east, and the inland
and the west, as well as reinforcing the coastal margin/
interior contrasts.   As a general rule in attempting to
type the climatic environment, *frequencies* of specified
conditions give more guidance than statements of *extremes*.

The importance of limiting climatic conditions,
temperature especially, for human comfort and efficiency and
building considerations receive attention elsewhere in this
volume, especially in chapters 16 and 17, but also in
chapters 18 and 19.

The climatic discussion will focus upon the nature of
the rainfall patterns.   Discussion of climatic mechanisms
appeared in chapter 3 and will not be examined.   The strong
seasonality of the rainfall with a relatively short summer
"Wet" (December to March) and a long predominantly rainless
period over the rest of the year is a dominant quality of
the climate.   Actual rainfall totals must be assessed in terms
of their effectiveness.   High radiation totals provide the
energy for large potential evapotranspiration losses.   The
further north, within the study area, the greater the
percentage of the annual rainfall which falls in the short
summer wet season.   Towards and beyond the Tropic the more
chance there is (though an irregular one) of rain in winter
from upper level systems within the high altitude westerlies.
The structure and the texture of the soils and the nature and
completeness of the cover afforded by the vegetation introduce
a finer scale of detail in the evaluation of rainfall effectiveness so that general maps on a sub-continental scale
provide but part of the information.   Internal differentiation
of climatic potential must be expected, therefore, for several
reasons over the whole area.

Australia introduces a compact landmass into the sub-
tropical and higher latitude tropical areas which are
particularly subject to the variations in the intensity and
location of the subsident atmospheric circulation associated
with the sub-tropical anticyclone.   Even allowing for the

ten or more degrees of latitudinal shift of the high pressure
cells towards the equator in winter and poleward in summer,
most of northern Australia is likely to receive only fluctuating incursions of the humid equatorial north-westerlies
from the summer monsoon and the related zone of convergence
of the Inter-Tropical Convergence. Even during the wet
season unstable air cannot be guaranteed to penetrate far
from the north coast and over Cape York south from Thursday
Island. Tropical cyclones can be important rain-producers
(as well as causing destructive winds on a narrow coastal
fringe). The circumstances favouring their development and
subsequent landward movement vary widely and their contributions to the summer rainfall, as they move west from the Coral
Sea, south from the Gulf of Carpentaria or south-west across
the north-west coast, are highly variable and irregular.
Heavy rain can on rare occasions reach well into the interior
dry regions producing brief floods. As well as these major
rain producing event convective instability is highly
sensitive to small alterations in the vertical changes in
atmospheric temperature and humidity. Heavy showers can
fall on one area leaving other places but a few kilometres
away, drought affected.

It is apparent that rainfall can be very variable from
one wet season to another, as well as within an individual
wet season, and that sharp spatial variations can occur over
quite small distances. The main source of moisture is
derived from the oceans to the east and north. The further
from the coast the less the prospect of rain with decreasing
humidity and a more stable atmosphere. Over much of northern
Australia the prevailing surface circulation is from the
east or south-east. Rainfalls decline sharply away from the
coast, especially the east coast where there is the added
obstacle of the Eastern Uplands. Continental conditions of
drought are carried towards the west and north-west coasts
which are on the leeward side of the land mass.

High ground usually introduces wetter regions into a
given rainfall region but unless the land height is considerable this effect is minimised where anticyclonic conditions
prevail. The Eastern Uplands give rise to the wettest area
in Australia, but only in a localised region where the high
ground and the coastal orientation are particularly favourable. The high ground effects are very restricted and
elsewhere in northern Australia the orographic effects on
rainfall amounts are far less, the more so in the interior.
Nevertheless the hydrographic patterns around some of the
more elevated inland areas show some height effects on
rainfall.

For much of the north, annual median falls below 500 mm
and to below 200 mm in interior lands in the centre and west.
This aridity is accentuated by high variability, low reliability and high intensities which result in excessive run-off
losses.

VIII.   HYDROLOGICAL RESPONSES

Few rivers and creeks in the north have more than
ephemeral flow, except in the wettest north-eastern and
eastern margins.   Rivers fill to dangerous flood levels in
a few hours, remain as raging torrents for little longer and
decline back to a trickle or a series of unconnected pools
in a matter of days or less.   In some cases a more
continuous water movement persists in the river gravels
below the surface.   In the inland areas most of the drainage
is local and disorganised and water rapidly loses itself in
short distances from the source of the flow.   One exotic
source of water of importance is that which feeds the Channel
Country of south-west Queensland from high rainfall sources
from the western slopes of the Eastern Uplands via the
Diamantina and other rivers towards Lake Eyre (see chapter 11
following).
Between about 120°E and 136°E and south of about 17°S
surface water is not normally available.   Outside this area
water is usually available on the surface with variable
reliability, except near the north and east coasts where a
perennial regime is possible.   In most areas water holes dry
up by July or August unless a good wet season is supplemented
by winter rainfall.   Some of the arid regions can draw upon
artesian supplies from sedimentary basins.   The most
important underground water is drawn in central and western
Queensland and adjacent parts of the Northern Territory from
the Great Artesian Basin.   A number of smaller basins include
the Georgina, Daly and Wiss Basins, the Amadeus Basin, the
Bonaparte-Ord Basin, the Canning-Fitzroy and the Camaroon
Basins.   These have been little or only partially developed
and their safe yields are not fully known.

IX.   VEGETATION AND GRAZING POTENTIAL

A number of different vegetation patterns have been described, each with an emphasis upon different criteria. The decline of effective moisture towards the interior is the most significant factor, but a number of other influences must be considered and will contribute to a more complex differentiation. It may be noted that, as late as 1981, Jones remarked, "there is little understanding of the dynamics of the various plant communities and of the factors which result in their natural distribution".

In this discussion it has been chosen to adopt a relatively simple differentiation of the vegetation of northern Australia which emphasizes the grazing potential. Nine divisions were presented in 1981 by Mott, Tothill and Weston and they are discussed by Perry and Mott in Chapter 9 following so they are not listed here.

Rainforest does not appear in this classification. This rapidly contracting vegetation type is one of major importance, but now remains in its best condition only in small less accessible and wetter parts, mostly on the Eastern Uplands. Some of the rainforest has been cleared for tropical pasture. Apart from any argument that all existing rainforest should be conserved, areas converted to pasture need careful management to prevent their deterioration as the initial store of soil nutrients becomes exhausted.

A number of factors linked to the environmental characteristics discussed in earlier sections have contributed to the present features of the vegetation apart from a purely climatic relationship. Edaphic factors have an increasing significance in the arid and semi-arid areas. Floristic elements in the vegetation associations owe some of their characteristics to past climates and plant migrations. The present state of the plant cover also reflects disturbance due to fire, grazing and clearance, while the effects of indigenous fauna, the kangaroo in particular and feral animals, cannot be neglected. Unwise management of grazing including overstocking and its effects, especially in times of drought, has resulted in the deterioration of the natural plant cover some of which seems almost irreversible. Unpalatable species have replaced palatable species, annuals have increased at the expense of perennials and areas exposed by the destruction of a protective plant cover have suffered soil deterioration and erosion. In addition in such localities a more rigorous micro-climate is produced thus impeding recovery to the initial condition of the vegetation.

Pasture improvement and increased carrying capacity are
of major importance to land-use in most of the area being
discussed, except for the relatively limited area in. the north
and east suitable for cropping and for the western interior
lands which are devoid even of adequate grazing, except
immediately after a rare rainfall. Considerable pasture
improvement in the east and north followed the successful
introduction of the Townsville stylo (see chapter 1).
Considerable effort and experimentation is being directed to
finding new species from other tropical environments that
will increase the rate and extent of pasture improvements.
The consequence will be, of course, a change in the ecosystem.
For the very reason that all the natural sub-systems reflect
a delicate balance and complex relationship between their
component elements, changes that affect the plant composition,
or the soils or microclimate, need to be well understood
and cautiously proceeded with. This caution is made even
more necessary in such a large area as we are dealing with,
in which a great deal of environmental data still need to be
assembled and scientifically interpreted.

X. OTHER ASPECTS

Insect pests and plant and animal diseases are integral
parts of the different ecosystems. Introduction of new
plants, or changes in the growth pattern of existing plants,
introductions of pastoral animals or the effects of changes
in the population of native fauna, feral animals such as the
water buffalo or the pig or the problems caused by such
animals as the dingo or rabbit emphasize the significance of
changes of this nature when reviewing the characteristics of
different areas. Such considerations open up a subject to
which a chapter by itself could be devoted. It must
suffice here to indicate that such relationships are also a
significant aspect of natural sub-systems.
One other area will be omitted from this discussion. The
geological history of northern Australia has endowed many
areas with valuable and major mineral and energy resources
(see chapter 14 by Ollier and chapters 18 and 19 related to
mining towns). These resources are part of the physical
environment, particularly from man/s point of view. They
have played an important role in the European development of
the north. They will undoubtedly continue to have as great
if not greater significance. Their exploitation are a
potential source of disturbance and modification of the other
aspects of the natural environment.

XI. CONCLUDING COMMENTS

The complex interaction between different elements which are associated in the natural sub-systems that can be identified in northern Australia has been only partially illustrated in this chapter. Almost all the interrelationships could be analysed more deeply, some have been passed by with little or no reference and the patterns or systems identified under the specific headings express only some of the contrasts and interconnections. Patterns displayed by natural phenomena can be differentiated according to a variety of criteria and, as was indicated in the introductory remarks, at different levels of detail. At least, however, the very complexity of the natural systems provides a justification for seeking to appreciate the nature and significance of the 'arenas of life'. The successful acting out of human lives in such arenas places a heavy emphasis on the need for understanding as best we can the prospects and the problems that any form of action will cause.

REFERENCES

Courtenay, P.P. 1982. *Northern Australia: Patterns and Problems of Tropical Development in an Advanced Country,* Longman, Melbourne.

Heathcote, R.L. 1975. *Australia,* Longman, London.

Isbell, R.F. 1981a. Soils, in *Rural Research in Northern Australia: A Report to the Commonwealth Council for Rural Research and Extension,* Australian Government Publishing Service, Canberra, pp.12-17.

Isbell, R.F. 1981b. Soils and Land Use. Paper presented to a seminar *CSIRO in North Australia* held at the Davies Laboratory, Townsville.

Jones, R.J. 1981. Pasture Research. Paper presented to a seminar *CSIRO in North Australia* held at the Davies Laboratory, Townsville.

McCown, R.L., Jones, R.K. and Peake, D.C.I. 1980. Short-term benefits of zero- tillage in tropical grain production, *Australian Agronom Conference Proceedings,* Lawes, April, p.220.

Mott, J.J., Tothill, J.C. and Weston, E. 1981. The native woodlands and grazing lands of northern Australia as a grazing resource, *Journal of the Australian Institute of Agricultural Science,* 47, pp.132-41.

5

THE ENERGY RESOURCES OF NORTHERN AUSTRALIA

C. D. ELLYETT

Emeritus Professor of Physics
University of Newcastle
New South Wales

I. INTRODUCTION

Approximately three-quarters of Queensland and about two thirds of Western Australia lies north of the 26th parallel. The whole of the Northern Territory is north of this parallel. All three areas in the past several decades have become, or now have the potential to become, enormous suppliers of energy. Tropical Queensland is synonymous with coal, although there are also significant uranium and oil shale deposits. The Northern Territory has - on world standards - huge resources of uranium. It also has both oil and natural gas. In addition, in the very recently discovered Jabiru oil deposit in the Timor Sea 300 km off the Northern Territory coast north and west of Darwin, a vast new field has possibly been discovered comparable with the Bass Strait fields. The north of Western Australia has enormous supplies of natural gas at sea in the North West Shelf.

All told, although distances are vast, with indigenous supplies there should now be adequate non-renewable energy resources within or near the tropics of Australia, except where distances to isolated communities make renewable resources, such as solar energy, a small scale economically viable alternative. There should also be much energy becoming available either for export, or for transport to *The South* of Australia. Which energy a particular State will use will tend to be predetermined by what lies within its own borders, but this has limitations. For example, almost all road transport requires either a liquid or a compressed gas. Hence coal is a commodity either for factories, power stations, or export.

NORTHERN AUSTRALIA
ISBN 0 12 545080 X

The need for liquid fuels raised many supply problems until recently, but all three northern regions are rapidly becoming self-sufficient in either oil or natural gas, or both, although the Northern Territory in particular has not yet reached that position of self-sufficiency.   There is a tendency to buy energy fuels interstate.

It is proposed to present for both of the northern States, and The Territory, an outline for their principal non-renewable fuels, under the headings of *coal, oil and natural gas,* and *uranium.*   A fairly brief statement will then be made for each area on its progress with renewable resource development.   In general these renewable resources will be found to play a quite minor role at present, and will probably continue to do so for the rest of this century, owing especially to the abundance of renewable resources as alternative energy sources.

Which type of fuel to use in any particular application is a complex mixture of engineering needs, coupled with the costs of winning and transporting the fuel.   These constraints vary with time and from place to place.   Problems of cost variability and locational advantage will not be considered in this chapter, nor will the implications of political actions.   The chapter focusses on a description of the distribution and the size of the various resources.

II.   STATE RESOURCES

A.   *Queensland*

1.   *Coal.*   This is the principal energy resource in Queensland.   Figure 1 shows the vast Bowen Basin reserves, which include more than half Queensland's total coal resource area.   Another major region is the Galilee Basin, to the west of the Bowen.   Considerable further reserves are found south of the 26th parallel. There is an estimated 5.52 x $(10)^6$ tonnes of coking coal and 8.31 x $(10)^6$ tonnes of lesser quality steaming coal, in economically recoverable quantities, at present (1983) prices (Queensland Energy Advisory Council 1983).   Measured and indicated *first class* reserves throughout Queensland stand at 14.2 billion (billion is 1,000 million) tonnes of coking coal and 15.1 billion tonnes of steaming coal.   Fifteen per cent of the coking coal and 35% of the steaming coal should be available by open cut (Queensland Government Mining Journal 1983).

The northern portion of the Bowen Basin contains almost all of the State's coking coal and about 70% of its proven

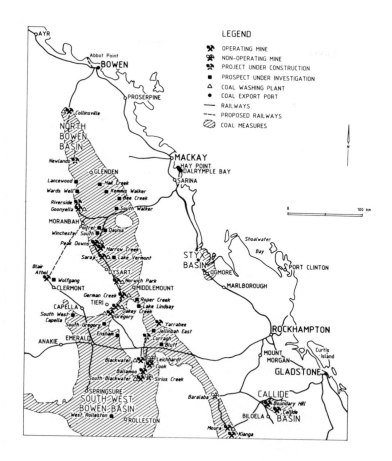

FIGURE 1. The north-eastern portion of Queensland's coal resources[a].

[a](Adapted from Queensland Government Mining Journal, Dec 1982)

reserves of steaming coal. Most exports are of coking coal.
     In the far north of the Bowen Basin, the Collinsville
mine exports coking coal through the port of Bowen, and also
supplies the State coke works. Steaming coal from this mine
is used for power generation at Collinsville, and to the south
a steaming coal project is under construction at Newlands.
The increased output from these mines will be shipped through
a new port under construction at Abbot Point, less than 20
kilometres north of Bowen. Coking coal from a number of both
open cut and underground mines is exported through Hay Point,
about 15 kilometres south of Mackay.

Developments must be planned many years ahead of the first shipments and there is presently (1983) a surplus capacity in Queensland due to the effects of the world recession and in particular an over-estimate of requirements by the Japanese. Four big draglines are held on a care and maintenance basis and five new coking coal mines coming on stream will easily meet any short term increase in demand.

2. *Oil and natural gas.* There are several petroleum fields near Rolleston and between Rolleston and Injune. There is also one further west at Gilmore. All lie within 2° north of the 26th parallel (Allen 1983). The majority of the fields, however, are in the south of the State, but this is obviously a case where supplies can to some extent be spread north into the tropical regions. Oil production in Queensland is still small, especially north of the 26th parallel.

A longer term potential source of oil lies in the oil shale deposits. These oil shale deposits extend over a considerable region of northern Queensland. One zone lies in coastal areas near Bowen, Rockhampton and Gladstone, but the most extensive deposits are in the Julia Creek region (Plate 1 shows the landscape of the area. Its scale and homogeneity over large distances can be gauged from the fact that these two pictures were taken, one hour apart while travelling at 120 km. per hour on the road between Mt. Isa and Julia Creek : Parkes September 1983)

PLATE 1. *The Julia Creek region of north west Queensland where the oil shales are located. (Photographs Parkes, D.N. September 1983).*

Feasibility studies for open-cut exploitation, followed by retort processing, are being carried out in a number of the coastal areas. Recovery appears to be too costly in the short term, compared with conventional petroleum exploitation.   It is also the case that Queensland's oil shale is generally of low grade, but total State reserves are in excess of 3.5 M megalitres.

3.   Uranium.   Exports over the past seven years have been, perhaps surprisingly in the present political climate of uncertainty about uranium mining in Australia, as much as about one-third of the coal energy exports, by value.
Uranium occurs in the north of the State, in the Cairns-Townsville hinterland, and in the north west, in the Mt. Isa region. Total reserves of uranium oxide stand at 29,700 tonnes (Allen 1983).   A $100 M uranium mine has been proposed for Ben Lomond, near Townsville, and an environmental statement has been submitted to the Federal Government.   In the present political circumstances (October-November 1983), it seems unlikely that export licences will be approved and the project will therefore not be able to proceed.   At the time of writing the Federal Government's uranium policy is still not clear, even within the Labor party which is in government.   The logic of allowing one large mine to proceed in South Australia;   but not others, particularly in the Northern Territory, is open to question.   However at the moment it must be assumed that the present stated policy will prevail and northern Queensland's uranium resources will not leave the ground or the country.

4.   Renewable Resources in northern Queensland.   Solar energy is of course the primary renewable resource, and may be used as such, either for the direct production of heat or for electricity.   In addition there are many subsidiary resources, including the processing of *plants,* the use of *water* stored naturally at high levels for the generation of electricity and the harnessing of the *wind* and the *waves,* all of which stem directly from a solar input.
Throughout northern Australia, direct solar energy use is still an insignificant fraction of the total energy usage.   It is however becoming increasingly important to isolated home-steads. *Hydro-electricity,* a subsidiary resource, is based in the north-eastern parts of the State but represents only 4.5% of the total energy generated.   Significant amounts could be produced from the Herbert (west of Cairns) and the Burdekin (west of Townsville) rivers, if the energy potential which they hold were to be harnessed.   Expenditure on these projects has not been authorised. *Ethanol,* derived from cane sugar, is

technically feasible as a fuel extender in petrol but is at
present not cost competitive.

Clearly northern Queensland's energy resources are based
heavily on coal and this seems likely to continue for some
considerable time.

B.   *Northern Territory*

1.   *Coal*. For the Northern Territory this section is
very simple: There is no commercially viable source of coal.
Although the average load demand for electricity in Darwin is
only about 600MW at the present time (1983), capacity is
expected to be inadequate by about 1986. The Northern
Territory Government has decided to build a new 300MW coal-
fired power station, to be commissioned at Channel Island in
Darwin Harbour by 1987 (Northern Territory Energy Policy 1982).

2.   *Oil and natural gas*. The Northern Territory is at
present in a weak position, both financially and politically,
in that all its primary energy, for both transport and
electricity production, is liquid petroleum imported from
Singapore. However, oil, condensate, and gas reserves now
exist both on and off shore, and are associated with the
Amadeus Basin to the south, the Browse Basin to the north, and
the Bonaparte Basin also to the north. In fact oil reserves
now substantially exceed any State except Victoria, and are
equal to 7.8% of the June 1982 estimate of total Australian
reserves.

The Amadeus Basin contains the Mereenie oil/gas reserves
and the Palm Valley gas reserve, south and west of Alice
Springs. Reports are widely divergent on the size of the
'proven' Mereenie oil reserves, ranging from 37 to 330 million
barrels, together with 34 million barrels of condensate and
liquified petroleum gas (LPG). There is also thought to be 25
billion cu m of gas recoverable (Northern Territory Energy
Policy 1982). The field is about to be developed and oil
should be flowing at 1,000 barrels/day by mid-1984. Eventually
it should be piped the 250 km to Alice Springs, one of the
fastest growing areas in northern Australia (see chapter 6,
Parkes).

An oil discovery announced in the Australian press on 2nd
September, 1983, may considerably increase the oil resources
of the Northern Territory. This announcement related to the
BHP's Jabiru 1A well in the Timor Sea, 300 km off the coast of
the Northern Territory. At the moment it has a potential of
200 M barrels of recoverable oil. Never before has a 57m oil
column been discovered in Australia outside the prolific
Gippsland Basin, where most of the country's oil is produced.

Discovery of further oil zones, deeper in the well, could take
this figure to over 550M barrels. It is conceivable that
there are other near-by oil fields, but it will take a major
exploration program over the next two years to evaluate the
area's full potential. Water depths of only 119 m mean development using conventional technology will be possible.
Cyclone related hazards will however require that some
additional costs are incurred.

   3.   *Uranium.* Although the Northern Territory uses no
uranium and is too small in population to even contemplate a
nuclear reactor, it does possess 13% of the western world's
uranium. The four main sources of supply, or potential
supply, are:

Ranger (existing)	124,000
Jabiluka	207,380
Nabarlek (existing)	12,000
Koongarra	13,300.

There are also many other potential sources in the South
Alligator Valley.
   The Federal Government is allowing the two existing mines
to proceed, but in October 1983, blocked the other two from
starting. Jabiluka, which has the richest supply in the
world, had obtained a Northern Territory mining lease, had
completed arrangements for the payment of royalties to the
aboriginal people, had secured contracts, and hoped to start
mining at the rate of 4,500 tonnes/year. Australian uranium
supplies are earmarked for commercial nuclear reactors for the
generation of electricity in many overseas countries. Ranger
earns approximately $300M per annum. All of Nabarlek has been
mined and is being processed at the rate of 1,500 tonnes per
annum.

   4.   *Renewable Resources.* The Territory has a near
vertical sun for much of the year. There is, however, a well
defined 'wet' season, with resultant cloud cover, in the Top
End of the Territory, which reduces solar collector efficiency
for this part of the year (See the chapter by Lee and Neal
above). *Solar* water heaters are used in the majority of
domestic residences in the Darwin area, and the capital outlay
is repaid in about four years. An experimental *solar pond* in
Alice Springs shows some promise for the generation of electricity in remote areas which also have a suitable water
supply available, but the technique has not as yet achieved
commercial viability in Australia.
   The present use of fuel oil and distillate by the Northern Territory Electricity Commission, the small scale grid

systems, scattered throughout the Territory, and the absence
of economies of scale, result in extremely high cost electricity generation. However, because of this condition, although
electricity generation by solar energy systems is not yet
competitive, it may first become so in the Northern Territory.
The Commonwealth Government provides a subsidy to the
Northern Territory Electricity Commission, in order to
'equalise' electricity prices with those in other States.
Tariffs are tied to those of consumers in the northern Queensland regional zone, resulting in an annual subsidy of $50M to
the Territory at the present time (1983).

Nowhere in the Northern Territory provides a sufficient
head of water to allow significant *hydro-electric* power
generation.

Wind power is another subsidiary, renewable resource. To
be effective a wind power density in excess of 100W/m is
required. The Northern Territory does not appear to have any
significant or sustained wind power, except possibly at
Tennant Creek (Edwards 1983).

The Northern Territory, where possible is moving away
from imported oil to coal obtained from other Australian
States, and is starting to use its own natural gas, and
presently, oil. These latter resources should grow in use
significantly over the decade ahead. Uranium remains purely
as an exportable commodity: if political favour permits it.

C.   *Western Australia*

1.   *Coal.* Although no coal is found north of the 26th
parallel, it is necessary to mention briefly its growing
importance in the south of the State to obtain an overall
assessment of the energy situation in northern Western
Australia. Although Western Australia has less than 1% of
Australia's coal, the Collie coalfield south of Perth now has
400M tonnes of extractable reserves, and probably more. There
is a new field at Eneabba, north of Perth, and much low grade
but useable coal has been found north of Esperence.

2.   *Oil and natural gas.* Oil has been found on the
continent near Barrow Island and in shallow water north of the
island, which will take over from the declining Barrow Island
reserves. Appraisal wells at South Pepper and North Herald,
near Barrow Island and in shallow water, are likely to lead to
producing wells, giving 10,000 barrels a day by September
1984, and 22,000 barrels/day by October 1985. Harriet No. 1,
in only 22 metres of water, 17 km north of Barrow Island has
just been declared commercial and should be producing by early
1985. Its reserves, now only 12M barrels, are expected to

increase substantially with further exploration. The Bambra
oil discovery is only 6.5km north of Harriet. Thus a quite
active oil producing area is emerging.

Oil discoveries however pale into insignificance
compared with the *North West Shelf* natural gas discoveries,
which will, in consequence, be described in rather more
detail.

Woodside Oil Co. NL started in a very small way in
Gippsland, Victoria, in 1953. By 1958 it was almost bankrupt
and just survived for several years. By 1961 the company
realised it must merge with overseas capital and expertise.
In 1962 it obtained large offshore exploration permits from
the Western Australian Government, and from the Commonwealth
Government for offshore areas adjoining the Northern
Territory. The area being explored is shown in Figure 2.

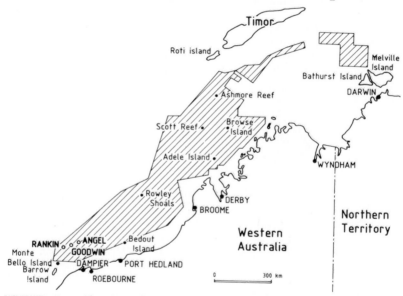

FIGURE 2.   The North West Shelf natural gas region

At one stage Woodside's permits covered 367,000 sq km, or
more than half the total shelf area of 600,000 sq.km. By 1963
a Company Consortium was established. After several changes
the current joint venture group, for the onshore part of the
liquified natural gas project, is

Woodside Petroleum Development P/L	13.333%
Shell Development (Australia)	16.666%
BHP	16.666%

B.P. Co. Ltd.   16.666%
California Asiatic Oil Co.   16.666%
Woodside is likely to sell a further 16.666% to the Japanese
joint venture Mitsubishi and Mitsui, but intends to retain 50%
shares of the gasfield and offshore facilities (*The Australian*
December 8th 1983).

Drilling was started in October 1967 at Ashmore Reef
(Figure 2). The results were negative. In 1971 large gas
reservoirs were found at Scott Reef, followed in the same year
by immense gas and condensate discoveries at North Rankin,
quickly followed by Angel and Goodwin. At present the North
Rankin field has 245 billion cu m of recoverable reserves, the
Goodwin Field has 73 billion cu m, and the Angel field has 40
billion cu m of recoverable reserves. This gives a total of
358 billion cu m of natural gas, trapped in layers of porous
sandstone, more than 3km below the sea floor. Its composition
is approximately 3.26% CO_2; 0.62% N_2; 88.58% CH_4; 5.69%
C_2H_6; 1.25% C_3H_8 and traces of higher paraffins.

The North Rankin A platform, sitting in 125m of water is
to be followed shortly by a second platform. Rankin A was
completed in 1982, weighs 50,000 tonnes and cost $600M. Both
platforms will have more than 30 production wells direction-
ally drilled. A platform at Goodwin will be built in the
early 1990's (Woodside Petroleum Ltd. Reports). Sand and
water will be removed on the platform before the raw gas, at
123psi (or 8580kPa) and the condensate will then travel 134km
to shore by a 40" (1016mm) seabed pipeline already in place.
A processing plant has been built at Withnell Bay, near
Dampier and the purified gas will then travel a further 1,300
km by pipeline to Perth and other towns. The State Energy
Commission will buy 10.9M cu m/day for 20 years. Estimated
costs for the natural gas project are $10.6 billion. Local
gas should commence to flow in July 1984. Export of liquified
natural gas (LNG) to Japan has been arranged in order to
create an economically viable base load. The first exports of
some 2M tonnes should commence in the latter part of 1988;
rising perhaps to 6M tonnes/year by the mid 1990's (*The
Australian*, December 8th 1983). Peak export earnings should
be greater than the whole Australian wool clip or the total
earnings of the iron ore industry.

In addition to all the above a very large North Scott
Reef No. 1 gas field was found in the Browse Basin (Figure 2)
in 1982, and in 1983 gas, rich in condensate was discovered
in the Dampier Sub-Basin (Woodside Petroleum 1983).

An innovation applicable to all of outback Australia, but
being studied particularly in Western Australia, is the conversion of remote diesel engines to operate either on diesel
or on natural gas. Away from pipelines compressed natural gas

can be used. A considerable reduction in operating costs
results (State Energy Commission of Western Australia 1982).

   3.   Uranium. Central Western Australia has one large
potential uranium mine at Yeelirie but this also will
probably not be allowed to start. (See chapter 14 by Ollier
for more detailed discussion of the mining industry).

   4.   Renewable Resources. Much *solar* research and
development is being done in Western Australia (for details
see the Solar Energy Research Institute of Western Australia
(SERIWA) Annual Report 1981/82). Solar resources can now
obviously be used successfully on a small scale, Outback. An
advantage of course, is the high level of solar radiation
received in much of the State. A remote homestead power
supply using a windmill combined with an array of flat plate
solar cells recently supplied 7.8kW, and such combined systems
will probably be adopted more generally. A *hydro-electric*
station, which is a secondary solar resource, is being considered for the Ord River Dam, generating 2 x 25 MW, and
feeding from Lake Argyle. It will supply Wyndham, Kununurra,
and the new diamond mine, but it is of course, in industrial
terms a small unit.
   Many possibilities are being looked at in Western
Australia for obtaining fuel from the biomass, including
particularly *ethanol* from cane sugar (Energy Policy Western
Australia 1982, State Energy Commission W.A. 1981). There are
potentially large cropping areas in the Ord and Fitzroy River
regions, which could yield 2.2M kl/year. Any biomass source,
however, would take a long time to develop to signifiance,
would have the inconveniences of planting, fertilising,
cropping and transportation to a processing point, and would
in consequence be more expensive at present than oil products.
A better route to obtain more transport fuel would probably be
the conversion of natural gas to petrol.
   Finally Kimberley *tidal* power is a renewable resource
with some potential. Western Australia has nearly all of
Australia's tidal potential. Studies in 1974 cut construction
costs from those calculated in a much earlier study, and
reduced the lead time, but it is still not economically
attractive (Saunders 1976). Lack of a nearby energy market is
a major problem. Tidal power is non-uniform and a retiming
mechanism must be used, including pumped storage and double
basin schemes. In general much generating capacity must be
installed to obtain a comparatively small energy output, and
at the present time this is not cost effective.
   In general then the renewable energy source options,
while valuable in small or isolated situations, pale into

insignificance against the vast natural gas and condensate resources soon to be available in tropical Western Australia, supported now by some oil, and coal in the south.

III.   CONCLUDING COMMENTS

The energy picture emerging from this short resource survey is one of exciting promise. For decades the north was regarded as a barren area of little consequence to the rest of Australia. Now the two States and the Territory, thanks largely to newly developed scientific techniques for natural resource surveys, have burst forth with enormous resources. North Queensland is synonymous with *coal*, both coking coal and steaming coal, in great quantities . The Territory is synonymous with *uranium*, but the cloud here is the present Federal Government policy, fraught with inconsistencies and uncertainties. Fortunately oil and natural gas are both showing great potential in the Territory, so its future on this basis alone seems assured. Western Australia has a great future with natural gas.

Throughout northern Australia, and particularly with the emphasis being given to it by Western Australia, the renewable energy options are proving valuable for the generation of small amounts of power, and for isolated positions.

For the foreseeable future however it is the non-renewable resources, in vast quantities, which are going to be the energy source for the various ecosystems of northern Australia. As discussed later by Saini (Chapter 15), Auliciems and Dedear (Chapter 17) and Szokolay (Chapter 16), efficient use of this finite resource pool nevertheless will ultimately depend on improved building design and development of new material technology and cooling systems.

REFERENCES

Allen, R.J.   1983.   *Queensland Government Mining Journal,* pp.147,150.

Edwards, P.R.   1983.   A wind power survey of the Northern Territory of Australia, *Solar World Congress,* Perth, (August).

Energy Policy Western Australia.   1982.   *Government of Western Australia,* November.

Northern Territory Energy Policy.   1982.   *Energy Division, Department of Mines and Energy, N.T. Government,* Darwin, pp.9,13.

Queensland Energy Advisory Council.   1983.   *Annual Review and Energy Statistics,* Brisbane.

Saunders, D.M.   1976.   State Energy Commission of Western Australia, July.

State Energy Commission of Western Australia.   1981. *Research Priorities for Transport Fuels,* No. 4, Perth.

State Energy Commission of Western Australia.   1982. (2nd July).

Woodside Petroleum Ltd.   1983.   *Chairman's Address to Shareholders,* May 4.

6

THE HUMAN POPULATION AND SOME DIMENSIONS OF ECOLOGICAL STRUCTURE IN NORTHERN AUSTRALIA

DON PARKES

Department of Geography
University of Newcastle
New South Wales

I. INTRODUCTION

This chapter discusses features of the distribution and ecological structure of the population in northern Australia. The data are drawn from computer tape records of the 1981 Australian census. All values discussed relate to population counts on June 30, 1981. The Local Government Area is the unit of observation. Most of the ratios used here have been calculated specifically for this discussion and are not generally available in published catalogues, produced by the Commonwealth, State or Territory Divisions of the Bureau of Statistics.

Variables used relate to population and housing. They allow description and comparison of *local areas* in terms of their values on each of the selected variables. The data, and the results of analysis upon them, are described as *ecological*. They refer to the variable characteristics of a *population* within a bounded area, as well as to associated aspects of housing or *habitat*. They do not refer to individuals.

The considerable volume of data which is available for each local area has to be generalised in some way if a comprehensible indication of structure or order is to be obtained. Therefore, the discussion which follows is predicated on a *model*; a generalisation derived from the application of certain rules of procedure, to a set of data.

NORTHERN AUSTRALIA
ISBN 0 12 545080 X

The outcome is a description of the regionally variable characteristics of population and of 'habitat'.

One hundred and eighteen local government areas, from now on usually called *local areas*, have been used in this study. For the Darwin region, local areas within the City of Darwin have been amalgamated. This amalgamated area is also joined with two contiguous statistical subdivisions, known as Balance of 1945 Area and 1973 Acquisition, to produce a single local area here called urban Darwin. All other areas are statutory local government areas or areas treated as such by the Australian census.

The use of local government areas, as a basis for discussion of population distribution and structure has at least two disadvantages. The most important disadvantage is that the distribution of persons *within* each area is not revealed. Therefore, distribution must be assumed to be either even over the area or concentrated at a single point; for instance at the geographical centre or *centroid*. For mapping purposes the latter has been assumed. Distortion is greatest in the extensive territories away from the urban concentrations and especially when there is an urban place located within a larger statutory area, as for instance for Alice Springs, Tennant Creek and Katherine in the Northern Territory. The other disadvantage arises when local government boundaries cut across well populated territory, as for instance in the more densely populated areas of eastern Queensland. The boundary between Townsville and the growing sub-urban shire of Thuringowa to the south and west provides one example. Smaller scale census districts are not appropriate.

A number of Australian geographers have published studies of the distribution of population, Australia-wide, but apart from density criteria or estimates of the potential of the population there has been no attempt to evaluate the ecological composition of the population (Gale 1978, Holmes 1972, 1978, 1981, Burnley 1981, Parkes 1983). Courtenay (1978, 1982) has presented the most recent, thorough study of northern Australia focussing on problems of tropical development in an advanced country, however population data relates, inevitably, to 1976 and uses the much coarser regional scale of statistical divisions. Courtenay's book (1982) and the other studies referred to above do however present a volume of interesting and complementary data and argument to enhance the discussion which follows. Gale's study of Aborigine population is a particularly valueable reference, used in conjunction with Burnley's shorter chapter, following.

II.   POPULATION AND 'HABITAT'

A.   *The Arena: One Variable at a Time*

Figure 1 shows the pattern of local areas in northern Australia.   These are the polygons within which, and to which, the analyses and discussion which follows, are related. Where possible, given the small scale, names have been included.   The small local areas such as Weipa, Katherine, Tennant Creek and Alice Springs are presented by an enhanced symbol.

The median size of local areas is 7,775.7 square kilometres, but the average size is 34,820 square kilometres.

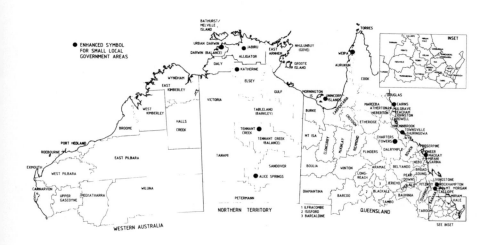

FIGURE 1.   Local government areas of northern Australia used as local area polygons[a]

[a]*This map is based on a computer plot prepared for the author by the Division of National Mapping in Canberra.   A program prepared by P.J. Young, University of Newcastle, excluded those local areas with centroids south of the 26th parallel.*

This demonstrates the wide range of sizes involved and the
large number of relatively small areas; most of them in
Queensland, to the east, and the south east of the region.
The largest of the local areas is East Pilbara (380,397.3
square kilometres), an area the size of the state of
Victoria, but with a population of only 9,751 people.   The
mining town of Newman, in the west of the area is the major
concentration of people (5,466).   The smallest local area is
the bauxite mining area, Weipa on the north west coast of the
Cape York Peninsula, with an area of 5.52 square kilometres
and a population of 2,433.

Based on the set of local areas shown in Figure 1, the
population of northern Australia was 908,370 people in 1981.
These figures relate to census counts and not to usual place
of residence.   South of the 26th parallel, about 14 million
people occupy 3.5 million square kilometres.   Northern
Australia, as defined in Figure 1, has an area of 4.1 million
square kilometres.   Figure 2 shows the distribution of
population across the arena.

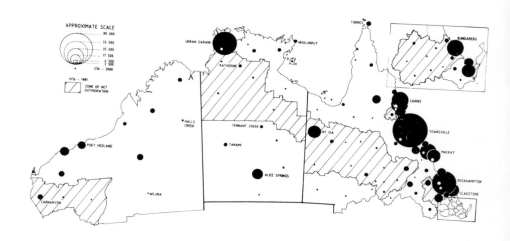

FIGURE 2.   Population distribution in northern
Australia 1981[a]

[a]The net migration pattern is generalised from Hugo (1983),
and has been added to this map.

The actual population for each local area in 1981 will be found in Table 1. Local areas are ranked according to size: associated population change between 1976 and 1981 is also shown.

TABLE 1. *Local Area Population Size 1981 in Rank Order and Population Change 1976-1981*[a]

RANK	AREA	1981 POP.	1976-81 CHANGE	RANK	AREA	1981 POP.	1976-81 CHANGE
1	TOWNSVILLE	81172	807	60	CLONCURRY	3651	-385
	DARWIN URBAN	56482	13138		FITZROY B	3382	477
	ROCKHAMPTON	52383	1250		DALRYMPLE	3338	752
	CAIRNS	39096	4239		MONTO	3299	71
	BUNDABERG	30937	481		CARPENTARIA	3273	1184
	PIONEER A	24887	5611		TAROOM	3237	134
	MOUNT ISA	24390	-2106		MOUNT MORGAN	3136	-331
	MULGRAVE A	23327	9431		TENNANT CK	3118	882
	GLADSTONE	22712	3764		BAUHINIA	3086	714
10	MARYBOROUGH	21530	3		EXMOUTH	2899	78
	MACKAY	20664	440	70	GAYNDAH	2859	45
	AYR	18477	62		FLINDERS	2740	-135
	ALICE SPRINGS	18395	4246		HALLS CREEK	2541	607
	JOHNSTONE	17438	662		MUNDUBBE	2481	86
	HERVEY BAY	16402	6098		WEIPA	2433	143
	LIVINGSTONE	15711	4077		KOLAN	2358	-326
	ROEBOURNE	15173	3939		ELSEY	2296	90
	BANANA	14519	350		SANDOVER	2263	431
	MAREEBA	14003	1867		GROOTE EYLANDT	2230	171
20	HINCHINBROOK	13683	-291		BLACKALL	2223	63
	BOWEN	13645	2353	80	VICTORIA	2157	-3
	PORT HEDLAND	13490	1711		TENNANT CK B	2089	345
	THURINGOWA A	13432	5396		TIARO	2066	191
	PROSERPINE	10833	3087		WINTON	1995	57
	EAST PILBARA	9751	1570		MIRIAM VALE	1971	495
	WEST PILBARA	8870	1377		PEAK DOWNS	1958	719
	PIONEER B	8845	1183		DALY	1830	124
	CALLIOPE	8700	3645		BARCALDINE	1783	3
	EMERALD	8435	2411		WOONGARRA B	1600	36
30	DUARINGA	8329	636		BATHURST	1586	149
	WOONGARRA A	8265	1038	90	PETERMANN	1523	228
	MULGRAVE B	8008	-1121		MCKINLAY	1477	9
	CARDWELL	7853	1375		WILUNA	1476	597
	BELYANDO	7700	490		BIGGENDEN	1411	-121
	ATHERTON	7501	1261		RICHMOND	1383	-59
	CARNARVON	7215	490		BURKE	1328	191
	SARINA	6922	1070		MEEKATHARRA	1281	-4
	BROADSOUND	6908	3529		FITZROY A	1263	727
	CHARTERS TWRS	6823	-1091		EIDSVOLD	1256	24
40	WEST KIMBERLEY	6627	1581		JERICHO	1177	-43
	TORRES	6131	130	100	GULF	1115	228
	DOUGLAS	5957	1211		TABLELAND	1115	-26
	WYNDHAM	5259	1188		ARAMAC	1082	23
	DARWIN BALANCE	4930	1619		JABIRU	1022	1022
	BROOME	4869	790		ETHERIDGE	1010	70
	MIRANI	4739	-150		NEBO	914	114
	GOOBURRUM B	4725	-3		AURUKUN	791	171
	ALLIGATOR	4481	1127		MORNINGTON	773	371
	WOOCOO	4456	1314		TAMBO	762	94
50	COOK	4456	1238		BARCOO	711	54
	EAST ARNHEM	4320	868	110	UNINCORP ISL.	675	322
	THURINGOWA B	4296	1418		BOULIA	660	25
	EACHAM	4137	704		ISISFORD	605	174
	TANAMI	4058	529		GOOBURRUM A	536	37
	ISIS	4023	97		ILFRACOMBE	460	32
	NHULUNBUY	3879	326		DIAMANTINA	366	111
	LONGREACH	3846	-206		PERRY	309	5
	KATHERINE	3737	610		CROYDON	255	33
59	HERBERTON	3688	9	118	UPPER GASCOYNE	236	-10

[a]*Indented labels are areas with same name, usually part (A) and (B) of a single shire. Tenant Creek B is balance of Tennant Creek Town.*

1. *Northern Western Australia* has a population of
79,687 people, occupying an area of 1,470,843 square kilometres, more than 50% of the entire state.   83% of this
population live in urban centres and bounded rural localities.   Urban centres, according to the census classification, have 1,000 or more people and 'include known holiday
resorts of less population if they contain 250 or more
dwellings of which at least 100 are occupied on census
night'.   Bounded rural localities have between 200 and 999
people.   Thus, because these pockets of population in
northern Western Australia occupy a very small area, a mere
13,483 people are distributed over an area of about 1,400,000
square kilometres, about 18% of the entire continent.   By way
of comparison, the North American state of Texas will fit
into this area about five times and has a population of
12,000,000:   a population nearly 1000 times larger on about
20% of the area!   The northern area of Western Australia
occupies about 33% of the northern Australia arena and has
8.7% of the population.

2.   *The Northern Territory* extends from the Timor and
Arafura Seas in the north to the 26th parallel in the south.
There were 123,324 people occupying an area of 1,343,675
square kilometres, in 1981:   an area almost exactly the same
size as northern Western Australia.   Because the Northern
Territory, as a whole, lies within the defined area of northern Australia it is possible to use census under-enumeration
estimates.   For the Northern Territory the estimate is 4.98%
(Statistician for the Northern Territory 1983), suggesting an
'actual location' population of 129,366.   Because it is not
possible to use under-enumeration adjusted figures for the
other two states, or for the local areas, all figures used in
the discussion which follows, relate to unadjusted census
counts.
     Of the enumerated population of 123,324;   84.5% resided
in 34 urban centres and bounded localities, of which only ten
centres accommodated a total of 91,479 people: 74.2% of the
total population of the Northern Territory.   This already
high concentration is increasing.   In 1976 only 66.4% of the
population lived in urban centres.   The remaining 19,074
people are distributed over an area of about 1,300,000 square
kilometres.   This dispersed population is distributed in the
ratio of one person per 70 square kilometres.   The equivalent
ratio for Western Australia, north of the 26th parallel is
one person for every 109 square kilometres.

3.   *Northern Queensland* has a population of 705,305

people: 77.6% of the total population of northern Australia,
distributed over an area of 1,294,244 square kilometres.   In
northern Queensland, 511,305 people live in urban centres and
bounded localities.   This is 72.5% of the total population,
and is a marginally smaller proportion than occurs in the
northern region of Western Australia or in the Northern
Territory.   However there are many more centres involved and
whereas a mere 13,483 people in northern Western Australia
and 19,074 in the Northern Territory were dispersed over the
remainder of their respective areas, in Queensland, north of
the 26th parallel, 194,050 people were involved.   Queensland
thus demonstrates much closer settlement than the other two
major regions of northern Australia.   The overall density
being one person per 7 square kilometres.   None-the-less
there are vast areas of 'empty' territory especially in the
Cook, Carpentaria and Mareeba areas of Cape York and in the
south west, in the Channel Country of Diamantina and
Windorah, and in Boulia immediately to the north and
extending to the Northern Territory border.   In a general way
we can say that northern Queensland is more 'closely settled'
by a factor of 10 than the Northern Territory and more
closely settled by a factor of 15 than is northern Western
Australia.

B.   *Attributes of Population and 'Habitat' in Local Areas*

A summary of the percentages and the absolute numbers of
persons, households and housing related characteristics is
shown in Table 2.   Northern Australia is treated as a single
region.   These values are not elsewhere available in this
form because northern Australia is not defined as a region
for official enumeration purposes.

Variable attributes of population and habitat, which are
considered in more detail in the next section, are summarised
in Table 3.   The structure of the ratios used, is also
defined in the table.   Unlike Table 2, the values here are
based on the mean of the values for each local area: thus
the value of 16.7% for Aborigines and Islanders is the
average value over all of the 118 local areas, whereas in
Table 2 it is the percentage based on total numbers for the
entire region of northern Australia.

Most local areas in northern Australia experienced an
increase of population between 1976 and 1981 as can be seen
from Table 1.   The greatest decrease was in Mt. Isa in north
western Queensland.   Mt. Isa is sometimes described as the
largest 'city' in the world.   This is misleading!   The

TABLE 2.   *Summary of Selected Features of Northern*
             Australia

FEATURES	Region Mean %	Absolute Number
Australian born as % of total population	85.0	772846
Aborigine and Islanders as % of total population	8.5	76863
Unemployed as % of working age [>15 <61 for f. <66 m.]	4.0	23471
Never married adults as % aged over 15 yrs.	25.1	183989
Separated and divorced as % aged over 15 yrs.	6.3	41339
Lived in same LGA in 1976 as % total population	48.7	442445
Fertility ratio - under 5 as % females aged 15–49	36.3	80618
Youth dependency ratio - 5-14 as % over 15 yrs.	26.4	174006
Old age dependency ratio - over 65 as % of work age	11.5	67418
Fulltime education as % over 15 yrs.	4.3	27791
First and higher degrees as % over 15 yrs.	2.8	18020
No qualifications as % over 15 yrs.	65.4	427297
Household income <$12 000 as % all households	32.5	77808
Worked >35hrs./week as % of working age	48.3	283333
Household med/high density housing as % all households	9.9	23744
Household in improvised dwellings as % all households	1.6	3790
Household own/buy dwellings as % all households	54.3	130085
Tenant of housing authority as % all households	5.0	12062
Tenants of private dwellings as % all households	28.1	67423
Overseas born as % of Australian born	17.5	135357
TOTAL POPULATION OF NORTHERN AUSTRALIA	100.0	908370

description merely refers to the fact that the constituted
local government area, associated with the 'chartered' city
of Mt. Isa, has an area of 41,190.56 square kilometres,
extending to the Northern Territory border, 200 kilometres to
the west and including the small pastoral service town of
Camooweal: at the other end of the same 'Main Street'.
However the urban centre of Mt. Isa did experience an
absolute population decline of 1,698 people, or 6.7% during
the period 1976-1981.   The greatest increase was in Darwin,
13,138 people or 23.2%.

Selected aspects of the data presented in Tables 2 and
3 will now be discussed before turning to discussion of the
ecological structure of local areas.

76,863 Aborigines formed 8.5% of the population of
northern Australia.   Because under-enumeration is usually
higher among Aborigines than among the rest of the population,
it is possible that their numbers could be as high as 80,000,
if an under-enumeration value of 5% is used.   The median
value for all local areas was only 3.9%:   the mean as
mentioned above, was 16.7%.   The highest proportions were
on Bathurst/Melville Islands in the Timor Sea, off the north
west coast of Darwin;   the population was 1,475, or 93% of
the total population.   Burnley discusses some general
demographic aspects of the Aborigine population of northern
Australia in the next chapter.

TABLE 3. Variables and Summary Statistics

V A R I A B L E S	$\overline{x}$	S.D	C.V.%	MIN. VALUE	MAX. VALUE	CODE USED ELSEWHERE
% Regional population	0.8	1.3	151.0	0.02	8.9	REGPOP
% Aborigine & Islanders	16.7	25.5	152.2	0.00	93.0	ABORTSI
% Unemployed	3.6	2.6	72.0	0.00	26.7	UNEP
% Never married adults	29.4	5.9	20.0	16.00	53.0	NMA
% Separated & divorced	5.8	1.7	28.7	1.80	9.8	SEP&DEV
% Same LGA in 1976	47.5	13.4	28.1	1.90	73.2	SAMELGA
Fertility Ratio as %	40.3	9.3	23.1	0.00	78.4	FERTILITY
Youth dependency ratio %	28.6	6.5	22.7	0.00	54.2	YOUTHRAT
Old age dependency ratio %	10.0	6.1	60.9	0.70	32.5	OLDRAT
% Fulltime education	3.8	1.6	40.2	0.00	8.8	FULLTIED
% Degrees	2.2	1.1	52.2	0.00	5.7	DEGREES
% No qualifications	67.3	10.1	14.9	5.80	88.0	NOQUALIF
% <$12,000 income/house	33.7	12.0	35.5	0.00	57.4	LD12000
% Work >35 hours/week	47.7	11.7	24.4	7.0	71.0	WORKO35H
% Med/high density house	4.6	5.7	123.3	0.00	23.0	MEDHIDEN
% Improvised dwellings	5.8	12.3	212.9	0.00	67.8	IMPROVIS
% Own/buy house	45.5	24.0	52.8	0.00	81.4	OWNBUY
% Tenants of house/auth	3.9	7.6	193.7	0.00	36.1	TENHOSAU
% Tenants private dwell	27.3	19.2	70.2	0.00	90.2	TENPRIDW
% Dwell sale empty let	1.8	1.4	78.5	0.00	8.7	SALETEMP
% Born overseas	12.1	10.2	84.6	0.80	52.6	BORNOSEA
% Pop change 1976-1981	19.5	27.5	141.0	-13.79	150.0*	% CHANGE

* The value 150.0 relates to Jabiru. Between 1976-1981 the population increased
   from zero to 1022. It was necessary to set a smaller ratio so that the effect
   of an extreme outlier would not occur in the multivariate analyses.

The ratio of *unemployed* to the total population of
persons of working age was 4% for the northern Australia
region, and 85% of local areas reported a value less than 5%.
In the 1981 census people were asked to answer the question,
'Did the person look for work last week?' Looking for work
was defined in various ways including registration with the
Commonwealth Employment Service. The highest reported level
of unemployment was in Aurukun, in the Archer River District
of northern Queensland, with a value of 26.7%. Other above
average figures were reported at Petermann in the south west
of the Northern Territory (7.4%), in Meekatharra in Western
Australia (7.32%) and a number of other areas, usually with
above average proportions of Aborigines.

The high *mobility* of the population in northern
Australia, also referred to by Bauer in Chapter 1, is
indicated by the low median value of 48% resident in the same
local area in 1976. The median *fertility ratio*, children
under the age of five to women in the age group 15 to 49,
was 39.1%. The mean value was 40.3% indicating a distribution with very little skew. The highest values were in
those areas where the aboriginal population exceeded 60%, for
instance in Daly in the Northern Territory (63.4%). The
highest value was in the Upper Gascoyne region of Western

Australia, south of the Pilbara (78.4%). However there were
only 236 people in this region of 56,989.84 square kilometres,
of whom only 45 were Aborigine and only 51 were women in the
age group 15 to 49.
     A youth *dependency ratio* between children aged 5 to 14
years and the total population aged over 15 years, had a
median value of 28.3%, and a mean of 28.6%. Once again the
population is distributed with very little skew, (0.8).
Excluding the unincorporated islands of Queensland, which had
a total population of 675 people and no children under the
age of 15 years, the youth dependency ratio changed between
12.0% and 54.2%. Apart from the Upper Gascoyne, which is an
empty and desolate area with the lowest local area population
in northern Australia, other areas with low youth dependency
included Exmouth, East and West Pilbara (WA) with their iron
ore mining towns and Etheridge, Cairns, Hinchinbrook,
Townsville, Proserpine and Charters Towners in Queensland.
In general those local areas with Aborigine proportions in
the upper quartile (top 25%), also have upper quartile values
for youth dependency. In many of the 'outback' towns of
northern Australia, children in the age group 12 to 18 are
sent away to boarding schools and this does affect the count
of the number of children in the youth dependency ratio, by
how much precisely it is not possible to estimate here. The
aboriginal areas with relatively high youth dependency ratios
are likely to experience some aggravation of the unemployment
levels during the next decade or so, but increasing life
expectancy will probably reduce the youth dependency level
during the same period. *Old age dependency*, the ratio of
people over the age of 65 years to those of working age, has
a median value of 8.8% and a mean value of 10.0%. Low ratios
of less than 5 persons over 65 years of age per hundred
persons of working age were found in 21 (18%) of the local
areas. Many of these were mining areas, including Tennant
Creek (Township) with 3.7%, Nhullunbuy the bauxite mining
town in Arnhem Land (1.4%), Groote Eylandt (1.2%), East
Arnhem (Balance) 1.7%, the uranium mining town of Jabiru
(2.%), West Pilbara including the iron ore mining settlements
of Tom Price, Paraburdoo and Pannawonica (2.7%), Roebourne
including the Robe River Iron company's port town of Wickham
(7.2%) and East Pilbara, with Newman as the dominant settlement (2.5%).
     The proportions of persons aged 15 years and over with
higher *tertiary qualifications*, that is degrees and higher
degrees, was highest in the urban Darwin region, with a ratio
of 5.7%. The median value for northern Australia was 2.05%
and the mean was 2.2%. Alice Springs was second highest with

a value of 5.2%, followed by the uranium mining settlements of Jabiru (5.1%), Nhullunbuy (4.8%), and Groote Eylandt (4.6%), all in the Northern Territory. At the other end of the education level indicator, the ratio of those with *no qualifications*, who were also aged over 15 years, to the total population aged 15 years and over, had a median value of 68.5%; the mean value was 67.3%. Wiluna in Western Australia had the highest proportion of persons with no qualifications (88%). The lowest value, excluding the unincorporated islands of Queensland was at Jabiru, 36.5%.

The number of households with *annual incomes* equal to or less than $12,000, as a ratio of all households, had a median value of 35.0% and a mean of 33.7%. Mount Morgan, 30 kilometres south and west of Rockhampton in Queensland, is a copper smelting and refining town, linked to the distant Peko-Wallsend ore deposits at Warrego, 50 kilometres north and west of Tennant Creek in the Northern Territory. The Shire of Mount Morgan, within which lies the Mount Morgan urban centre (population 2,974 of a total Shire population of 3,136) had the highest proportion of households on incomes less than $12,000 per year, in 1981. This local area apart, however, it is in the extensive areas of high aboriginal populations, where there is no supporting mining operation or tourist industry, that the highest proportions of low income people are found. It can only be inferred that these people are also Aborigine. Some examples include Tanami (54.1%), Sandover (51.8%), Gulf (45.0%), all in the Northern Territory; Halls Creek (49.0%) and Wiluna (51.2%) in Western Australia. It is worth noting however that across all the 118 local areas there is no correlation (r=0.05) between the proportion of households with incomes less than or equal to $12,000 and the proportion of Aborigines. This must however be further assessed in terms of the cultural significance of the European notion of household and of the behavioural characteristics of Aborigine people in relation to the sharing of 'wealth'. The *lowest* proportions of low income households are associated with the tourist and mining areas; for instance Alice Springs (21.1%) and Katherine (23.7%) have aboriginal populations of 16.4% and 16.1% respectively: the mean for northern Australia was 16.7%. However it is in the newer mining areas of Western Australia's Pilbara and the Top End of the Northern Territory that low income proportions are at a minimum. East Pilbara, dominated by Newman has a value of 10.8%, Port Hedland in Western Australia (12.0%), Roebourne (8.1%) and West Pilbara dominated by the Hamersley mining towns of Paraburdoo and Tom Price 7.3%). Nhullunbuy (7.5%) and Jabiru (3.7%) in the Northern Territory, and Weipa

in Queensland (3.8%) all indicate the high income levels and
real purchasing power of the new mining regions.   Until
recently, and in the case of Paraburdoo, until July 1983,
many of the newer mining towns have been *closed* company
towns.   Residence in these towns was dependent on employment
by the company or by the various service and contract
companies with leased accommodation in the town.   Such
contracts and other services were closely tied to the ecologically dominant mining function.   For northern Australia as a
whole the proportion of households with incomes less than or
equal to $12,000 was 32.5%, as shown in Table 2.

   Migrants from 'southern cities' and from overseas form a
high proportion of the population in northern Australia.   In
the East Pilbara iron ore mining region of Western Australia
the ratio of overseas born to Australian born was 52.6%:
every other person met was likely to have been born overseas.
The lowest was on Bathurst/Melville Islands, in the Timor Sea
north of Darwin, with a mere 0.8%.   The median value for
northern Australia was 8.5%, the mean was 12.1%.   Throughout
the mining towns, especially those developed during the past
15 years, the proportion of overseas born exceeds one third
and is usually closer to 40%.   Alice Springs, dependent on
public service and tourism to an increasing extent, has 23.7%,
including a substantial population from the United States,
associated with the Pine Gap communication facility, 20
kilometres to the south and west.   Tennant Creek, an older
mining centre 500 kilometres north of Alice Springs has one
overseas born resident to every five Australian born.   Many
of the Australian born are however children of the overseas
born.   It is possible that these high levels of transience,
which are characteristic of small urban settlements 'out
back', would be even higher if the proportion of overseas
migrants was not so high.

   Our discussion now shifts to a consideration of the
ecological structure of the local areas in northern
Australia.   Whereas the present and previous section has
dealt with a number of attributes of the population, treated
as separate features such as fertility or old age dependency,
in the next section 22 variable characteristics of population
and 'habitat' will be treated in terms of the manner in which
they vary together, from area to area.   As mentioned in the
Introduction, it is from a 'model' of the northern Australia
population, its distribution and ecological structure, that
we are able to gain some preliminary insight into not only
how many people occupy this vast arena, or their distribution
over it, but also into the manner in which certain selected
attributes associate with one another.   In some local areas

a set of attributes can be observed to occur in high proportions, while others are absent or weakly represented.   In
other areas there is an obverse condition.   By correlating a
number of variables with each other we can identify
systemmatic changes in their variation, from place to place.
This systemmatic variation can then be studied by various
statistical models in order to derive an indication of the
underlying structure of the variation, suggested by the
correlations.   This search for the underlying structure is
here undertaken as a 'model' known as principal components
analysis.   From this model it is possible to allocate
scores, new values, to each of the 118 local areas and then
to group and map them so that a picture of the northern
Australia arena, in terms of all of the variables used, can
be produced.   The identification of similarities and
differences, using objective methods which can be replicated
by other researchers, for other regions, at other times is
an important part of the process of 'scientific' enquiry:
an ongoing process open to refinement and magnification of
detail relates to the use of other statistical models,
different sets of variables but possibly still including all
or most of those used here.

The choice of variables is a function of the availability of data and the objectives of the study, as well as the
time and money costs of preparation, analysis and interpretation.   There are also constraints related to the mathematics
of the method.   In the selection which follows I have tried to
keep the number as small as possible (some might say it is
already too large because the ratio of variables (22) to
observations (local areas 118) should ideally be about 1:10).
Those selected are designed to act as sensitive indicators of
population and habitat in northern Australia.   No variables
related to occupation types or industry classes have been used
because it was thought likely, that in the northern Australia
environment, where there is rather little diversity of occupation and industry, relative to the range of possible categories, there would be rather a severe overemphasis of these
categories.   This overemphasis would reduce the sensitivity of
the demographic and housing related attributes which have been
selected.

III.   ECOLOGICAL STRUCTURE AND POPULATION GROUPINGS

A.   *Independent Dimensions of Structure*

As we have seen local area populations in northern Australia vary between a few hundred people, and over eighty thousand people. Each, within its statutory boundary, is a *container* for a range of characteristics which combine in various ways.

The 22 variables and codes which are used in this study are the same as those which appear in Table 3. The first step in the building of a model is the calculation of correlation coefficients between all 22 variables. These coefficients appear in Table 4.

From these correlations, the rotated and sorted components, also known as dimensions, are derived. The first six dimensions are shown in Table 5. These six dimensions account for 74% of the total variation among the original 22 variables. Only the first four dimensions are considered in detail here.

Dimension 1: Ecological Change orders the local areas of northern Australia in a manner which distinguishes the recent mining and other growth zones from local areas which have been established for a longer period of time. The characteristics which contribute most to this distinction are a relatively small proportion of people who were also resident in 1976, a

TABLE 4.   *The correlation Matrix*[a]

	1	2	3	4	5	6	7	8	9	10	11	12	13	14	15	16	17	18	19	20	21	22
1 REGPOP	1.0																					
2 ABORTSI	-21	1.0																				
3 UNEMP	11	28	1.0																			
4 NMA	-14	37	-03	1.0																		
5 SEP&DEV	20	15	14	27	1.0																	
6 SAMELGA	05	29	15	-04	-50	1.0																
7 FERTILIT	-28	57	05	19	-02	14	1.0															
8 YOUTHRAT	-18	54	19	04	-28	47	51	1.0														
9 OLDRAT	18	-31	09	-44	-28	38	-34	-22	1.0													
10 FULLTIED	16	-27	06	12	-24	17	-30	09	16	1.0												
11 DEGREES	33	-03	02	03	55	-49	-03	-11	-28	01	1.0											
12 NOQUALIF	-12	02	-04	06	-46	56	24	33	31	14	-50	1.0										
13 LD12000	-10	05	03	-15	-34	48	-03	05	68	13	-43	56	1.0									
14 WORKO35H	-02	-69	-42	08	-13	-18	-13	-26	00	23	03	30	-08	1.0								
15 MEDHIDEN	59	-18	09	-11	39	-19	-22	-19	00	06	51	-36	-38	02	1.0							
16 IMPROVIS	-20	73	26	09	16	16	34	29	-18	-27	00	00	25	-58	-21	1.0						
17 OWNBUYHO	16	-58	03	-43	-42	34	-38	-08	66	39	-35	42	55	34	-14	-41	1.0					
18 TENHOSAU	11	02	04	01	43	-34	00	-09	-19	-23	47	-25	-38	04	49	-06	-39	1.0				
19 TENPRIDW	06	27	03	17	18	-13	36	19	-44	-14	38	-27	-64	-12	35	-04	-66	19	1.0			
20 SA:ETEMP	-04	-31	-18	-04	-20	05	-08	-01	11	14	-13	15	01	36	03	-30	21	00	-05	1.0		
21 BORNOSEA	24	-25	-02	-14	42	-58	-23	-36	-26	-12	58	-64	-62	02	58	-19	-37	46	43	-05	1.0	
22 % CHANGE	-06	-02	-06	-11	15	-40	02	-08	-18	-15	26	-33	-28	-09	01	-01	-20	08	28	-16	24	1.0

[a]*The correlation coefficients are printed without the leading zero BEFORE the decimal point. The decimal point has also been omitted in order to allow the entire lower triangle of the matrix to be printed. Thus the value -21 would be read as -0.21.*

TABLE 5.   Varimax Rotated and Sorted Component Loadings[a]

	D1	D2	D3	D4	D5	D6
SAMELGA	0.78					
% CHANGE	-0.74					
NOQUALIF	0.74	-0.37				
LD12000	0.64	-0.38		-0.46		
MEDHIDEN		0.89				
REGPOP		0.68				0.32
DEGREES	-0.39	0.65				
TENHOSAU		0.63				-0.47
BORNOSEA	-0.60	0.62				
SEP&DEV	-0.34	0.51	0.26		0.45	-0.26
WORKO35H			-0.87			
IMPROVIS			0.74			-0.33
ABORTSI			0.72	0.47	0.31	
UNEMP			0.60			
SALETEMP			-0.58			
TENPRIDW	-0.31	0.36		0.74		
YOUTHRAT	0.31		0.27	0.71		
FERTILIT				0.66		
NMA					0.91	-0.37
OLDRAT	0.48			-0.53	-0.53	
FULLTIED						0.85
OWNBUYHO	0.38			-0.48	-0.48	0.41
VP	3.518	3.406	2.951	2.615	2.007	1.742

[a]*The above component loading matrix has been rearranged so that the columns appear in decreasing order of variance explained by the components. The rows have been re-arranged so that for each successive component, loadings greater than 0.50 appear first. Loadings less than 0.25 have been replaced by blanks. VP in the final row is the eigenvalue or sum of the squared loadings and it indicates the absolute amount of the total variance which is explained by the dimension. Thus the first dimension D1 accounts for 3.518 out of a total of 22.0: the number of variables.*

relatively high level of population increase, relatively few people with no qualifications, and low proportions of households with incomes less than $12,000. Where population increase was high the proportion of overseas born was also high, old age ratios were low, there were relatively high proportions of persons with higher tertiary qualifications and the proportion of people resident in the local area who were separated or divorced was also relatively high. Tenancy of private dwellings, rather than owner occupation, and relatively low youth dependency ratios, *together define one of the structural components of population and habitat*

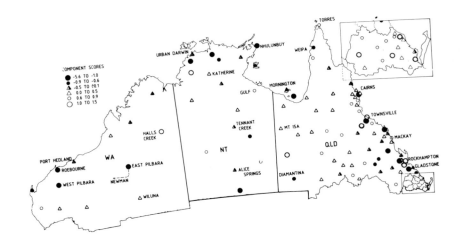

FIGURE 3.   Dimension 1:   Ecological Change.

differences in northern Australia.   The map, Figure 3, shows
the geographical pattern in *generalised* form.

On the first dimension of *ecological space* in northern
Australia, older areas such as Biggenden, Charters Towers,
Eidsvold, Mount Morgan, Maryborough, and Bundaberg all with
positive scores, differ most from areas such as Jabiru,
Broadsound, Fitzroy, West Pilbara, Darwin (Balance), Calliope
and Nhullunbuy.

Dimension 2:   Non-Urban/Urban order the local area in a
manner which distinguishes the built up areas with relatively
high proportions of medium and high density housing, a
relatively high proportion of the total population of northern Australia, higher proportions of people with tertiary
qualifications, housing authority tendancies, overseas born
and separated and divorced individuals, from local areas with
relatively low proportions on each of these items.   The
dimension is not simply structured in terms of population
size.   The newer mining areas of East Pilbara, Nhullunbuy and
Weipa for instance appear to have a similar profile to other
'urban' areas, with much larger populations, such as Mackay,
Mount Isa, Alice Springs, Townsville and Darwin.   Figure 4
shows the geographical pattern, in generalised form.   While
this dimension succeeds in distinguishing the urbanized local
areas rather well, their positive scores generally being at a
greater distance from the average than is the case for the
negative scores, it does not distinguish well between the

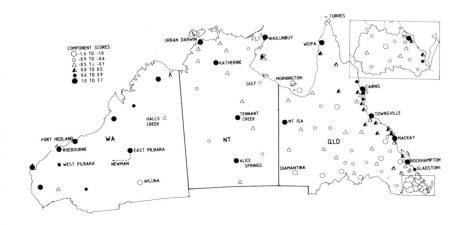

FIGURE 4.   Dimension 2:   Non-Urban/Urban.

great bulk of local areas in northern Australia.   This accords
reasonably well with prior expectations however, as most of
the local areas in northern Australia are still associated
with rather extensive ecosystems, with low populations and
essentially pastoral, rangeland related infrastructure.
Although the 'model' we have used to identify the dimensions
of population ecological structure, generates independent
dimensions, study of the local area score distributions does
suggest that the items involved in ecological change;   the
first dimension: are also assicated with the nascent urbanized zones which have recently developed as a result of mining
enterprise, for instance in the East and West Pilbara, Weipa,
Groote Eylandt, Tennant Creek, Port Hedland.

Dimension 3:   Race Ecology (Figure 5) identifies
variations in local area ecological structure, especially in
terms of Aborigine, work and housing related items.   Five
variables load highly on this dimension, but none of them
appear on either of the first two dimensions. (Table 5).   The
most discriminatory variable is the proportion of persons who
worked more than 35 hours per week.   The dimension indicates
that in local areas where the proportion of Aborigines (and
Islanders) is relatively high, there is also a relatively
high proportion of improvised dwellings, unemployed people
and, along with lower proportions working over 35 hours per
week, there are relatively few houses for sale, or to let;
indicating the operation of a distinct housing market.

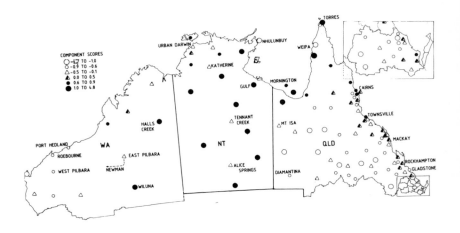

FIGURE 5.   Dimension 3:   Race Ecology

Because these are *ecological correlatives* it is not possible
to interpret them in terms of individual associations.   That
is to say we cannot state that Aborigines live in improvised
dwellings, are unemployed, and do not work for more than 35
hours per week.   However the ecological structure of this
dimension indicates that some local areas have high coincidence of these attributes and other do not.

Demographic characteristics such as fertility, youth and
old age dependency ratios are either not represented at all
on this dimension, or only rather weakly, for instance youth
dependency appears, but with a very low value of 0.27.   The
mining areas tend to have either low negative scores or else
are set about the average, with scores between + or - 0.5.
In other words these local areas tend to be distant, in
ecological space, from the extensive areas of Aborigine
'concentration', such as Aurukun, Wiluna, Tanami and
Petermann, with high positive scores.   These often include
large tracts of land which are either owned by the Aborigine
people or which include substantial reserves with which the
Aborigine people are associated.

Dimension 4:   Demographic/Race (Figure 6) reveals further
structural aspects of ecological variability in northern
Australia.   Like the third dimension, this one also teases out
systemmatic variations which include the Aborigines.   However
in this case the *racial* element is somewhat subordinate to
the demographic elements and to housing tenure.   The latter
show an inverse relationship between areas with high

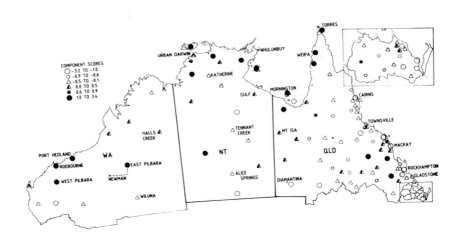

FIGURE 6. Dimension 4: Demographic/Race

proportions of privately tenanted dwellings; a typical condition in the newer mining areas and areas where owner-buyers
are in relatively high proportions.

Most of the local areas with positive scores greater
than 1.0 are either aboriginal concentrations such as Bathurst/
Melville Islands, East Arnhem, Burke and Mornington, or are
recent mining areas with new mining towns within them, such
as West Pilbara (Paraburdoo, Tom Price, Pannawonnica), East
Pilbara (Newman), Nhullunbuy and Weipa. These areas, though
differing on other dimensions have similar demographic and
housing (habitat) profiles. On this dimension of ecological
space, low proportions of household incomes below $12,000 per
year are inversely related to fertility ratios, youth dependency ratios, Aborigines and privately tenanted dwellings but
positively related to old age dependency ratios and relatively
high proportions of owner-buyers. The majority of areas are
only weakly distinguished from one another.

The discussion now concludes with a description of the
clustering or grouping of local areas into ten categories.
This grouping is derived by a method known as *k-means
clustering* (Engelman and Hartigan 1981) in which a set of
observations, in this case 118 local areas are partitioned
into a specified number of groups, within which there is
greater similarity among the members than with non-members.
Various metrics may be used to determine the similarity of
observations, and in this case the Euclidean distance has
been used. The six dimensions (Table 5) from which scores

108 Don Parkes

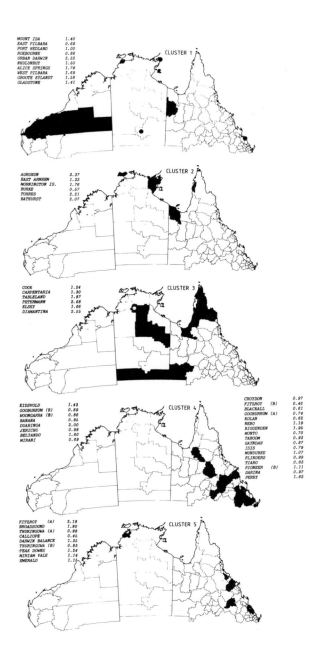

FIGURE 7.   Clusters of Local Areas on six dimensions.

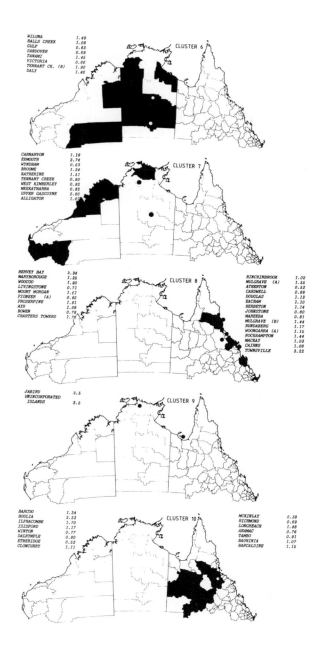

```
WILUNA            1.49
BALLS CREEK       1.08
GULF              0.83
SANDOVER          0.69
TANAMI            1.45
VICTORIA          0.66
TENNANT CK. (B)   1.90
DALY              1.46
```
CLUSTER 6

```
CARNARVON         1.19
EXMOUTH           2.74
WINDHAM           0.63
BROOME            1.24
KATHERINE         1.51
TENNANT CREEK     0.80
WEST KIMBERLEY    0.92
MEEKATHARRA       0.82
UPPER GASCOYNE    2.60
ALLIGATOR         1.6
```
CLUSTER 7

```
HERVEY BAY        2.34
MARYBOROUGH       1.25
WOOCOO            1.90
LIVINGSTONE       0.71
MOUNT MORGAN      1.61
PIONEER   (A)     0.86
PROSERPINE        1.81
AYR               1.08
BOWEN             0.75
CHARTERS TOWERS   1.76
```
CLUSTER 8
```
HINCHINBROOK      1.02
MULGRAVE  (A)     1.55
ATHERTON          0.53
CARDWELL          0.68
DOUGLAS           1.13
EACHAM            1.10
HERBERTON         1.14
JOHNSTONE         0.60
MAREEBA           0.81
MULGRAVE  (B)     1.44
BUNDABERG         1.17
WOONGARRA (A)     1.15
ROCKHAMPTON       1.44
MACKAY            1.03
CAIRNS            1.06
TOWNSVILLE        3.22
```

```
JABIRU            2.5
UNINCORPORATED
    ISLANDS       2.5
```
CLUSTER 9

```
BARCOO            1.24
BOULIA            2.23
ILFRACOMBE        1.70
ISISFORD          1.17
WINTON            0.77
DALRYMPLE         0.80
ETHERIDGE         0.52
CLONCURRY         1.11
```
CLUSTER 10
```
MCKINLAY          0.38
RICHMOND          0.69
LONGREACH         1.48
ARAMAC            0.76
TAMBO             0.91
BAUHINIA          1.07
BARCALDINE        1.16
```

FIGURE 7 continued

were calculated are the basis for partitioning the initial set
of 118 local areas into ten clusters of local areas which are
most like each other local area within a cluster.

B.   *Clusters of Ecologically Similar Local Areas*

Unless there are theoretical or other empirical reasons
to determine the number of clusters to be formed from a
larger set of observations, the decision has to depend on
heuristic methods and other considerations such as how the
results are to be presented. Separate analyses were undertaken using five, seven and ten clusters. The results from
the ten-cluster analysis are presented in Figure 7.

The clustering procedure used here begins with 118,
single area clusters. In other words the initial set of 118
local areas are treated as belonging to distinct clusters.
In the strictest sense of course this is the most precise
condition, given the initial set of observation units. The
method proceeds by first grouping the two most similar areas
and continuing to group new areas, sometimes to other single
areas but progressively to the average of the values in some
existing group. The logical limit to the method is to end up
with a single cluster: in this case composed of all the
local areas in northern Australia. However the clustering is
terminated here with ten clusters.

Whereas our discussion in the previous section
distinguished among local areas in terms of their position on
each of six dimensions: four in some detail: we now
consider each area in terms of scores on all six dimensions
together. Without a mathematical model to establish
consistent and objective rules for deciding which areas most
probably belong in which groups, it would be quite impossible
to cope with a six-dimensional structure. Resort to
intuitive or other subjective criteria have their place in
some studies but they do not allow for replication or
refinement of the same or similar data sets at some time in
the future.

The local area members of each cluster, together with
Euclidean distance from the centre of the cluster to which
they belong are also presented. The smaller the value on the
Euclidean distance scale, the closer the area is to the
centre or mean value of its cluster. Thus for instance in
cluster 1, urban Darwin with a value of 2.25 is the most
'peripheral' member of the cluster. East Pilbara (0.69) and
Roebourne (0.88) are the most similar on the six dimensional
profile. In cluster 2, Aurukun, on the west coast of the
Cape York Peninsula has a relatively large value of 3.37 but

its inclusion in the same cluster as East Arnhem, Mornington, Burke, Torres and Bathurst/Melville Islands is, at least for the moment intuitively acceptable. Other clusters can be studied from the maps in Figure 7. Clusters 4, 5, 8 and 10 are dominated by Queensland local areas. Clusters 3 and 6 accommodate most of the Northern Territory, except for the climatically and physiographically distinct region known as The Top End. Clusters 1 and 7 accommodate all of the areas in western Australia.

This clustering of areas within each state to a marked degree is interesting in itself because there were no locational criteria used in the initial set of 22 variables. Western Australia has 13 local areas and the Northern Territory, 20 local areas. On the basis of a random allocation to each of the ten clusters, each cluster should have at least one Western Australia local area in it and two Northern Territory local areas. For Queensland local areas there should be at least eight in each cluster. Clearly the allocation is not random in terms of proportional allocations. Northern Australia's population and habitat appears to be structured, both in terms of *ecological* space and *geographical* space.[1]

IV.   CONCLUDING COMMENTS

Northern Australia's vast arena has an exceptionally small population; less than 10% of the entire nation, on an area covering half the continent. This population is distributed as an increasing mass towards the east coast of Queensland. This general trend to the population surface is structured in terms of demographic and other attributes of the population and some underlying *ecological* dimensions of structure have been suggested. Further refinement of methods and tests of the stability of the structure should be undertaken so that an ecological model of the northern Australian population can be developed and used for longitudinal studies in the future.

NOTES

[1] A stepwise multiple discriminant analysis using the same 22 variables, with local areas initially grouped in their State regions and using the 13 local areas of the Hunter Region of N.S.W. as a control group, correctly reallocated 89% of all local areas into their original grouping. All Western Australian local areas were correctly reallocated. Three of the thirteen local areas in the Hunter were more like areas in

the northern Queensland region than like other areas in the Hunter, however in two cases the probabilities were only 0.52 and 0.51, the remaining 48% and 49% respecitvely being the posterior probability for membership of the Hunter. One of these areas is the Great Lakes Shire which has many of the characteristics of retirement and resort facilities found in the Hervey Bay, Prosperpine region of Queensland. The other two, Musswellbrook and Merriwa share coalmining and cropped pasture features with areas in the Biggenden, Kolan to Bowen region of south east northern Queensland. The majority of areas, 10/13 however are distinctly different. The implications are that the two states and the Northern Territory are containers for local areas which are distinctive of these regions. Also that northern Australia appears to have a distinct population ecology.

REFERENCES

Burnley, I.H.   1981.   Population change and Social Inequalities in Sparsely Populated Regions of Australia. In *Settlement Systems in Sparsely Populated Regions: The United States and Australia*, (eds.) Richard E. Londsdale and John H. Holmes, Pergamon, New York, pp.105-124.

Courtenay, P.P.   1978.   Tropical Australia.   In *Australia: a Geography*, (ed.) D.N. Jeans, Routledge & Kegan Paul, London, pp.289-315.

Courtenay, P.P.   1982.   *Northern Australia: Patterns and Problems of Tropical Development in an Advanced Country*, Longman Cheshire, Sydney.

Engelman, L. and Hartigan, J.A.   1981.   K-Means Clustering. In *BMDP Statistical Software*, (eds.) W.J. Dixon *et al.*, University of California Press, Berkeley, pp.464-473.

Gale, F.   1978.   A Social Geography of Aboriginal Australia. In *Australia: a Geography*, (ed.) D.N. Jeans, Routledge & Kegan Paul, London, pp.354-365.

Holmes, J.H.   1972.   Macrogeographic Analysis of Population Distribution in Australian States.   Conference Paper: Australian Institute of Geographers, Canberra, *(mimeo)*.

Holmes, J.H.   1978.   Population. In *Australia: a Geography*, (ed.) D.N. Jeans, Routledge & Kegan Paul, London, pp.331-353.

Holmes, J.H.   1981.   Sparsely Populated Regions in Australia. In *Settlement Systems in Sparsely Populated Regions: The United States and Australia*, (eds.) Richard E. Lonsdale and John H. Holmes, Pergamon, New York, pp.70-104.

Hugo, G. 1983. Macro Demographic and Social Trends in Australia: Their Implications for Urban Development and Management. *Proceedings of the 16th Annual Conference of the Australian Institute of Urban Studies,* October 27-28, Brisbane, Australian Institute of Urban Studies, Canberra.

Parkes, D.N. 1981. Living in the Southern Centre of the North, *Transactions of the Menzies Foundation,* 2, pp.95-111.

Parkes, D.N. 1983a. "Hello Alice!": Aspects of the Human Ecology of an Arid Zone Settlement, *Research Papers in Geography, No. 26, University of Newcastle, N.S.W. (ISBN   ),* pp.1-55.

Parkes, D.N. 1983b. The Future Development of Human Settlement. In *The Northern Territory of Australia: Present Indicative and Future Conditional,* (ed.) D. Giese, Northern Territory University Planning Authority, Darwin, pp.23-34.
A selection of papers prepared for the 7th International Meeting of the Commonwealth Human Ecology Council to have been held in Darwin, Alice Springs and Armidale in conjunction with the University of New England, February, 1983. *(ISBN 07245-0460-5)*

7

ABORIGINES IN NORTHERN AUSTRALIA

I. H. BURNLEY

Department of Geography
University of New South Wales
Kensington, New South Wales

I. INTRODUCTION

It is in northern Australia, especially in areas geo-
graphically remote from European settlement that the more
traditional ways of life of Aborigines have been preserved.
These areas are Arnhem Land, the Kimberleys, Cape York
Peninsula and the Pilbara. It is also in these northern
locales that there was less decimation of the aboriginal
population through smallpox, measles and other contagious
illnesses, although these nevertheless took their toll (See
Stanley, chapter 20). Whereas Radcliffe-Brown (1930)
estimated that there were 100,000 Aborigines in Queensland in
1788, by the 1921 census there were only 15,450, little more
than one seventh of the original number. In the Northern
Territory, too, the aboriginal population fell from about
35,000 in 1788, to 18,000 in 1921.

By 1981 the Aborigine population of Western Australia
had recovered to 30,750, that of the Northern Territory to
28,680, and that in Queensland to 34,000. If Radcliffe-
Brown's figures were correct, the Aborigine population in the
Northern Territory has recovered to 82 percent of the 1788
estimate and in Queensland to 34 percent. For comparison,
the total aboriginal population of Australia is estimated to
be about 170,000. Possible underenumeration in 1981 in the
southeastern states present a problem. However the Northern
Territory, Western Australia and Queensland counts are
considered to be reliable, since the census was taken through
personal interviews with Aborigines whereas with the general
population the census schedules were left with households and

115

the filled in schedules were collected later by a census
collector.

It must be mentioned here that recent research by N.
Butlin (forthcoming) estimated that the aboriginal population
of New South Wales and Victoria was considerably larger
(over 250,000 in N.S.W. alone) in 1788. This has been done by
'reverse surviving' 19th Century aboriginal populations, and
while the actual size of the pre 1788 aboriginal population
at its peak is still debatable, it almost certainly was
larger than Radcliffe-Brown's estimate. This means that the
extent of decimation and aboriginal population decline was
more comparable to that of native born groups in North and
South America and many parts of the Pacific.

It is difficult to determine the rate of aboriginal
population growth in northern, as in other parts of Australia
using census data. Information on the aboriginal population
was only collected in detail in 1966, 1971, 1976 and 1981.
From 1971, the nature of the 'race' question changed, with
the emphasis being on whether persons identified themselves
as Aborigines or Torres Strait Islanders. Further, it appears
that there was some over-enumeration in 1976.

However the census populations for LGAs north of the
Tropic of Capricorn yield an average annual (compound) rate
of population growth between 2.2 and 2.5 percent, virtually
all of which was from natural increase. This is more than
double the natural increase rate of the total Australian
population. Indeed in 1971, the Gross reproduction Rate in
the aboriginal population of the Northern Territory was 3.0
and the Net Reproduction Rate was over 2.0. That is, 3
daughters were being born to replace every aboriginal mother
although 2 were surviving to the reproductive age range.
These indexes illustrate two things: first the higher
aboriginal fertility (the total Australian Net Reproduction
rate was 0.9 in 1981-83), and the high infant mortality in
1971. The Aboriginal Infant Mortality Rate in the Northern
Territory in 1971 was in excess of 140 per 1,000 live births;
by 1981 it had fallen to 50 per 1,000. This reflects health
innovations, special hospital care such as that provided at
the Alice Springs Base and Katherine Hospitals, the
improvement of itinerent nursing services, improvement of the
Royal Flying Doctor Services, and the success of some
Aborigine-initiated health advisory services, although it
should be noted that there are no aboriginal doctors in the
region who practice as qualified 'Western' practitioners.
The infant mortality rate, although much improved, is still
four times the Australian rate.

While overall health conditions of the Aborigines in
northern Australia have been improving, the life expectancy
is still low: it was 55 for men (at birth) in 1981 and 59
for women in the Northern Territory.   While appropriate
mortality statistics are not available for northern
Queensland and northern Western Australia, it is believed
that the life expectancy in these regions would be no better
than that in the Northern Territory.

II.   REGIONAL ASPECTS OF ABORIGINAL POPULATION STRUCTURE

Consideration of the age-sex profiles of aboriginal sub-
populations in the major areas of concentration in northern
Australia at the 1981 census illustrate the demographic
potential and aspects of mortality trends.   Table 1 indicates
that the variations between Queensland, Western Australia and
the Northern Territory in the age profile of the aboriginal
population were not significantly different, although the
proportion of persons over 65 was a little higher in Western
Australia than in Queensland.   The overall low proportions
of elderly reflect higher mortality among the Aborigines than
in the total population: proportions over 65 were generally
under 3.7 compared to over 10 percent among the total
Australian population.   This and other demographic features
are evident in the ecological structures discussed in the
previous chapter by Parkes.
With over 40 percent aged under 15, the higher fertility
of the aboriginal population is clearly evident.   The age
profile of the aboriginal population in northern Australia
indicates a population that has begun to modernise, but has
not yet passed through *demographic transition*.   That is, the
mortality rate and the fertility rate are high;   although
almost certainly both are declining (Lancaster Jones, 1969;
Smith, 1980).   The result is considerable natural increase
since the crude death rates while high by Australian and
other *modernised* countries' standards, are significantly
lower than the crude birth rates.   As infant mortality has
been declining it is likely that fertility, and thus natural
increase will eventually fall, although they are likely to
remain significantly above the Australian national levels
for at least a generation.
Urbanisation may be one important cause of the partial
modernisation of the aboriginal population.   If this is so,
communities near, or partially integrated with urban centres
would exhibit a population structure with lower mortality

TABLE 1.   Proportions of the Aboriginal Population in Major Concentration Areas in Northern Australia, in Broad Age Groups in 1981.

Age	QUEENSLAND TOWNSVILLE	ROCKHAMPTON	COOK	CARPENTARIA	MULGRAVE	MT. ISA	TOTAL QUEENSLAND
Under Age 15	41.3	44.3	38.4	37.9	42.0	45.4	43.7
15 - 44	47.2	43.7	44.7	44.6	46.3	44.4	44.0
45 - 64	9.7	9.7	13.3	13.4	9.2	7.8	9.7
65 +	1.8	2.4	3.7	4.2	2.4	2.4	2.6
Total Percent	100.0	100.1	100.1	100.1	99.9	100.0	100.0
Total Number	2642	1005	1682	1514	2024	1636	33,967

Age	NORTHERN TERRITORY BATHURST -MELLVILLE	ALLIGATOR -BALANCE	EAST-ARNHEM -BALANCE	ELSEY -BALANCE	ALICE SPRINGS	DARWIN	TOTAL NORTHERN TERRITORY
Under Age 15	41.6	42.1	46.1	41.5	41.9	37.4	42.4
15 - 44	43.7	41.0	44.1	42.0	45.7	51.1	43.7
45 - 64	11.9	13.5	9.0	13.4	8.2	9.6	10.5
65 +	2.9	3.3	0.9	3.1	4.2	1.9	3.4
Total Percent	100.1	99.9	100.1	100.0	100.0	100.0	100.0
Total Number	1473	2357	3784	1194	2974	779	28,681

Age	WESTERN AUSTRALIA WYNDHAM EAST KIMBERLEY	WEST KIMBERLEY	WIL UNA	TOTAL WESTERN AUSTRALIA
Under Age 15	39.6	39.0	35.2	42.5
15 - 44	43.9	39.6	46.4	43.6
45 - 64	11.5	15.7	13.6	10.1
65 +	4.9	5.7	4.9	3.7
Total Percent	99.9	100.0	100.1	99.9
Total Number	1418	2760	1230	30,748

Source: 1981 Census: Table LAB01

and lower fertility. Table 1 indicates, however, that with
the partial exception of Darwin and Alice Springs, this was
not the case. Even in Darwin, the proportion of the aboriginal population under 15 was 37.4, and the proportion of
elderly was very low. This, of course, could partly reflect
age-selective net migration. Meanwhile the proportion of
elderly in Alice Springs was above average. However differences in population structure between the remote Western
Australian communities and the Northern Territory and
Queensland areas were greater than "urban" and "rural"
differences generally, suggesting, possibly, more population
modernisation in Western Australia. In Queensland, two of
the remote areas where Aborigines form a large proportion of
the population - Cook and Carpentaria - had lower proportions
of young persons and higher proportions of elderly, although
regional differences were not marked. It could be that in
remote areas where traditional folkways have been maintained,
diet is better, and mortality is less severe.

III.   LANGUAGE AND EDUCATIONAL VARIATIONS

Table 2 shows variations in the proportion of the
aboriginal population having never attended school and speaking English only, or have limited proficiency in English.
These figures include the Torres Strait Islanders; it was
not possible to obtain this information for each group
separately at the local level. The main area where Torres
Strait Islanders affect the interpretation is in Bathurst-
Melville in the Northern Territory, and Rockhampton and
Townsville to a slight extent.
The proportion who never had attended school was
relatively low in Queensland. This was partly because of the
institution of *Reserves* in Queensland, these normally having
schools attached. In West Kimberley and Wiluna in Western
Australia, over a third and two fifths respectively of
persons over age 15 had never attended school, and 26 percent
of those in Wyndham-East Kimberley. In the Northern
Territory, 29 percent in Aligator (Balance) and Elsey
(Balance) had never been to school. These are remote areas,
and as in East Arnhem, traditional life styles are maintained
to a degree. Most of these remote communities are not
associated with urban places. The Aborigines who are now
'adult' have not been able to be integrated in to the
Western Australian and Northern Territory school systems.

TABLE 2. *Proportions of the Aboriginal Population Aged Over 15 who had Never Attended School* and English Language Proficiency**, 1981: Selected Areas in Northern Australia*

QUEENSLAND

	Never Attended School %	Uses English Only %	Uses Other Languages Speaks English: Not Well %	Uses Other Languages Speaks English: Not at All %
Townsville	2.5	84.5	1.7	0.1
Rockhampton	1.5	95.9	0.3	-
Cook	8.1	55.0	2.0	0.2
Carpentaria	16.1	77.5	3.9	0.4
Mulgrave	1.7	94.8	0.6	-
Mt. Isa	11.1	88.6	0.6	-
Total Queensland	4.6	82.1	2.1	0.3

NORTHERN TERRITORY

Bathurst-Mellville	10.2	2.3	15.6	1.0
Aligator-Balance	29.0	3.5	37.3	3.6
East-Arnhem Balance	16.4	1.4	33.5	16.3
Elsey-Balance	29.7	9.3	25.4	3.1
Alice Springs	17.3	47.2	8.9	1.3
Darwin	14.8	54.3	3.5	-
Total Northern Territory	25.1	24.7	27.4	4.8

WESTERN AUSTRALIA

Wyndham- East Kimberley	26.0	67.2	9.1	0.5
West Kimberley	34.6	30.3	20.3	2.3
	41.2	3.2	36.2	16.6
Total Western Australia	18.4	69.4	8.7	1.5

*Refers to Aborigines and Torres Strait Islanders together
**English Language Proficiency applies to Aborigines and Torres Strait Islanders age over 5.

Source: (1) Tables LAB07]
         (2) Tables LAB05]   1981 Census

Proximity to urban places appears to have been influential to an extent: the number who never attended school was low in Townsville and Rockhampton and the proportion using only English was relatively high. In the Northern Territory, the proportion using only English, although less than half, was very much higher in the Alice Springs and Darwin areas than in the other areas of concentration. It will be noted that even in the remotest areas, the proportion speaking no English at all was relatively low although a third in the Alligator, East Arnhem and Wiluna areas spoke poor English and 20 percent in West Kimberley. On the one hand, significant proportions spoke little English and had never attended school in remote areas of the Northern Territory and the north of Western Australia. There are some important implications for future potential for absorption into the *outback* or the *urban economy*, given the relatively rapid aboriginal population growth. On the other hand, significant numbers knew some English, and spoke it well.

In the Northern Territory, Aborigines have dominated the workforce in the cattle industry (Stevens, 1970;   1974). This industry did not require much knowledge of English but it allowed Aborigines to develop new skills as stockmen to advantage. Changes in the industry may displace some, however, and in conjunction with rapid population growth, there may be problems of sustenance in the future. Urbanisation or long distance migration do not appear to be solutions.

This raises the question of the granting of land rights to Aborigines. The most extensive areas so far granted on a *freehold* basis are the Northern Territory, especially in east Arnhem land where major concentrations of Aborigines exist, and in Central Desert, Lake Mackay, Haasts Bluff and Petermann which occupy about 25 million ha. west and north-west of Alice Springs. Arnhem Land east contains some 9 million ha.   In Western Australia there are some smaller *leases* in the Kimberleys, in a number of separate locations at Forrest River (1.3 million ha.), Violet Valley, Beagle Bay, Pantijan and Noonkanbah, with leases also near Wiluna. There are also a number of freehold areas in the western Pilbara.   There is by no means a perfect concordance between aboriginal population concentrations of substantial size in Western Australia and *land rights areas*.

It is not possible to readily relate to migration data (place of residence one year and five years previously) from the 1981 census to the aboriginal land areas. Since in northern Australia particular Aborigine peoples tend to identify with particular areas, it is not to be expected that long distance or inter-state migration would be involved. The only exception to this is aboriginal movement to and from Adelaide and the Northern Territory, and much of this is referral for health reasons although there has been some urbanization migration to Adelaide, and also return migration from Adelaide (Gale 1974).

In Alice-Springs, almost 40 percent of Aborigines did not change residence during 1976–81; 22 percent migrated in from elsewhere in the Northern Territory. It is not possible from these data to estimate out-migration. In East Arnhem, 75 percent did not move between 1976–81; there was very little in-migration from elsewhere, either in the Northern Territory, or interstate. In Cook two thirds did not change residence at all, and less than 8 percent migrated in from other areas of Queensland. In Carpentaria less than 9 percent moved in from elsewhere in Queensland; 65 percent did not change residence and less than a sixth changed residence within the same LGA. In Mt. Isa, on the other hand, while over half did not move at all, about 15 percent moved in from elsewhere in Queensland, and 3 percent from the Northern Territory. In Mulgrave, over half did not move and 11 percent moved in from elsewhere in Queensland, mainly adjacent areas. In Rockhampton 25 percent moved in from elsewhere in Queensland.

The trends in the Kimberleys were very similar to those in the Northern Territory; there was very little in-movement from outside the local region. Furthermore much of the apparent urbanization migration in Queensland, and Darwin, was almost certainly exchange migration with almost as many moving back home. However an urbanization trend appears to be underway in Coastal Queensland, and in the Northern Territory to an extent. Nevertheless the educational levels, language problems, and lack of urban skills preclude any substantial urbanization and it is difficult to see large scale incorporation into modern cities such as Alice Springs or Townsville, other than in public housing or in fringe camps. Given the undoubted rapidity of population growth, however, it must be questioned whether larger aboriginal communities can be supported in the desert areas, although in Arnhem Land, larger populations may be supportable.

REFERENCES

Commonwealth of Australia 1982. *Aboriginal Land and Population*, Division of National Mapping, Canberra.

Gale, Fay 1974. *Urban Aborigines*, Australian National University Press, Canberra.

Jones, F. Lancaster, 1970. *The Structure and Growth of Australia's Aboriginal Population*, Australian National University Press, Canberra.

Radcliffe-Brown, A.T. 1930. Estimates, *Commonwealth Yearbook* 23, pp.687-696.

Smith, L. 1980. *The Aboriginal Population of Australia*, Australian National University Press, Canberra.

Stevens, F. 1970. The Aborigines. In *Australian Society*, (eds.) S. Encel and A.F. Davies, Second Edition, Cheshire, Melbourne.

Stevens, F. 1974. *The Aborigines in the Northern Territory Cattle Industry*, Australian National University Press, Canberra.

8

AUSTRALIAN TROPICAL MARINE ECOSYSTEMS

J. S. BUNT

Australian Institute of Marine Science
Townsville, Queensland

I. INTRODUCTION

As an island continent directly linked with three of
the world's major oceans, Australia the national entity, is
both affected by and dependent upon its surrounding seas and
their resources in many ways. This dependence and its
economic significance may be expected to intensify with the
passage of time. Yet our existing knowledge of that marine
domain falls considerably short of present needs. We
especially lack comprehension of the diverse, complex and,
in some ways, highly fragile ecosystems of our marine
tropics. This account attempts to sketch current knowledge
of those systems and to indicate how some of their living
resources have been managed in terms of mariculture.
References have been provided as a guide to the relevant
literature.

II. PHYSICAL ENVIRONMENT

To the limits of the declared EEZ, or Extended Economic
Zone, the Australian marine tropics extends from
approximately 110°E long. in the Indian Ocean to 160°E long.
in the Coral Sea and from around 9°S lat. in the Arafura Sea
to 26°S lat. on the western and eastern sides of the
continent. The area bounded by the real coastal and
arbitrary offshore boundaries of this region occupies
roughly 3 million square kilometers and its land-sea

NORTHERN AUSTRALIA
ISBN 0 12 545080 X

interface, on the basis of 0.7 km intervals on 1:250,000
maps, extends a total of nearly 20,000 km. Measuring the
Australian coastline has been considered by Galloway & Bahr
(1979) on which the preceding estimate is based.

The literature on Australian geography is extensive.
However, the volume edited by Jeans (1978) is a valuable
source and within it a chapter on the coast by Davies (1978)
is especially pertinent. A useful more detailed treatment
of coastal geomorphology may be found in Bird (1976). An
inventory of Australia's coastal lands has been prepared by
Galloway (1981). Accounts such as these are difficult to
summarize as a background to the following account. It must
suffice to say that the greater part of the Australian
tropical coastal lands is low-lying although much of the
coastal plain in the east is narrow and backed by the Great
Dividing Range. The most prominent extent of rocky coast
occurs in the Kimberley area of the northwest while, in the
north and elsewhere, the development of tidal plains is
extensive. Open sandy beaches occur most extensively along,
but are not limited to, the western coastlines.

Within this broad setting, irregularities in the
continental margin have created a wide range of major and
minor embayments, all of which provide settings for
distinctive living communities and ecological processes.
Inshore environments are further influenced by drainage from
the land. No less than 98 watersheds, widely diverse in
character, discharge to the sea within the region
(Australian Water Resources Council, 1976). Deliveries of
fresh water vary widely from watershed to watershed, from
year to year and over annual cycles. The annual discharge
of the Ord River in N.W. Australia, for example, ranges from
24% to 232% of the average, while, for the Burdekin River in
Queensland, the range of 2% and 332% of the average is even
more dramatic. Features of this kind reflect patterns in
rainfall ranging from as little as 200 mm to over 2,400 mm
per annum with a generally strong bias to the summer
monsoons. Cyclones ranging in annual frequency from 0.7 to
1.3 within 160 km of the coasts (Davies, 1978) are an
important influence. Detailed information on tropical
Australian cyclones over the period 1909 to 1975 may be found
in Lourensz (1977), see also Lee and Neal in this volume.

A substantial proportion of the territorial sea is
associated with continental shelves although its boundaries
do encompass a segment of the N.E. Indian Ocean, part of the
Coral Sea Basin and the northernmost extensions of the
Tasman Sea. It is important to note, in these terms, that
most of the Timor Sea and virtually all of the Arafura Sea

including the Gulf of Carpentaria are less than 200 m
deep. The N.W. shelf to the same depth is also a prominent
feature and the same may be said for the Queensland shelf,
more particularly towards its northern and southern limits,
and especially since it serves as foundation for the Great
Barrier Reef. Extending some 2,200 km and containing at
least 2,500 individual reefs, that living system and its
geological foundations are of global significance. The
lagoon behind it affects the entire N.E. coastline and its
ecological character.

Our understanding of the circulation and general
movements of the waters around tropical Australia is
generally rather limited both at large and small scales in
time and space. For the eastern coast and western Coral
Sea, recent overviews have been provided by Pickard (1977,
1983). A broad account of circulation in the Gulf of
Carpentaria is available in Forbes & Church (1982).
References to the Indian Ocean N.W. of Western Australia may
be found in Andrews (1977). In fact, however, the nature of
the large scale circulation in that sector has received
relatively little attention. Nevertheless, for ecological
purposes in general, greater interest relates to meso- and
microscale events at and over the continental shelf and at
the land-sea interface including embayments, estuaries and
the lower reaches of rivers under influence from the sea.
To my knowledge, a synthesis of information at these scales
for the Australian tropics has never been attempted,
although a compendium by Radok (1976) provides useful
information on tides, mean sea levels and wind fields. An
early account of Australian estuarine hydrology by Rochford
(1951) deals almost exclusively with non-tropical coasts.

Immediate driving forces of concern include the tides,
temporal wind patterns, river flood discharges, cyclonic
events and salinity gradients created by evapotranspiration
especially at the land-sea interface. The consequences in
terms of water movement, circulation and exchange are
complicated by coastline configuration and general
bathymetry. Tides in the region vary from as little as 1 to
as much as 11 m in amplitude at springs. Short term
barometric effects can affect predicted tidal behaviour
considerably. This is notably true of the shallow Gulf of
Carpentaria. The complexities of tidal behaviour in
northern estuaries and riverine systems have been little
studied although upstream tidal data are presently being
collected at some locations on the coast of Queensland
(Boto, pers. comm.) and in the Northern Territory (Chappell,
pers. comm.). Wolanski et al. (1980) have explored the

hydrodynamics of a tidal channel on Hinchinbrook Island,
showing that asymmetry between ebb and flood currents aids
in maintenance of channel configuration. Wolanski (1983),
has also undertaken studies of tides on the northern Great
Barrier Reef continental shelf. The scope for expanding
investigations of this kind to other areas of interest is
considerable.

     Meso- and microscale processes of water movement at and
over the Queensland shelf break, little understood until
recently, are now being revealed in studies reported by
Andrews (1983) and others at a conference on the Great
Barrier Reef. The documentation of upwelling phenomena,
longshore current patterns, shelf waves, topographically-
induced eddies, the patterns of shallow water flow over reef
flats, dispersion of river flood plumes and related
phenomena is now active and directly relevant to studies of
associated ecosystems. Attention to the physical
oceanography of the N.W. shelf remains limited although
there has been a good deal of private study given to design
parameters for offshore structures. Only two papers given
at the Australian Physical Oceanographers' Conference held
at Flinders University in 1983 dealt with that region. None
of the papers given at the same conference considered waters
off far northern coasts. In view of the link between the
Indian and Pacific Oceans via the Arafura Sea and Torres
Strait, along with major fresh water inputs from West Irian
and Papua New Guinea, the general neglect of this area is
surprising.

III.   THE ECOSYSTEMS AND THEIR PRODUCTIVITY

There are five principal classes of marine ecosystem in the
Australian tropics. These include the mangroves, seagrass
beds, coral reefs, benthic systems in general other than
coral reefs, and plankton-based communities of the open sea,
nearshore waters and estuaries. At the upper intertidal
margin, it would be reasonable to include as well the
distinctive communities of the saltmarshes. Each will be
treated separately.

A.   *The Saltmarshes*

     There is little resemblance between the so-called
saltmarsh communities of tropical Australia and those of

Europe or North America. According to Saenger *et al.* (1977)
only 7 plant species are represented and the most prominent
are low succulents. These communities, in effect, are an
extension of those which are most fully developed in the
temperate south. Indeed, their existence, although quite
extensive in the upper parts of the tropical intertidal
zone, is marginal, leaving vast areas entirely bare and
commonly encrusted with salt deposits, especially during the
winter dry season. Where conditions are less forbidding,
however, vegetal ground cover is often complete and, under
most favourable conditions, e.g. parts of the wetter
Queensland coast, succulents such as *Sueda* and *Arthrocnemum*
may be replaced by dense stands of sedge, sometimes co-
existent with mangroves and *Melaleuca*. At the other
extreme, suitably adapted species are rarely exposed to
fresh water influence and are inundated by seawater only on
extreme tides. Since, as far as I am aware, saltmarsh
productivity has never been investigated in the Australian
tropics, no further detail will be provided on these
communities.

B.   *The Mangroves*

Worldwide, the literature on the mangroves is
substantial. A bibliography by Rollet (1981) contains over
5,000 references. Yet our knowledge of those forests as
productive ecosystems is not especially well-developed.
This is certainly the case for Australia. Even so, national
interest in the mangroves has intensified to such an extent
in recent years that our understanding is showing rapid
improvement (e.g. see Clough, 1982).
Galloway (1982) has estimated that the Australian
mangroves occupy a total area of some 12,000 km^2, largely in
the tropics although outliers extend as far south as
Westernport Bay in Victoria, Bunbury in Western Australia,
and parts of the coast of South Australia. Their
development and floristic diversity is greatest along the
wetter coasts of Queensland from which Bunt *et al.* (1982)
have listed well over 30 species. Among them, Bunt and
Williams (1982) found it necessary to recognize at least 30
distinctive vegetational associations. Such complexity had
not previously been recognized by authors such as Macnae
(1966) who based his account on the existence of only 5
vegetational zones from the upper tidal limit to open
water. While such clear and limited zonation is sometimes
evident, it has proven an oversimplification which, in many

respects, neglects the consequences of environmental variability induced by tidal pattern, salinity regime and other
factors as discussed by Bunt and Williams (1981) for
Queensland, Kenneally (1982) for Western Australia and Wells
(1982) for the farther northern coasts. A synthesis of
information on the vegetational character and environmental
relations of the mangroves throughout their northern
distribution has not yet been attempted but should soon be
feasible. At the same time, the need still exists for
further ground survey since it has been found that even
recently compiled floristic lists have sometimes overlooked
species of interest. It is also known that some species
have irregular or discontinuous distributions, the reasons
for which are not presently clear.

The physical structure of the forests, like their
floristic composition, is also rather variable and this is
true as well for the range in development to be seen in
individual species. Suffice to say that, under conditions
of stress, stunting can be so acute, that normally large
trees can be reduced to the size of small shrubs. On the
other hand, under favourable circumstances, it is not
unusual to view extensive forest canopies well over 30 m
high. The effects of salt and water stress are general in
western Australia but variable along the northern and
eastern coasts. Luxuriant growth is always associated with
fresh water influence, either directly from high local
rainfall or indirectly through exposure to prolonged and
reliable fresh water drainage. Topography, which provides
regular and extensive tidal inundation, is also important in
sustaining well developed forests. Under such conditions,
the boundary between the mangrove and adjoining terrestrial
forests may be blurred or even obscure. Otherwise, the
boundaries are normally quite clear and, indeed, over large
areas the mangroves may lie separated from the nearest
terrestrial vegetation by wide expanses of highly saline
bare mud.

Anywhere in the world, the study of mangroves in
ecosystem terms and especially in relation to their
productivity is comparatively recent. In a large sense, the
stimulus to do so arose from investigations in Florida by
Odum and Heald (1972) whose results indicated that the
mangroves represent a significant natural resource with
important influences on marine animal food chains especially
in nearshore environments. Subsequent studies have
demonstrated that, at least in terms of their primary
productivity, the mangroves of Australia equal or rival
those of the Americas. Methods used have varied from

sophisticated direct measurements of photosynthetic activity in mangrove leaves to less direct but more convenient survey techniques based on recording the attenuation of light through forest canopies, but more generally have depended on the simple collection of litter fall through time, itself only a partial indication of the total new yield.

Up to the present, data on litter fall relate only to collections made in mangroves on Hinchinbrook Island close by the coast of northern Queensland. There, rates of total annual litter fall on a daily basis over a large number of sites ranged from slightly less than 1.0 to 7.7 $g.d.w.m^{-2}$ (Bunt, 1982b). The ranges of values for individual vegetational associations were found to span a substantial proportion of the total range between associations. Duke *et al*. (1981) found that, in general, leaves were the most important component of collected litter fall. Williams *et al*. (1981) reported marked seasonal periodicity in the delivery of litter with peaks over the summer period. Patterns of litter fall varied distinctively from species to species. Boto and Bunt (1981) have shown that at sites in Queensland, and probably more generally, most litter is exported to coastal waters by tidal action. Its ultimate fate, however, remains to be established.

Using less direct methods, also on Hinchinbrook Island, Bunt *et al*. (1979) obtained estimates of net forest primary production ranging between 16 and 26 kg C ha^{-1} day^{-1}, suggesting that litter fall alone may account for only 50 to 78% of total production. Subsequent measurements at other locations around the northern coast by Boto (pers. comm.) have been found generally compatible with the findings from Hinchinbrook. Direct measurements of photosynthetic gas exchange as influenced by atmospheric environmental conditions have been made with several of the more common northern Australian mangroves by Andrews *et al*. (in press) to whom the reader must be referred for detail.

With respect to mangrove sediments, their nutrient status and physicochemical characters, Boto (1982) in a recent review was forced to conclude that little is known of the conditions along Australian tropical coasts although the situation is changing. For example, Boto and Wellington (1983, in press) have demonstrated that phosphorus levels are generally low and that, under some conditions, levels of combined nitrogen may be limiting for growth as well. Studies currently underway by Boto (pers. comm.) indicate that, notwithstanding generally reducing conditions, nitrate is the necessary source of nitrogen. Its supply may be

assured by aerobic conditions sustained in the immediate
vicinity of the mangrove rhizosphere.

In summary, while knowledge of the vegetational
character, productivity and foundation ecosystem properties
of the Australian mangroves is now steadily improving, the
nature of the interaction between forest production and
dependent consumers as well as their trophic relationships
remains much less understood. That this is so is reflected
in a short review by Redfield (1982) and in earlier accounts
e.g. by Macnae (1966). At the same time, the fact that
commercially important species such as mud crabs, barramundi
and prawns, part of a much larger fauna (Hutchings and
Recher, 1981) are intimately associated with the mangroves
and their environment indicates the urgent need for detailed
studies.

C.   Sea Grasses

Globally, the sea grasses have long attracted
scientific attention, some of it intensive over the last
decade or so and couched directly in ecosystem terms (e.g.
McRoy and Helfferich, 1977). Within that context, a
relatively small number of workers has been active in
Australia. As an overview by Larkum (1977) and a more
recent account by McComb et al. (1981) show, however,
Australian investigations have been centred almost entirely
in temperate environments. Personal observations and
informed advice from others suggest that, although the
seagrasses of the Australian marine tropics are widespread
and although they are known to be floristically diverse (Den
Hartog, 1970), in general they do not produce luxuriant
stands of the kinds well known elsewhere in the tropics.
Nonetheless, grass beds of one kind or another are to be
found in a variety of situations from mud flats fronting
mangroves, to sandy embayments and as components of reef
communities. Accordingly, their total contribution as
primary producers must be considerable but the magnitude of
that production remains to be assessed. However, in
Cleveland Bay near Townsville in Queensland, Heinsohn (see
Hutchings, 1981) estimated that as little as 3.5 ha of bed
would be adequate to support one dugong, a grazing marine
mammal, for one year. Elsewhere, dugongs have been observed
to eat 50-55 kg of seagrass/day (Jones, 1967). If the
estimates by Heinsohn were proved correct and if the
magnitude and diversity of northern seagrass faunas as a
whole are at all comparable with those reported by Hutchings

(1981) under more temperate conditions, then our ignorance
in this field may be considered a matter of serious
neglect. Data assembled from a number of locations
worldwide by Thorhaug (1982) indicate that seagrass
productivity ranges from as little as 0.4 to as much as
93 kg ha^{-1} day^{-1}, the upper limit exceeding the known
productivity of the Australian mangroves.

D.   *Coral Reefs*

Coral reefs exist at various locations around the
tropical coasts of Australia although their most outstanding
development is along the coast of Queensland in the Great
Barrier Reef where most research effort has been
concentrated. The much smaller coral reefs of Western
Australia and the reef communities of the Northern Territory
remain little studied although that situation is now
changing.

The Great Barrier Reef extends a total distance of some
2,000 km and includes around 2,500 individual reefs.
Reference is usually made to Maxwell (1968) for background
on modern thinking on the geology of the system although an
account by Hopley (1982) has since become more relevant,
based on research over the intervening decade or so.
Presently emerging views on the antecedent surface, the
origins of the entire structure and its evolution have been
summarized in a review by Davies (1983).

As an environmental entity and as an ecosystem, the
Great Barrier Reef began its present development with the
most recent post-glacial rise in sea level which came to
relative stability over the period of roughly 20,000-6,000
years B.P. A general bibliography on the province prepared
by Frankel (1978) reflects a considerable history of
scientific interest. Modern ecosystem studies, however, are
only now gathering momentum. Indeed, the foundations for
such studies are still being laid. For example, a
comprehensive taxonomic treatment of the corals by Veron and
Pichon (1976, 1979, 1982) and Veron *et al.* (1977) has only
recently been completed and many other important groups such
as the sponges (Wilkinson, pers. comm.) are very poorly
known. Nonetheless, a considerable background of
information and some understanding of reef community ecology
is now emerging. Direct observations are also now available
on rates of reef growth and productivity, on the processes
of nutrient cycling in the reef environment and on aspects

of community trophodynamics. Even so, a coherent account of
Australian coral reefs in ecosystem terms is scarcely yet
practical because so many gaps remain to be filled and
because available observations are, in many respects,
fragmentary in time and space. In a necessarily brief
account, it is possible only to touch on recent advances.
For a convenient assembly of more detailed background, the
reader is referred to Barnes (1983a) and to various papers
in Baker *et al.* (1983).
    In a system as vast as the Great Barrier Reef, a single
description of the living reef community would be
inappropriate and entirely misleading. Indeed, studies by
Done (1982) on the corals and by Williams and Hatcher (1983)
on the fish over a set of reefs across the central section
of the province have revealed distinctive differences in
community structure at several scales. The matter of
zonation across individual reefs has been reviewed by Done
(1983). Suffice to say that the diversity of community
pattern in space is considerable and that patterns are
subject to changes in time, sometimes dramatically in face
of environmental extremes such as cyclones or following
heavy predation by the notorious crown-of-thorns starfish,
Acanthaster planci. The relative importance of the many
environmental and biological factors likely to be involved in
influencing community structure remains a matter of active
debate. Physically, however, differences in wave climate,
water sediment load, current fields and tide behaviour are
all important. Identifying biological influences is rather
more difficult. Apart from the direct effects of predator-
prey interactions e.g. *Acanthaster*-coral, various forms of
competition for space are recognized stemming from earlier
work by Lang (1971; 1973) either by direct inter-species
aggression or by release of toxic products e.g. by soft
corals (Sammarco *et al.*, 1983). Further, the boring
activities of sponges are attracting attention (see Done,
1983 and Wilkinson, 1983) and territorial behaviour, e.g. by
damselfishes has been demonstrated by Sammarco (1983) to
affect reef algal communities in a complicated fashion.
Processes of recruitment, reproductive strategies,
demographic behaviour and other biologically-based controls
on the development and modification of community structure
might also be considered.
    The matter of gross material fluxes including the
production and consumption of organic matter and $CaCO_3$
budgets in coral reefs has been reviewed concisely by Kinsey
(1983a) whose earlier and ongoing work has been influential
in these studies. Barnes (1983b) has now introduced

instrumental techniques to facilitate such investigations and is using these tools in a systematic fashion over reefs in the central section of the Great Barrier Reef. Data gathered by that author have shown how net productivity varies across reef flats and with incident light intensity. In collaboration with Chalker (pers. comm.), Barnes (1983c) found gross production on one study reef in late March to be 8.8 g C m^{-2} day^{-1} and respiration 7.6 g C m^{-2} day^{-1}, giving a P/R ratio of $1.16:1$. Net deposition of $CaCO_3$ was found to be 3.6 kg m^{-2} yr^{-1}. Kinsey (1983b) has begun comparable flux studies in reef lagoons.

Under what may be termed the general banner of reef metabolism, various more detailed studies are in progress. The primary productivity of turf and crustose algal communities has been measured by Borowitzka et al. (1983). Drew and Abel (1983) have examined the growth of the carbonate-depositing alga, Halimeda, a major contributor to reef sediments. Processes and rates of biological nitrogen fixation have been studied by Wilkinson et al. (1983) and others, and a number of investigators are directly measuring rates of growth of individual coral species. Hatcher (1983) has considered the trophic significance of detritus in coral reef ecosystems. Once again, the reader will find useful reviews of many of these topics in Barnes (1983a).

The manner in which coral reefs sustain such high levels of production in what seem to be characteristically oligotrophic waters has been a matter of great interest for some time. Certainly in situ nitrogen fixation is important. How though, is reef growth enabled to continue and how is biomass to be sustained, even with tight resource cycling, without a substantial input of primary nutrients such as phosphorus? Two potential sources exist in the Great Barrier Reef, one in drainage from the land and the other introduction from the open sea. In the latter regard, observations at the shelf break by Andrews and Gentien (1982) have demonstrated physical processes which deliver relatively cold nutrient-rich water from below the thermocline up the continental slope to and beyond the shelf break in the central section of the Great Barrier Reef. The phenomenon occurs with a period around 90 days and, in terms of nitrogen alone, has been estimated to represent an onshore flux of 20 µg at l^{-1} throughout the water column in a reef zone some 50 km wide. The physical processes involved have been studied further by Wolanski and Pickard (1983). Little is yet known about nutrient fluxes from landward sources although aspects of the problem are now receiving attention by Kinsey and collaborators (pers.comm.).

E.   Other Benthic Communities

This field has been rather neglected in the past,
probably because of logistic requirements or because many
potential study sites are difficult of access.

A relatively early study by Endean *et al.* (1956) paid
attention to the intertidal rocky shores of Queensland north
to 16°39'S. North of 25°S, it was recorded that an oyster
zone dominated the upper intertidal with barnacles below
that. The significance of wave action and salinity and
various other environmental factors were considered in
relation to the zonation. In a separate report, Endean *et
al.* (1956) found similar patterns in a series of continental
islands off the Queensland coast although the intrusion of
corals and zoanthids was noted. Attention to the benthic
algal communities has also been limited although Ngau and
Price (1980) have described the distribution of 144 algal
taxa of intertidal sites in the vicinity of Townsville.

On the tropical coasts of Western Australia and at
other locations, George and his colleagues (pers. comm.)
have undertaken important studies with individual groups of
invertebrates such as the Fiddler and other mangrove-
associated crabs with respect both to basic taxonomy,
habitat preferences and biogeography. Wells (1983) has
studied the ecology of mangrove molluscs both in the Exmouth
Gulf area and in the Kimberleys. Nearshore benthic
community surveys are in progress as well in the Dampier
Archipelago by W.A. Museum researchers and some of this work
has been extended to include elements of the benthic fauna
at Rowley Shoals.

Further offshore, a major survey by C.S.I.R.O. is in
progress on the benthic communities and sediment substrates
of the N.W. Shelf. A comparable investigation is underway
by Birtles and colleagues (pers. comm.) across the
continental shelf of Queensland near Townsville. The Darwin
Museum is investigating benthic ecology at locations off the
coast of the Northern Territory. Until this information is
published, however, there is little more than can be said
except that these and further studies of the tropical
benthos of Australia along lines recently reported by Fry *et
al.* (1983) are an outstanding need.

F.   Pelagic and Related Nearshore Communities

Unlike marine ecosystems such as coral reefs and
mangroves whose physical attributes are readily apparent and

firmly associated with a substrate, the existence and
behaviour of those based on the generally microscopic
plankton and subject to the fluid motions of the sea are
much less obvious. Their significance, nonetheless, is
well-recognised. Unfortunately, the attention they have
received in Australia's northern waters has been limited
with a concentration of effort on particular species of
interest in commercial fishing. Concern, in any event, has
generally and naturally been biased towards questions of
productivity and attention here will be limited to that
topic.

In these terms, it is useful to consider, at first,
current knowledge on the nutrient status of the Australian
tropical sea. A valuable overview of that subject by
Rochford (1980) reveals, in fact, that the primary
nutrients, N and P, are generally low. Indeed, inorganic
phosphates exceed 0.2 μg at.1^{-1} at large scales only in the
Arafura Sea, including parts of the Gulf of Carpentaria and
on segments of the N.W. Shelf. Similarly, nitrates at the
surface exceed 0.5 μg at.1^{-1} only in part of the N.W. and
southern Coral Sea and, to a very limited extent, off N.W.
Cape, Western Australia. Information on nutrient inputs
from the land is quite limited. Boto (pers. comm.) found
inorganic phosphate levels in a series of north Queensland
rivers to lie in the range 0.06-0.33 μg at.1^{-1} and nitrate
levels generally, although not always, <1.0 μg at.1^{-1}.
Concentrations of ammonium ion were similar to those for
nitrate. Nutrient levels as recorded by the Department of
Transport and Works in the Alligator Rivers region of the
Northern Territory are also low. Data is available for the
rivers of Western Australia. However, Chittleborough (pers.
comm.) has suggested that the values for nutrients such as
nitrate may not be reliable. In any event, Rochford's
(1980) view that such inputs are generally likely to be low,
partly because Australian soils are nutrient deficient by
world standards, at least for phosphates, is probably
correct.

On the basis of extensive phytoplankton surveys, Wood
(1954; 1964) recognized several ocean regions in the
Australian tropics. More recent studies by Revelante et al.
(1982), however, suggest that temporal differences in
phytoplankton assemblages are more important than spatial.
At the same time, evidence was obtained of a diatom-rich
flora close to the coast, at least in the Great Barrier Reef
lagoon. Unfortunately, virtually no attention has been
given to the productivity of close inshore waters or the
estuaries of northern coasts.

For the open sea, taking into account available information, Bunt (1982a) made an estimate, emphasized as tentative, that primary production to the limits of the Australian 200 nm EEZ could lie between 365-662 x 10^6 t C annually. Within the tropical part of that province, annual rates might lie btween 160-310 x 10^6 t C. It would be premature to consider how productivity may vary within the region although the N.W. Shelf is presently receiving detailed attention by C.S.I.R.O. because of its apparent fertility and Motoda et al. (1978) have recorded isolated instances of high production in the Gulf of Carpentaria. Planktonic primary production associated with upwelling events along sections of the Great Barrier Reef is being studied by the Australian Institute of Marine Science.

Against such a fragile background, it is known that the tropical zone supports important stocks of prawns, squid, scallops and a wide variety of harvestable fish species which depend wholly or in part on planktonic sources of primary production. According to data for 1977 summarized by Radway Allen (1978), total Australian fishery landings amounted to 114,000 tonnes. It would be impractical here to attempt a summary of past and current research relating to those resources. It is also impossible to establish exactly the proportion of that harvest of tropical origin. Nonetheless, it can be said that, by reasonable estimates in carbon equivalents, total Australian landings reported represent only 1.6-3.0 x 10^{-5} of the speculated primary production from planktonic sources. Conventional trophic considerations would suggest that the yield could be at very least an order of magnitude higher. Clearly, in these terms alone, our understanding of pelagic and associated ecosystems is not adequate and this is particularly the case for tropical waters. The background to this need, Australia wide, has been treated by Jeffrey et al. (1982) and the requirements directly in terms of fisheries by Radway Allen (1978).

IV. POTENTIALS FOR MARICULTURE IN THE AUSTRALIAN TROPICS

In principle, at least, there is no question that opportunities exist to develop mariculture in the Australian tropics as has been done over long periods, for example, in S.E. Asia with expansion and sophistication of those activities continuing. Indeed, there has been a considerable history of interest in such industry in this country, even though the success of particular enterprises has not always been achieved. The culture of pearls, of course, must be counted as a valid aspect of mariculture and the skills involved have now spread from centres such as Broome in Western Australia to an undertaking in the Escape River of Cape York. More recent, if chequered developments, include endeavours to culture species as diverse as prawn, turtles and oysters, while another with real promise involves an aboriginal enterprise to farm crocodiles. Current interests now extend to giant clams and to the heavily fished and highly prized barramundi as well as the equally attractive mud crab (*Scylla serrata*). Other possibilities include the culture of ornamental shells and corals, the latter already a practice under essentially natural conditions by licenced collectors.

With such a long and environmentally varied coastline, much of it sparsely or totally uninhabited and relatively pristine and with a number of species available of commercial value, the attractions in developing various forms of mariculture in tropical Australia would appear obvious. On the other hand, it is not possible to evade a number of possible difficulties some of which have already been experienced by pioneers in the field. Many of the problems were debated at a recent symposium sponsored by the Queensland Department of Primary Industries in Brisbane during March, 1983. The obstacles range from the frequent need for basic biological understanding through matters of arranging appropriate plant design to legislative and regulatory requirements and not least, the subtleties of modern marketing practice. In the latter regard, attention to the economics of production in relation to anticipated return is crucial to such an extent that investment in overcoming technological difficulties is considered pointless unless the economic and marketing aspects have first been researched with care.

Prospects appear highest for products of high commercial value and, indeed, experience has proven that mariculture as a means to produce cheap animal protein simply is not viable in the economic climate of developed countries. Unfortunately, detailed information on initiatives taken in this field within tropical Australia is difficult to obtain and often not publicly reported. For general background on the subject, however, a text by Bardach *et al.* (1972) is valuable.

V.   CONCLUDING COMMENTS

For a province of such magnitude, such diversity and such intrinsic natural and resource value, it may seem remarkable that so little should be known of Australia's tropical marine environments, living communities and often unique and important life forms. Only with the expansion of a still small population, with modern facilities in communication and travel and through an awareness brought about through international initiative, particularly through the U.N., and its activities, has a genuine concern for the sea and its resources become evident. Declarations of the 200 nm EEZ has provided a particular focus for that concern. Within such a climate, it is to be expected that earlier neglects will steadily be made good.

As a foundation for more detailed study, there is a particular need to consider the influence of the land on the sea, notably the input of primary nutrients by river discharge to coastal zone ecosystems and the relative importance of those nutrients as a basis for marine organic production compared with detrital materials of terrestrial origin. These are questions of central importance which have been largely overlooked.

REFERENCES

Andrews, J.C. 1977. *Deep-Sea Research 24*, p.1133.
Andrews, J.C. 1983. In *Proceedings: Inaugural Great Barrier Reef Conference*, (eds.) J.T. Baker *et al*, James Cook University of North Queensland, Townsville, p.403.
Andrews, J.C. and Gentien, P. 1982. *Mar. Ecol. Prog. Ser. 8*, p.257.
Andrews, T.J., Clough, B.F. and Muller, G.J. (in press) *Proceedings Second International Symposium on Biology and Management of Mangroves*, W. Junk Publishing Company.
Australian Water Resources Council 1976. *Review of Australia's Water Resources 1975*, Australian Government Publishing Service, Canberra.
Baker, J.T., Carter, R.M., Sammarco, P.W. and Stark, K.P. 1983. In *Proceedings: Inaugural Great Barrier Reef Conference*, (eds.) J.T. Baker *et al*, James Cook University of North Queensland, Townsville.

Bardach, J.E., Ryther, J.H. and McLarney, W.O.   1972.
     *Aquaculture: The Farming and Husbandry of Freshwater
     and Marine Organisms*, Wiley-Interscience.
Barnes, D.J.   1983a.   (ed.)   *Perspectives on Coral Reefs*,
     Australian Institute of Marine Science and Brian
     Clouston, Publisher, Canberra.
Barnes, D.J.   1983b.   *J. Exp. Mar. Biol. Ecol. 66*, p.149.
Barnes, D.J.   1983c.   In *Proceedings: Inaugural Great
     Barrier Reef Conference*, (eds.) J.T. Baker *et al*, James
     Cook University of North Queensland, Townsville, p.275.
Bird, E.C.F.   1976.   *Coasts. An Introduction to Systematic
     Geomorphology*.   Australian National University Press,
     Canberra.
Borowitzka, M.A., Day, R. and Larkum, A.W.D.   1983.   In
     Proceedings: Inaugural Great Barrier Reef Conference,
     (eds.) J.T. Baker *et al*, James Cook University of North
     Queensland, Townsville, p.287.
Boto, K.G.   1982.   In *Mangrove Ecosystems in Australia*,
     (ed.) B.F. Clough, Australian Institute of Marine
     Science and Australian National University Press,
     Canberra, p.239.
Boto, K.G. and Bunt, J.S.   1981.   *Estuarine, Coastal and
     Shelf Sc. 13*, p.247.
Boto, K.G. and Wellington, J.T.   1983.   *Mar. Ecol. Prog. Ser.
     11*, p.63.
Boto, K.G. and Wellington, J.T.   (in press).   *Estuaries 7.*
Bunt, J.S.   1982a.   In *C.R.C. Handbook of Biosolar Resources
     Vol. 1, part 2*, (eds.) A. Mitsui and C.C. Black, C.R.C.
     Press, P.429.
Bunt, J.S.   1982b.   In *Mangrove Ecosystems in Australia*,
     (ed.) B.F. Clough, Australian Institute of Marine
     Science and Australian National University Press,
     Canberra, p.223.
Bunt, J.S., Boto, K.G. and Boto, G.   1979.   *Mar. Biol. 52*,
     p.123.
Bunt, J.S. and Williams, W.T.   1980.   *Aust. J. Ecol. 5*, p.385.
Bunt, J.S. and Williams, W.T.   1981.   *Mar. Ecol. Prog. Ser.
     4*, p.349.
Bunt, J.S., Williams, W.T. and Duke, N.C.   1982.   *J. Biogeogr.
     9*, p.111.
Clough, B.F.   1982.   (ed.) *Mangrove Ecosystems in Australia*,
     Australian Institute of Marine Science and Australian
     National University Press, Canberra.
Davies, J.L.   1978.   In *Australia, a Geography*, (ed.) D.N.
     Jeans, Routledge & Kegan Paul, London, p.134.

Davies, P.J.   1983.   In *Proceedings: Inaugural Great Barrier Reef Conference*, (eds.) J.T. Baker *et al*, James Cook University of North Queensland, Townsville, p.13.

Den Hartog, C.   1970.   *The Sea-grasses of the World*, North Holland, Amsterdam.

Done, T.J.   1982.   *Coral Reefs 1*, P.95.

Done, T.J.   1983.   In *Proceedings: Inaugural Great Barrier Reef Conference*, (eds.) J.T. Baker *et al*, James Cook University of North Queensland, Townsville, p.197.

Drew, E.A. and Abel, K.M.   1983.   In *Proceedings: Inaugural Great Barrier Reef Conference*, (eds.)   J.T. Baker *et al*, James Cook University of North Queensland, Townsville, p.299.

Duke, N.C., Bunt, J.S. and Williams, W.T.   1981.   *Aust. J. Bot. 29*, P.547.

Endean, R., Kenny, R. and Stephenson, W.   1956.   *Aust. J. Mar. Freshw. Res. 7*, p.88.

Endean, R., Stephenson, W. and Kenny, R.   1956.   *Aust. J. Mar. Freshw. Res. 7*, p.317.

Forbes, A.M.G. and Church, J.A.   1982.   *C.S.I.R.O. Marine Laboratories, Research Report 1979-81*, p.21.

Frankel, E.   1978.   *Bibliography of the Great Barrier Reef Province*, Australian Government Publishing Service, Canberra.

Fry, B., Scalan, R.S. and Parker, P.L.   1983.   *Aust. J. Mar. Freshw. Res. 34*, P.707.

Galloway, R.W. and Bahr, M.E.   1979.   *Aust. Geogr. 14*, p.244.

Galloway, R.W.   1981.   *Aust. Geogr. Studies 19*, P.107.

Galloway, R.W.   1982.   In *Mangrove Ecosystems in Australia*, (ed.) B.F. Clough, Australian Institute of Marine Science and Australian National University, Canberra, p.31.

Hatcher, B.G.   1983.   In *Proceedings: Inaugural Great Barrier Reef Conference*, (eds.) J.T. Baker *et al*, James Cook University of North Queensland, Townsville, p.317.

Hopley, D.   1982.   *The Geomorphology of the Great Barrier Reef*, John Wiley and Sons, New York.

Hutchings, P.A.   1981.   *Proc. Linn. Soc. N.S.W. 106*, p.181.

Hutchings, P.A. and Recher, H.F.   1981.   *Proc. Linn. Soc. N.S.W. 106*, P.83.

Jeans, D.N.   1978.   (ed.) *Australia, a Geography*, Routledge and Kegan Paul, London.

Jeffrey, S.W., Hallegraeff, G.M. and Heron, A.C.   1982, *C.S.I.R.O. Marine Laboratories Research Report 1979-81*, Cronulla.

Jones, S.   1967.   *International Zoo Yearbook 7*, p.215.

Kenneally, K.F.   1982.   In *Mangrove Ecosystems in Australia*, (ed.) B.F. Clough, Australian Institute of Marine Science and Australian National University, Canberra, p.95.

Kinsey, D.W.   1983a.   In *Perspectives on Coral Reefs*, (ed.) D.J. Barnes, Australian Institute of Marine Science and Brian Clouston, Publisher, Canberra, p.209.

Kinsey, D.W.   1983b.   In *Proceedings: Inaugural Great Barrier Reef Conference*, (eds.) J.T. Baker *et al*, James Cook University of North Queensland, Townsville, p.333.

Lang, J.C.   1971.   *Bull. Mar. Sci. 21*, p.952.

Lang, J.C.   1973.   *Bull. Mar. Sci. 23*, p.260.

Larkum, A.W.D.   1977.   In *Seagrass Ecosystems*, (eds.) C.P. McRoy and C. Helfferich, Marcel Dekker, Inc. New York, p.247.

Lourensz, R.S.   1977.   *Tropical cyclones in the Australian region, July 1909 to June 1975*, Australian Government Publishing Service, Canberra.

Macnae, W.   1966.   *Aust. J. Bot. 14*, p.67.

Maxwell, W.G.H.   1968.   *Atlas of the Great Barrier Reef*, Elsevier, Amsterdam.

McComb, A.J., Cambridge, M.C., Kirkman, H. and Kuo, J.   1981. In *Biology of Australian Native Plants*, (eds.) J. Pate and A.J. McComb. University of Western Australia Press P.258.

McRoy, C.P. and Helfferich, C.   1977.   (eds.) *Seagrass Ecosystems*, Marcell Dekker, Inc. New York.

Motoda, S., Kawamurra, T. and Taniguchi, A.   1978.   *Mar. Biol. 46*, p.93.

Ngau, Y. and Price, I.R.   1980.   *Aust. J. Mar. Freshw. Res. 31*, p.175.

Odum, W.E. and Heald, E.J.   1972.   *Bull. Mar. Sci. 22*, p.671.

Pickard, G.L.   1977.   A review of the physical oceanography of the Great Barrier Reef and Western Coral Sea, *Australian Institute of Marine Science Monograph Series Vol. 2*, Australian Government Publishing Service, Canberra.

Pickard, G.L.   1983.   In *Proceedings: Inaugural Great Barrier Reef Conference*, (eds.) J.T. Baker *et al*, James Cook University of North Queensland, Townsville, P.63.

Radok, R.   1976.   *Australia's Coast*, Rigby, Adelaide.

Radway Allen, K.   1978.   In *Australia's Offshore Resources*, (ed.) G.W.P. George, Australian Academy of Science, p.109

Redfield, J.   1982.   In *Mangrove Ecosystems in Australia*, (ed.) B.F. Clough, Australian Institute of Marine Science and Australian National University, Canberra, p.259.

144   J. S. Bunt

Revelante, N., Williams, W.T. and Bunt, J.S.   1982.   *J. Exper.
  Mar. Biol. Ecol. 63,* p.27.
Rochford, D.J.   1951.   *Aust. J. Mar. Freshw. Res. 2,* p.1.
Rochford, D.J.   1980.   *C.S.I.R.O. Fisheries and Oceanography
  Report, 1977-79,* p.9-20.
Rollet, B.   1981.   *Bibliography on mangrove research, 1600-
  1975,* UNESCO, Parish.
Saenger, P., Specht, M.M., Specht, R.L. and Chapman, V.J.
  1977.   In *Ecosystems of the World. I. Wet Coastal
  Ecosystems,* (ed.) V.T. Chapman, Elsevier, Amsterdam,
  p.293.
Sammarco, P.W.   1983.   *Mar. Ecol. Prog. Ser. 13,* p.1.
Sammarco, P.W., Coll, J., LaBarre, S. and Willis, B.   1983.
  Coral Reefs 1, p.173.
Thorhaug, A.   1982.   In *C.R.C. Handbook of Biosolar Resources,
  Vol. 1, part 2,* (eds.) A. Mitsui and C.C. Black,
  C.R.C. Press, P.471.
Veron, J.E.N. and Pichon, M.   1976.   Scleractinia of Eastern
  Australia, *Australian Institute of Marine Science
  Monograph Series, Vol. 1, Part I.* Australian Government
  Publishing Service, Canberra.
Vernon, J.E.N. and Pichon, M.   1979.   Scleractinia of Eastern
  Australia, *Australian Institute of Marine Science
  Monograph Series Vol. 4, Part III.* Australian
  Government Publishing Service, Canberra.
Veron, J.E.N. and Pichon, M.   1982.   Scleractinia of Eastern
  Australia, *Australian Institute of Marine Science
  Monograph Series, Vol. 5, Part IV.* Australian
  Government Publishing Service, Canberra.
Veron, J.E.N., Pichon, M. and Wijsman-Best, M.   1977.
  Scleractinia of Eastern Australia, *Australian Institute
  of Marine Science Monograph Series, Vol. 3, Part II.*
  Australian Government Publishing Service, Canberra.
Wells, A.G.   1982.   In *Mangrove Ecosystems in Australia,*
  (ed.) B.F. Clough, Australian Institute of Marine
  Science and Australian National University, Canberra,
  p.57.
Wells, F.E.   1983.   *Bull. Mar. Sci. 33,* p.736.
Wilkinson, C.R.   1983.   *Perspectives on Coral Reefs,*
  (ed.) D.J. Barnes, Australian Institute of Marine
  Science and Brian Clouston, Publisher, Canberra, p.263.
Wilkinson, C.R., Williams, D.McB., Sammarco, P.W., Hogg, R.W.
  and Trott, L.A.   1983.   In *Proceedings: Inaugural Great
  Barrier Reef Conference,* (eds.) J.T. Baker *et al,* James
  Cook University of North Queensland, Townsville, p.375.
Williams, D.McB, and Hatcher, A.I.   1983.   *Mar. Ecol. Prog.
  Ser. 10,* p.239.

Williams, W.T., Bunt, J.S. and Duke, N.C.   1981.   *Aust. J. Bot. 29*, p.555.

Wolanski, E.   1983.   *J. Geophys. Res. 88*, p.5953.

Wolanski, E., Jones, M. and Bunt, J.S.   1980.   *Aust. J. Mar. Freshw. Res. 31*, p.431.

Wolanski, E. and Pickard, G.   1983.   *Aust. J. Mar. Freshw. Res. 34*, p.65.

Wood, E.J.F.   1954.   *Aust. J. Mar. Freshw. Res. 5*, p.1.

Wood, E.J.F.   1965,   *Nova Hedwigia 8*, pp.5 and 453.

9

RANGELAND ECOSYSTEMS: FACTORS INFLUENCING THEIR PRODUCTIVITY

R. A. PERRY

CSIRO Division of Groundwater Research
Wembley, Western Australia

J. J. MOTT

CSIRO Division of Tropical Pastures
Cunningham Laboratories
St Lucia, Queensland

I. INTRODUCTION

The major land use in northern Australia is livestock grazing, mainly on native pastures. The nine main rangeland ecosystems are described. In the higher rainfall parts (see Chapter 2) the main factor limiting productivity is poor pasture quality particularly in the dry season but in the arid areas it is low pasture production. High freight costs due to the long distances from markets increase the costs of supplies and decrease returns.

This chapter draws heavily on previous papers by the authors (Perry, 1970; Mott, Tothill and Weston, 1981; Mott and Tothill, 1984) all of which have comprehensive bibliographies to which the reader is referred.

For more than 100 years the major land use in northern Australia has been livestock grazing, most of the area being used for beef cattle but with sheep important in central Queensland and minor in Western Australia. In the years 1977–80 livestock populations were 14 million cattle and 13 million sheep. The value of production fluctuates widely, e.g. in the 1977/78 financial year it was less than $400 million but in 1979/80 it exceeded $1,000 million due to both higher prices and higher turnoff.

NORTHERN AUSTRALIA
ISBN 0 12 545080 X

The general livestock husbandry system is yearlong set-stocking on large properties. Over most of the area the properties are held by individuals or companies under long term lease from the Governments of Queensland, Northern Territory and Western Australia but in the higher rainfall parts of Queensland the properties are freehold. Considerable areas of sandy deserts, mostly in Western Australia and Northern Territory remain as unalienated crown land. Properties vary greatly in size, those in the more arid parts averaging several thousand km^2 whereas in the better pastures of south-eastern Queensland 3000-8000 ha is normal.

Prior to the second world war the main control of livestock was through the relatively sparse network of watering points but since then most properties have been fenced and subdivided and many more watering points developed. Over most of the area watering points are either pumped bores or surface catchment tanks but in large areas of Queensland the bores are artesian and water is distributed from them in shallow bore drains.

Over most of the area livestock are grazed on native pastures. In some parts, particularly in Queensland, tree cover has been removed or reduced to improve native pasture production. Less than 1% of the area has been sown with improved pastures, with buffel grass being the main species sown.

II. THE ENVIRONMENT

The inland parts of the area are predominantly flat to undulating plains with either internal or unorganized drainage. The major soils are cracking clays, red sands or clayey sands, and red and yellow earths. The cracking clays are most extensive on the Mesozoic sediments of the Great Artesian Basin in Queensland but extend to the Barkly Tableland and Wave Hill districts of the Northern Territory and the Kimberley region of Western Australia. The red sands and clayey sands are the common soils on the sandplains and dunefields of the Simpson, Gibson and Great Sandy Deserts. Red and yellow earths occur on erosional and lateritic plains. Other soils such as calcareous earths, duplex soils, and lithosols occur to a lesser extent. In the higher rainfall areas extending inland from the coast for up to 300 km, the country is more dissected by streams flowing to the ocean. In these areas lithosols, red and yellow earths and solodics are the main soil groups.

Over the whole area the soils are infertile. Nitrogen
and phosphorus are almost universally deficient, and sulphur
and trace element deficiencies are probably widespread.
Physical properties are also poor; most of the soils have
poor water holding properties and many have hard setting
surfaces.

Climatically the whole area receives most of its rain in
summer and most of it has a short summer rainfall season and
a long dry winter. Towards the south, winter rainfall becomes
more important although it is low. In coastal Queensland some
high rainfall areas receive rain throughout the year.

III. PASTURE TYPES

In the following descriptions of the rangeland ecosystems
the same nine pasture types recognized by Mott, Tothill and
Weston (1981) are used. Christie (Chapter 10)discusses
aspects of the productivity of the grasslands in northern
Australia. The distribution of pasture types is shown in
Figure 1.

A.   *Tropical Tallgrass Lands*

These are the pastures of the far northern, medium to
high rainfall areas of Western Australia, Northern Territory
and Queensland. The grasses are various mixtures of perennial
tussock grasses (e.g. *Themeda australis, Sehima nervosum,*
perennial *Sorghum* spp, and *Chrysopogon fallax*) or tall annual
Sorghum spp. Shorter annuals (*Shizachyrium* spp. and *Aristida*
spp.) are important in many parts of the Cape York Peninsula.
Typically the pastures occur as the understorey of an open
canopy (10 to 15 m high) of trees such as *Eucalyptus
tetrodonta, E. dichromophloia, E. foelschiana,* and
Erythrophleum chlorostachys with scattered tall shrubs or low
trees including *Melaleuca viridiflora.*

The soils are infertile and even in the high rainfall
areas grass productivity is low (1100–140 kg ha^{-1} an^{-1} dry
matter) as is the nutritive value, particularly during the
long dry season. The carrying capacity is low (25–50 ha hd^{-1}),
calving percentages are only about 50%, and 5 to 7 years is
required for animals to reach a turnoff weight of 500 kg. The
area of approximately 32 million ha carried slightly more than
1 million cattle during 1977-80.

B.   *Black Speargrass Lands*

These pastures occur in eastern Queensland from the base
of Cape York Peninsula to the New South Wales border and
extending inland 100-300 km from the coast.   The pastures are
dominated by perennial tussock grass about 1 m or more high,
Heteropogon contortus being the most common but with
Bothriochloa spp. being important components particularly in
the drier westerly parts and on heavier soils.   The pastures
normally occur as an understorey in open forests or woodlands
in which *Eucalyptus crebra* and *E. melanophloia* are common
trees.   Thinning or clearing the trees leads to a 2 or 3 fold
increase in grass production and increased carrying capacity
in the southern parts but the response only occurs in below
average rainfall years in the northern parts.
These lands have a higher pasture productivity which is
nutritionally better than the Tropical Tallgrass Lands.
Stocking rates are 3-15 ha hd^{-1}, calving rates 55-60% and
cattle reach a turnoff weight of 500 kg in 4 to 5 years.   The
area of 29 million ha carried 3.5 million cattle in 1977-80.

C.   *Aristida-Bothriochloa Woodlands*

Most of these lands occur in a belt 100-200 km wide to
the west of the Black Speargrass lands in Queensland.   A
small area extends into the Northern Territory near the Gulf
of Carpentaria.   The pasture is predominantly perennial
tussock grasses less than 1 m high commonly *Aristida* spp.,
Bothriochloa spp. and *Chloris* spp. with *Chrysopogon* common in
the north and *Stipa* and *Danthonia* being included in the south.
The overstoreys over most of the area are 6-10 m high with
E. populnea, E. crebra, and *E. melanophloia* the most
important trees.   In the north various *Melaleuca* spp.
predominate in somewhat lower woodlands.
Pasture productivity is low as is the quality and carrying capacity is 10-40 ha hd $^{-1}$ of cattle or sheep to 2-3 ha.
The 34 million ha carried 1.7 million cattle and 1.9 million
sheep in 1977-80.

D.   *Brigalow Lands*

These lands occur in the same general location as the
southern half of *Aristida-Bothriochloa* lands.   In the natural
state the lands have a dense cover of brigalow (*Acacia
harpophylla*) trees 10-15 m high with only a sparse grass
understorey with a low livestock carrying capacity.   The soils

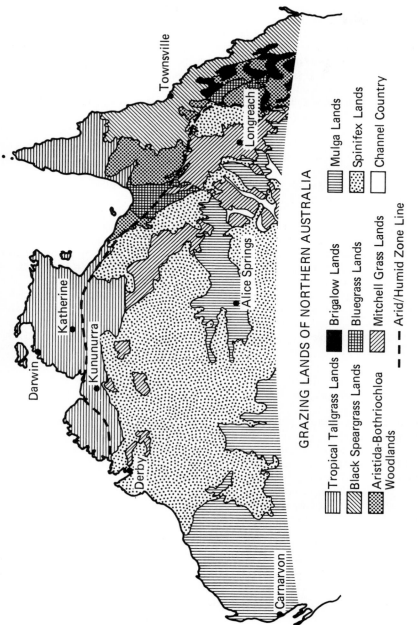

FIGURE 1. Grazing Lands of Northern Australia.
(After Mott, Tothill and Weston, 1981)

are reasonably fertile and with clearing the productivity of
the natural grasses increases and the carrying capacity rises
to 3-8 ha hd^{-1} of cattle or 1 sheep per hectare. However
brigalow regrows into dense stands unless the cleared land is
cultivated for several years. Much of the land has been
cleared and sown to introduced grasses such as buffel grass
and green panic which have a higher carrying capacity. The
9 million ha carried 1.8 million cattle and 1.4 million sheep
in 1977-80.

E.   *Bluegrass lands*

These lands occur in Queensland and are open grasslands
dominated by perennial tussock grasses about 1 m high among
which *Dichanthium* spp. are prominent. Two main types can be
recognized. In the north, near the Gulf of Carpentaria the
grasses are mainly *Dichanthium fecundum* and *Eulalia fulva*
growing on infertile clay soils and with fairly low production
and quality. The carrying capacity is about 15 ha hd^{-1}. In
central and southern Queensland the prominent grass is
Dichanthium sericeum growing on fertile soils. Both production and quality is higher and the carrying capacity is about
4 ha hd^{-1} although much of the area is used for cropping
rather than pastures. The total area of 9 million ha carried
9.4 million cattle and 60 thousand sheep in 1977-80.

F.   *Mitchell Grasslands*

These are open grasslands (or grasslands with scattered
low trees) growing on cracking clay soils under a mean annual
rainfall less than about 600 mm. They occur over large areas
of rolling downs in Queensland, on the Barkly Tableland and
Wave Hill districts of the Northern Territory and to a lesser
extent in northern Western Australia. The dominant grasses
are perennial tussock grasses 0.5-1 m high incorporating one
or more species of *Astrebla*. Shorter annual grasses,
particularly *Iseilema* spp., and forbs grow in the spaces
between the tussock grasses.

These are the most productive of Australia's arid zone
pastures with a carrying capacity of 10-15 ha hd^{-1} or a sheep
to 1-2 ha. The total area of 43 million ha carried 2.9
million cattle and 6.5 million sheep in 1977-80.

G.   *Mulga Lands*

These are shrublands (or low woodlands) 4-7 m high growing mostly on red earth soils and widespread across arid
Australia.   Over much of the area *Acacia aneura* is the
dominant but other *Acacia* spp. are characteristic in some
areas.   The tree density varies tremendously from a few per
hectare to thousands per hectare.   The understorey vegetation
is also very variable with scattered to dense low shrubs of
which *Cassia* spp. and *Eremophila* spp. are common and low
annual or short-lived perennial grasses or forbs.   Of the
grasses various species of *Eragrostis*, *Aristida*, *Thyridolopis*,
and *Monochather* are the most common.   Forbs include many
species particularly composites, legumes, crucifers, and
amaranths.   Herbaceous productivity is low and seasonally
variable but of reasonable quality.   Many of the shrub species
are grazed by stock in dry periods and so are a source of
drought forage.   Livestock carrying capacity is 25-50 ha hd^{-1}
of cattle or a sheep to 3-6 ha.   The area of over 60 million
ha carried 640 thousand cattle and 3 million sheep in 1977-80.

H.   *Spinifex Lands*

These lands are also widespread on the arid areas occurring on sandy deserts, rocky hillslopes and some other
habitats.   They are grasslands (or grasslands with a sparse
to moderately dense cover of low shrubs) dominated by one or
more species of *Triodia* or *Plectrachne* which are hard leaved
perennial tussock grasses of low palatability and nutritive
value.   Most of the area is not used for grazing and those
parts which are used have a very low carrying capacity.   In
1977-80 the 65 million ha carried about 600 thousand cattle
and 300 thousand sheep.

I.   *Channel Country*

These lands are the floodplains of the major rivers in
south western Queensland.   Following rains in the headwaters
these plains are flooded.   On recession of the floods a
luxuriant but short-lived pasture growth occurs.   In such
periods these lands are used for fattening cattle.   In 1977-80
that 5.4 million ha carried 130 thousand cattle.   The Channel
Country is discussed in more detail by Graetz, in Chapter 11 .

IV.   FACTORS INFLUENCING PRODUCTIVITY

In the higher rainfall areas in the northern and eastern parts of northern Australia the main factor affecting productivity is the poor quality of the pastures, particularly during the long dry season. The soils are infertile with nitrogen and phosphorus and probably other elements being deficient almost everywhere. With summer rainfall they produce a rapid growth of moderate quality which deteriorates rapidly at the end of the wet season. Thus pasture quality fluctuates seasonally within each year from moderate in the wet season to poor in the dry season.

In the arid areas it is quantity rather than quality which sets the limit on animal production. The rainfall has a summer maximum but towards the south there is a significant winter component. In years in which adequate rainfall occurs short pastures of reasonable quality are produced but in dry years little or no production occurs.

With climate limiting productivity in the arid areas there is little chance of increasing potential pasture production and the main consideration must be to maintain or improve rangeland condition to ensure sustained productivity. There is scope for investigating introduced species but the chances of success are not high. The situation is different in the higher rainfall areas where there are possibilities for increasing both quantity and quality of pasture productivity by clearing trees, applications of fertilizers and introduction of better plant species, particularly legumes.

Animal performance can be improved over the whole area by the use of more adapted livestock and by improved animal husbandry. The introduction of *Bos indicus* genes to the British breeds used traditionally provides greater resistance to cattle tick, greater resistance to heat stress and better forage use. Control of animal pests and diseases can also improve performance and reproduction.

Distance from markets has an important influence on production from the whole area. Costs of freight are high, increasing the costs of all supplies and decreasing returns from turnoff. The consequently lower margins of returns over costs reduce the options for improving animal and pasture production.

REFERENCES

Mott, J.J. and Tothill, J.C.   1984.   *Tropical and subtropical Woodlands.*

Mott, J.J., Tothill, J.C. and Weston, E.   1981.   The native Woodlands and Grasslands of northern Australia as a grazing resource for low cost animal production, *Journal of the Australian Institute for Agricultural Science 47:* pp.132-41.

Perry, R.A.   1970.   Arid Shrublands and Grasslands, *Australian Grasslands,* (ed.) R.M. Moore, Australian National University Press, Canberra, p.455.

10

PRODUCTION AND STABILITY OF SEMI-ARID GRASSLANDS

E. K. CHRISTIE

School of Australian Environmental Studies
Griffith University
Nathan, Queensland

I.   INTRODUCTION

The major semiarid grassland communities in northern Australia are the Mitchell (*Astrebla* spp.) tussock grasslands and the Spinifex hummock (*Triodia* spp.) grasslands. The resultant natural grassland following clearing in strips, or thinning of the mulga (*Acacia aneura*) shrublands is also a major livestock grazing system. The Mitchell grasslands are the most productive of the natural Australian arid and semi-arid vegetation systems and usually occur on treeless cracking clay soils, comprising a total area of around 450,000 km^2 (Orr 1975). The *Acacia aneura* (mulga) grazing lands occupy an area of 200,000 km^2 in semiarid Queensland alone and occur on red earths, stoney hillsides and on some clayey sands. Mulga itself, is principally used as a drought feed, stock mainly feeding on the perennial grass and forb ground-storey. The soils are infertile to highly infertile which limits seedling growth, and hence establishment of perennial plants, if plant populations are reduced by over-grazing. The spinifex grasslands are the most extensive and widespread community in arid and semiarid Australia occupying about one-third of this area. Spinifex communities occur on infertile red sands to rocky hillslopes. In their natural state most of the spinifex communities are grazed at very low stocking rates and vast areas are unoccupied crown land (Perry 1970).

NORTHERN AUSTRALIA
ISBN 0 12 545080 X

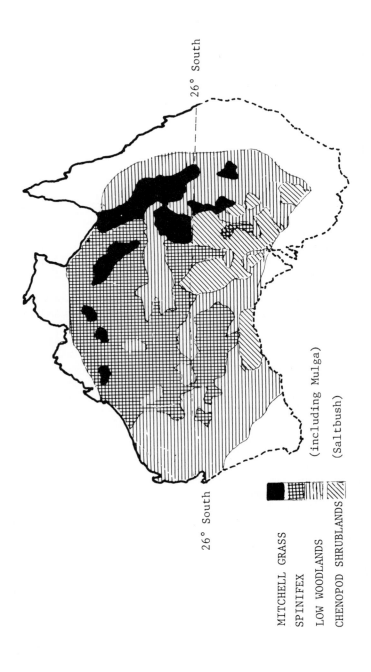

FIGURE 1. The distribution of the major Australian arid
and semi-arid plant communities[a]

[a]Based on CSIRO Division of Land Resources Management.

MITCHELL GRASS

SPINIFEX

LOW WOODLANDS   (including Mulga)

CHENOPOD SHRUBLANDS   (Saltbush)

26° South

26° South

Since the introduction of domestic livestock, available evidence indicates the presence of degradation in semiarid grazing lands of northern Australia (e.g. Burrows 1973; Dawson and Mawson 1973; Orr 1980). In *Acacia aneura* grazing lands, the evidence seems to be that the vegetation change is unidirectional and towards a greater preponderance of trees and shrubs. Furthermore, this change is not catacylysmic, but rather continuous, suggesting that the intermediate states are unstable. On the other hand, vegetation fluxes in the Mitchell grasslands may be regarded as locally stable points in systems which are characterized by having multiple-stable points (Burrows 1980). Although each of the natural grazing land systems can be contrasted in respect to soil fertility and vegetation flux patterns, each system does have one factor in common - that is in the control which precipitation input has on production processes. Consequently, there are enormous annual variations in net primary production. This is one of the central issues in their grazing management - how to best relate livestock numbers to the conditions of fluctuating herbage supply and still maintain stability of the soil and vegetation resource and cash flow!

Three complementary features are essential in any programme of management for semiarid grazing lands. Firstly, standards sensitive to the early stages of deterioration in grazed systems are required. Secondly, these condition standards should be used in a natural resource monitoring programme in order to detect direction of trend of the system under the existing land use practice and climate. Finally, management strategies aimed at maintaining long-term resource stability need to be developed. Although management strategies are numerous, the most fundamental is the estimation of animal carrying capacity. Any consideration of animal carrying capacity must be based on characterizing the seasonal and annual variation in net primary production of the plant community, for any defined state of condition of the resource, in relation to precipitation input. Moreover, for grazed systems, a level of herbage utilization by grazing animals, which gives near-maximal biological and economic animal production commensurate with soil and vegetation resource stability, must also be determined. Such an approach which integrates studies into net primary production, semi-arid grassland condition assessment and herbage utilization by livestock will be described for one major semiarid natural grassland system viz. the *Acacia aneura* natural grazing lands. Some implications of this work for grazing land management and administration will also be discussed.

II.   CONDITION ASSESSMENT STANDARDS FOR SEMI-ARID GRASSLANDS

Grazing land condition may be broadly defined as the status of a grazing site, with respect to its vegetation and soil characteristics, relative to its potential. If condition assessment is to be objective and meaningful, it must be based on measurable attributes which can be applied on a quantitative basis to a variety of sites (Roberts 1972). Roberts (1970) has proposed a simplified condition assessment scheme based on plant cover (or basal area), botanical

TABLE 1.   Criteria for Assessing the Condition Status of the Herbaceous Ground-Storey of the Mulga Grazing Lands of Semiarid Queensland.

A.   *BOTANICAL COMPOSITION*

Condition Standard	Botanical Composition *Proportion of Preferred Species[a] is,*
Excellent	> 75%
Good	50 - 75%
Medium	25 - 50%
Poor	< 25%

[a] based on dominant climax or pristine vegetation at the site

B.   *GROUND COVER*

Condition Standard	Basal Area[b] (%)	Relative Forage[b] Production Potential (% maximum)	Erosion[c]
Excellent	> 5.0	> 90	*Soil mantle stable and undisturbed Same condition as relict system.*
Good	3.0 - 5.0	60 - 90	*Up to 25% of soil mantle disturbed by erosive forces. Some surface crusting. Initial signs of sheet erosion. Surface run-off increases.*
Medium	1.5 - 3.0	30 - 60	*Up to 50% of soil mantle moderately affected by erosive forces. Definite signs of sheet or rill erosion. Soil mounds around obstructions.*
Poor	< 1.5	< 30	*50 to 100% of soil mantle very severely affected by erosive forces.*

[b] based on Christie (1978)

[c] based on Deming (1957)

a) Herbage Component
   Basal Area, 1.0%

   (ground-storey condition
    class: poor)'

b) Herbage Component
   Basal Area, 2.5%

   (ground-storey condition
    class: medium)

c) Herbage Component
   Basal Area, 4.0%

   (ground-storey condition
    class: good)

d) Herbage Component
   Basal Area, 5.5%

   (ground-storey condition
    class: excellent)

Plate 1.   Relationship between basal area of the herbaceous ground-storey and condition class of the Acacia aneura grazing lands (following Christie, 1978).

composition and soil surface condition for use in semiarid
grasslands.   Shifts in botanical composition, or the replacement of one type of plant cover with that of another,
represents the best means of detecting overgrazing, initially,
in grassland systems (Table 1A).   Botanical composition,
therefore, would generally be the most sensitive indicator of
deterioration particularly on fertile soils (e.g. the grey,
brown cracking clays/Mitchell grasslands) where plant cover
is normally always high.   This is because precipitation
and the time of year that major precipitation events
occur, will ultimately control the successional pattern on
degraded sites (Christie 1979).   However, on infertile soils
(e.g. red earth/mulga grazing lands) plant cover is low, so
plant cover would generally be more important than
composition, because re-establishment of any perennial plant,
preferred or undesirable, is a very slow, difficult process -
since it is both limited by water and nutrients.   Basal area
becomes one fundamental criterion for condition assessment
(Table 1B; Plate 1).   The central issue for managing *Acacia
aneura* grazing lands is to, at the very least, maintain cover
of the perennial plants and thus stability of the soil
resource in the long-term (Christie 1978).
     Annual monitoring of land condition at a number of
reference sites, for any one grassland system in a biological
region, or at the paddock level, allows the direction of
trend to be determined under the existing management practice
and climate.   It may be stable, downwards (deteriorating) or
upwards (improving) and thus provides a direct aid to land
managers.

III.   ECOLOGICAL CONCEPTS IN THE ESTIMATION OF ANIMAL CARRYING CAPACITY

     There is a real need to develop an ecologically based
method for assessing grazing capacity of Australia's semiarid natural grazing lands in order to maintain ecosystem
stability in the long-term.   Semiarid natural grazing lands
have a characteristic upper limit to net primary production.
This is the essential starting point in any consideration
of carrying capacity.   Net primary production for any one
grassland system must be quantified, both in relation to long-
term annual variation, as well as seasonal variation in the
plant community.   The next step is to determine how much of
this net primary production can be made available to stock
annually i.e. the degree of herbage utilization, and still

maintain ecosystem stability. Of course, any estimate of
animal carrying capacity will also be dependent on animal
intake rate, distribution of animals and the grazing management system - as well as the degree of herbage utilization
through consumption. However, in continuously grazed systems
where reasonably uniform distribution of animals exists, the
most important of these variables is the degree of utilization
of the herbage resource through animal consumption (Heady
1975).

A number of factors need to be considered in the
determination of a safe level of herbage utilization in semiarid areas. These include:

(a) Maintaining the long-term stability of the soil and
vegetation.

(b) Maintaining a desirable species balance (preferred
or key species: unpalatable plants).

(c) Ensuring that biological and economic animal
production is near-maximal.

(d) Ensuring some end of summer herbage carries over
into the following year.

(e) Allowing herbivores to selectively graze and hence
minimize nutritional limitations to growth and production.

Continued over-stocking, or over-utilization, leads to
land resource degradation. Over-use should not be simply
regarded as a decline in any one of these factors. Rather,
all of these factors are affected by over-stocking and their
interaction should be considered before making a decision on
resource use or abuse! All of these factors have been
considered in two existing semiarid grazing land studies in
south-west Queensland commenced in 1975 (Christie 1981) and
in 1978 (Burrows 1980)

A.   Net Primary Production of Semiarid Grasslands

For any one grassland system net primary production will
be dependent on the control exerted by the environmental
factors of water, nutrients and temperature. Arid and semiarid systems are generally regarded as water-controlled
systems because precipitation has the overriding control on
net primary production (Noy-Meir 1973).

A simulation model (Hughes and Christie 1982) which
estimates net primary production of the ground-storey
component of extensively grazed *Acacia aneura* grazing lands
of semiarid Queensland, Australia has been applied to the

assessment of grazing capacity of semiarid grasslands.
Historic climatic records for a continuous period of 25 years
were analyzed for Charleville, Queensland. Precipitation in
this environment occurs predominantly over the October-March
period and accounts for approximately 70% of the total
recorded over twelve months (Figure 2A). The limitations in
the use of precipitation as the sole determinant of net
primary production is evident from Figure 2 (A,B), where the
trend in actual length of the growing period was not always
well correlated with precipitation input. Hence, differences
in live biomass production (Figure 2C) were more closely
related to soil water balance output than to precipitation.
The total live herbage biomass crop produced over the October-
March growing season, for a ground-storey basal area site
value of 2.5% (Plate 1), varied annually, approximately five
fold - from 190 to 930 kg.ha^{-1} (Figure 2C). The long-term
mean value was 490 kg.ha^{-1}. yr^{-1} (Christie and Hughes 1983).
Derived values for long-term mean live herbage biomass
production over the October-March growing season, for sites
having a basal area of 1%, 4% and 5.5% (Plate 1) were around
130 (range 50 - 220), 690 (270 - 1360) and 730 (270-1450)
kg.ha^{-1}, respectively (Christie and Hughes *op. cit.*).

Although this analysis of net primary production dealt
with live biomass production only, another significant source
of biomass available for herbivores in semiarid environments
is the senescent plant material, or herbage carryover, from
year to year. Herbage carryover from year to year decreases
significantly as the level of herbage utilization through
animal consumption increases. At a conservative level of
annual utilization (e.g. 20%) herbage biomass carryover
provides an important reserve to supplement the smaller amount
of live biomass production in low rainfall years (Christie
and Hughes 1973). Ideally, grazing capacity should be
related to the total herbage biomass available to the grazing
animal at the end of each summer growing period and not simply
live herbage biomass (Figure 2C). As the annual level of
grazing use extends beyond 50%, carryover herbage biomass
becomes extremely small (Christie and Hughes *op cit.*).

B.   *Estimation of Grazing Capacity*

Many natural grazing land studies, such as Reid *et al.*
(1963), have shown a consistent relationship between the
number of animals grazed, the amount of herbage available
and the degree of utilization of the herbage biomass.

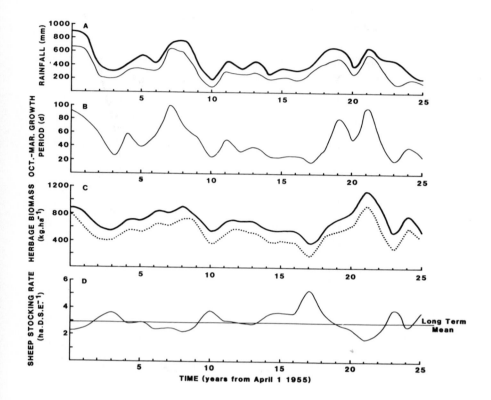

*FIGURE 2.   Analysis of historic rainfall records
for Charleville, Queensland, 1955-1980[a].*

[a]*Model output shows interrelationships between (A) Rainfall,
April-March (━━━━━) and October-March (────────);
(B) Length of October-March growing period (soil water
balance output);   (C) Live herbage biomass produced over the
October-March period (··············) and total herbage biomass
crop available at the end of this growing period (━━━━━)
and;   (D) Derived values for annual stocking rate.   Analysis
has been based on a constant ground-storey basal area of
2.5% (or medium condition class) and a value for the degree
of herbage utilization through sheep consumption, of 20%
(Christie and Hughes 1983).*

The following relationship was used by Christie, and Hughes (1983) to derive annual values for stocking rate,

$$S = c/UB \qquad (1)$$

where S is the annual stocking rate (ha. animal^{-1});   c is the annual herbage biomass consumption for a specified animal unit (kg.yr^{-1});   U is the percentage herbage utilization by animals (expressed as a decimal) and B the end-of-summer total peak herbage biomass crop (kg.ha^{-1}).

Stocking rate (S) was calculated on the basis of Dry Sheep Equivalents (D.S.E.) as the *Acacia aneura* grazing lands only have limited sheep breeding potential (Anson and Mawson 1969). Animal consumption for a specified sheep body-weight class (c) was estimated from the relationship of Vickery and Hedges (1972). Values for the degree of utilization (U) were based on two existing grazing studies in semiarid south-west Queensland (Burrows 1980; Christie 1981). Preliminary analysis of the data for the *Acacia aneura* grazing land study by Burrows (1980) suggests that utilization values of the herbaceous ground-storey component for this grazing system should be adjusted so that consumption does not exceed 20 per cent of the end-of-summer herbage production. This value was adopted by Christie and Hughes (1983) in their estimation of grazing capacity for the *Acacia aneura* grazing lands for sites in which ground-storey condition class varied. They also derived values for sheep carrying capacity, for each ground-storey condition class, from the following expression,

$$\overline{S} = (S_1 + S_2 \ldots . S_n)/N \qquad (2)$$

where $\overline{S}$ is the sheep carrying capacity (ha. D.S.E.$^{-1}$); S_1, S_2 .... S_n the annual stocking rate for year 1, year 2, ... and year n, respectively (ha. D.S.E.$^{-1}$) and N the time (years).

The estimated values for annual stocking rate for the 2.5% basal area site varied over time from 1.7 to 5.3 ha. D.S.E.$^{-1}$ (Figure 2D). The value derived for sheep carrying capacity, or long-term mean stocking rate, was 2.9 ha. D.S.E.$^{-1}$, for this site (Figure 2D). As basal area, and hence ground-storey condition class declined, end-of-summer peak herbage biomass crop and sheep carrying capacity decreased (Figure 3). There were only small changes in the values estimated over the excellent to good ground-storey condition classes. However, as condition class decreased from moderate to poor there was a pronounced decline in herbage

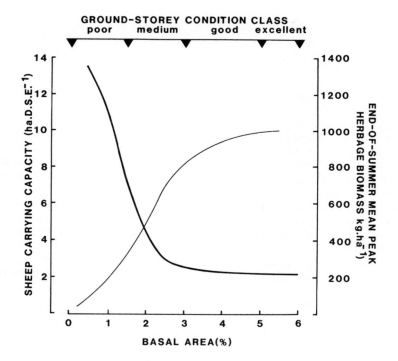

FIGURE 3.   Interrelationships between basal area of the ground-storey component, long-term mean value for end-of-summer total peak herbage biomass crop (————) and the derived value for sheep carrying capacity (▬▬▬▬). Analysis based on historic rainfall records for Charleville, Queensland, 1955-80 (Christie and Hughes 1983).

biomass production and sheep carrying capacity (Figure 3)
(Christie and Hughes 1983).

IV.   LAND ADMINISTRATION AND MANAGEMENT

These derived values for annual stocking rate are largely
theoretical, which may be extremely difficult, or impractical,
to achieve over time for a pastoral holding. However, the
derived value for animal carrying capacity should be best
regarded as providing the means for determining a nucleus
flock population to which animals could be added to, or
removed, at the end of each October-March growing period. As
Heady (1975) suggests, it is probably best, as a managerial
expediency, to combine fixed stocking of the nucleus flock
with flexible stocking of other animals to obtain the most
rapid possible improvement during the favourable years and
the least damage in poor years. Such a plan for the *Acacia
aneura* grazing lands would require annual stock adjustments
at the end of each October-March growing period when most of
the new live herbage biomass has been produced. Although the
simulation study described, only considered a continuous year-
round grazing system, there is no evidence for Australian
semi-arid natural grazing lands, that, rotational or other
grazing systems improve vegetation cover, or annual
production, over continuous grazing systems (Wilson 1977).
Ebersohn (1970) believes that the biggest single manip-
ulative practice possible for arid lands for their continued
productivity is an increase in the size of pastoral holdings.
Living area standards for the *Acacia aneura* grazing lands of
semiarid Queensland are now based on a maximum sheep
population of around 9,400 (Anon., 1971). If this value is
taken as the nucleus flock size, and using the derived values
for carrying capacity (Figure 3), then a viable size for a
pastoral holding for these grazing lands in the Charleville
region would vary from around 105,000 ha (for a 1% basal area
site) to 18,000 ha (5.5% basal area site). These derived
values are generally much larger than the current average
holding size most common in the area viz. 15,000 to 17,500 ha
(Mills unpublished data) - especially as ground-storey
condition deteriorates. Insufficient size of the pastoral
holding has contributed to overstocking and degradation of
these grazing lands (Weller 1973) so that holding size over
the poor to good condition class, remains a significant issue
for land administrators. Clearly, the size of pastoral

holding needs to reflect not only precipitation pattern but
also the land condition class.

V. CONCLUSION

Traditional methods for estimating grazing capacity have
been used solely on what a similar grazing land system has
carried in the past. Adjustments are made after a deficiency
has become obvious. Where an assessment has been too high,
an adjustment is not made until obvious deterioration has
occurred and, in many cases, deterioration has often been
critical and permanent (Newman 1974). The method of Christie
and Hughes (1983) represents an alternative method to the
traditional method. It is significant because it estimates
livestock numbers on the basis of herbage production per
unit area of land and not simply on a unit area of land basis.
Thus, the method integrates biotic and abiotic factors which
influence herbage production. In addition, the output from
the simulation model of Hughes and Christie (1982) has overcome the lack of long-term information on net primary
production dynamics for semiarid grazing lands. This has
been a major past limitation for vegetation management and
stocking policy. Consequently, the use of conventional
extension methods which try to inform pastoralists and
influence their management strategies have been relatively
unsuccessful in arid Australia, because of the lack of technical information suitable for extension (Young 1979). As
Burrows (1979) points out, almost all of the past management
decisions concerning Australia's semiarid natural grazing
lands have been based on powers of observation and ecological
intuition. For past practices to be altered and new ones
adopted, quantitative data linked to grazing management is
required for both land administrators and pastoralists. The
studies described in this chapter represent one such approach
for meeting these needs for one grazing land system in semiarid Australia. The scope for applying this approach to
other semiarid grazing land systems in northern Australia
warrants further investigation.

REFERENCES

Anon, 1971. Living area standards - sheep, *A. Rept. Land Admin. Commiss. (Qd.)* 1970-71, Dept. of Lands.

Anson, R.J. and Mawson, W.F. 1969. Ewes or wethers in the south-west, *Qd. Agric. J.* 95, pp.94-8.

Burrows, W.H. 1973. Studies in the dynamics and control of woody weeds in semiarid Queensland. *Eremophila I gilesii. Qd. J. Agric. Anim. Sci.* 30, pp.57-63.

Burrows, W.H. 1979. Vegetation management decisions in Queensland's semiarid sheeplands. In *Rangeland Ecosystem Evaluation,* (ed.) K.M. Howes, Aust. Rangel. Soc., Perth, pp.202-18.

Burrows, W.H. 1980. Range management in the dry tropics with special reference to Queensland. *Trop. Grasslds.* 14, pp.281-88.

Christie, E.K. 1978. Herbage condition assessment of an infertile range grassland based on site production potential. *Aust. Rangel. J.* 1, pp.87-94.

Christie, E.K. 1979. Eco-physiological studies of the semiarid grasses *Aristida leptopoda* and *Astrebla lappacea. Aust. J. Ecol.* 4, pp.223-28.

Christie, E.K. 1981. Ecosystem processes in natural grazing lands: their application in the estimation of animal carrying capacity. In *Desertification of Arid and Semiarid Natural Grazing Lands,* (ed.) E.K. Christie, Australian Development Assistance Bureau International Training Course, Brisbane, Australia, June 1981, pp.108-122.

Christie, E.K. and Hughes, P.G. 1983. Interrelationships between net primary production, ground-storey condition and grazing capacity of the *Acacia aneura* rangelands of semiarid Australia. *Agric. Systems,* 12, pp.191-211.

Dawson, N.M. and Mawson, W.Y.F. 1973. Utilization of arid lands of south-west Queensland. *Proceedings Third World Conference on Animal Production,* Melbourne, Australia, 1, pp. 2(d) 8-2(d).

Deming, M.H. 1957. *Range Studies,* USA Dept. of Interior B.L.M. Manual 9, Part 10.

Ebersohn, J.P. 1970. Vegetation resources of arid Australia. *Proceedings Australian Arid Zone Conferences,* Broken Hill, Australia, B, pp.1-11.

Heady, H.F. 1975. *Rangeland Management,* McGraw-Hill, New York.

Hughes, P.G. and Christie, E.K. 1983. Ecosystem processes in semiarid grasslands III A simulation model for net primary production. *Aust. J. Agric. Res.* 34 (in press)

Newman, J.C. 1974. Effects of past grazing in determining range management principles in Australia. In *Plant Morphogenesis as the Basis for Scientific Management of Range Resources*, Agricultural Research Service, Washington D.C., pp.197-206.

Noy-Meir, I. 1973. Desert ecosystems: environment and producers. *A. Rev. Ecol. Syst.* 4, pp.25-51.

Orr, D.M. 1975. A review of *Astrebla* (Mitchell grass) pastures in Australia. *Trop. Grasslds.* 9, pp.21-36.

Orr, D.M. 1980. Effects of sheep grazing *Astrebla* grassland in central western Queensland I Effects of grazing pressure and livestock distribution. *Aust. J. Agric. Res.* 31, pp.797-806.

Perry, R.A. 1970. Arid shrublands and grasslands. In *Australian Grasslands*, (ed.) R.M. Moore, Australian National University Press, Canberra, pp.246-59.

Reid, E.H., Kovner, J.L. and Martin, S.C. 1963. A proposed method of determining cattle numbers in range experiments. *J. Range Mgmt.* 16, pp.184-7.

Roberts, B.R. 1970. Assessment of veld condition and trend. *Proc. S. Afr. Grassld. Soc.* 5, pp.137-139.

Roberts, B.R. 1972. Ecological studies on pasture condition in semiarid Queensland. *Qd. Dept. Primary Industries Tech. Rept.*

Vickery, P.J. and Hedges, D.A. 1972. Mathematical relationships and computer routines for a productivity model of improved pasture grazed by merino sheep, *CSIRO Aust. Anim. Res. Lab. Tech. Pap. No. 4.*

Weller, M.D. 1973. Farming systems and practices. In *Murweh Shire Handbook*, (ed.) Anon, Queensland Department of Primary Industries, Brisbane, 6, pp.1-15.

Wilson, A.D. 1977. Grazing Management. In *Impact of Herbivores on Arid and Semiarid Rangeland Ecosystems*, (ed.) K.M. Howes, CSIRO, Melbourne, pp.83-92.

Young, M.D. 1979. Influencing land use in pastoral Australia. *J. Arid Envir.* 2, pp.279-88

11

THE UTILITY OF LANDSAT FOR MONITORING THE EPHEMERAL WATER AND HERBAGE RESOURCES OF ARID LANDS: AN EXAMPLE OF RANGELAND MANAGEMENT IN THE CHANNEL COUNTRY OF AUSTRALIA

DEAN GRAETZ
ROGER PECH

CSIRO Division of Wildlife & Rangelands Research
Deniliquin, New South Wales

I. INTRODUCTION

The semi-arid and arid lands of Australia cover a single vast central area of the continent extending to the western and southern coasts. They occupy some $5.3 \times 10^6 km^2$ or approximately 70% of the total land surface area of Australia and are second in size only to the North African – Middle East deserts. One-third of these arid lands are not used by European man on any permanent basis. The remaining area, equivalent to the total land area of India, is used as rangelands. All pastoral enterprise in the rangelands is based on using native vegetation for meat and/or wool production, supporting populations of some 30×10^6 sheep and 6×10^6 cattle – representing about 20% of the nation's flocks and herds.

The Australian rangelands are organized and managed on a very different basis to those in most other arid land areas of the world. Firstly, and of most importance, all the livestock enterprises are geared to a monetary, export market-oriented system which contrasts markedly with the subsistence systems of most of the world's arid lands. Secondly, the livestock industry is organized on an extensive, sedentary basis; there is no nomadism, there are no herders and virtually no seasonal grazing (Perry and Graetz, 1979).

NORTHERN AUSTRALIA
ISBN 0 12 545080 X

In this national context Landsat has been found to be readily applicable and of great value to rangeland resource management generally (Graetz *et al*, 1975; Graetz *et al*, 1980; Hacker, 1980).

II.   THE LAND AND WATER RESOURCES OF THE CHANNEL COUNTRY

The rangelands discussed in this chapter lie in the centre of Australia's arid lands and are remote and sparsely-settled though they carry appreciable herds of cattle. They are unique because they are based on the flood plains of three large ephemeral inland rivers, the Cooper, Diamantina and Georgina, that flow from areas of high summer rainfall into the arid heart of Australia. These three rivers are part of the larger Lake Eyre drainage basin. Figure 1 and Figure 2.

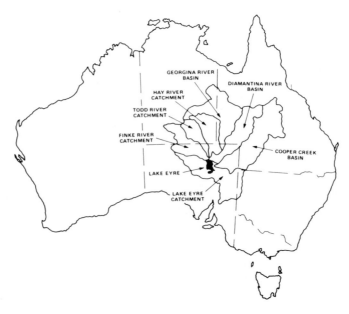

FIGURE 1. A location map of the Lake Eyre Drainage Basin within the arid lands of mainland Australia. The total area of the basin is some 1,170,000 km² or about 16% of the total area of the continent. The Drainage Basin is centred on the normally-dry salina, Lake Eyre, the true 'dead heart' of Australia.

Much of the floodplain is a complex braided network of small streams which give rise to its descriptive name - *The Channel Country*. They are also extensive (Figure 2) and are spread across two States (Table 1). In all, the area regarded as *Channel Country* in this chapter is bounded by Longitude 138°-146° and by Latitude 21°-29°, an area of approximately 660,000 km², most of which is run-off country. For simplicity it is sufficient to group the lands into just two types - the alluvial flood plains and the surrounding country. This

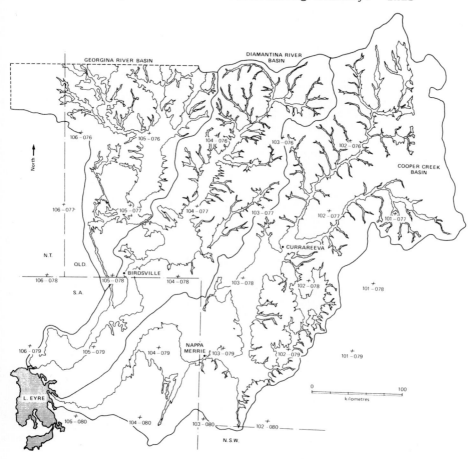

FIGURE 2. A map of the alluvial flood plains within the catchments of the Cooper, Dimantina and Georgina River systems. Superimposed are the locations and reference numbers of the relevant Landsat scene centres.

classification by landform type and genesis also groups
pasture according to the amount and type produced by rain
and its subsequent persistence. A much more detailed
description has been privided by the Australian Water
Resources Council (1975, 1978).

TABLE 1.   *A Summary of the Size (km²) and Distribution of*
the Alluvial Flood Plains of The Channel Country.   The figures
were determined from Figure 2 and % values of the total are
in brackets.

	Qld.	S.A.	Total	Total area in catchment
Cooper Creek	68 600 (66)	35 000 (34)	103 600	296 000
Diamantina	41 250 (75)	13 750 (25)	55 000	158 000
Georgina	49 000 (100)	–	49 000	242 000
Total	158 850	48 750	207 600	696 000

The flooding of *The Channel Country* results from monsoonal summer rainfall several hundreds of kilometres away
and can be very variable in size and occurrence.

Only one of the three drainage systems will be considered
in this chapter. A summary of all three is available elsewhere (Graetz, 1980). The Cooper Creek Basin, being the
easternmost, has the largest flood flows. It is also the
largest basin (296,000 km²) and its lands carry the most
intensive pastoral settlements for sheep production in the
central-northern reaches and cattle in the far south-west.

The Cooper Creek catchment flow record is the longest
available for the three river basins and is a reasonable
measure of the discharge from the two tributaries into the
wide braided stream network characteristic of The Channel
Country. The monthly flows are quite substantial;   the
maximum monthly record was 6,536 gl but during the great
floods of 1973/74 flows were larger, exceeding the rate curve
and so lost. By contrast, 55% of the monthly records are zero
flow. The average monthly discharges peak in the months of
January, February and March and fall off rapidly into the
winter months (Table 2).

A detailed summary of the existing flow records of all
rivers involved is presented elsewhere (Graetz, 1980).

*TABLE 2.   A Summary of the Monthly Flows (gl)[α] for Cooper
Creek Gauged at Currareeva (Lat. 25°20'S, Long. 142°44'E).
The values are averages over 37 years.*

	Oct	Nov	Dec	Jan	Feb	Mar	Apr	May	Jun	Jul	Aug	Sep	Annual
Mean	34	36	97	605	708	874	616	101	166	47	57	37	39

[α]*A giga-litre (gl)* $= 10^9$ *litres.*

III.   LAND MANAGEMENT IN THE CHANNEL COUNTRY

Within the structure of the national pastoral industry,
the management and the production characteristics of this
area reflect the unique nature of the land itself.
Most reviews of the beef cattle industry of Australia single
this area out for special mention and note the fattening or
finishing characteristics of the pastures (Kelly, 1971;
Yates and Schmidt, 1974).   However, they also list the
problems associated with the continuing use of this country,
the remoteness and its attendant penalties of high freight
and production costs, and the management of land where the
stock carrying capacity can vary enormously over short periods
of time, and because of its over-riding aridity, can be
subject to prolonged periods of drought (Allen, 1972; Dawson
and Boyland, 1974;   Dawson *et al.* 1975).
During a century of pastoral occupation of these lands
by European man various financially-successful stock
management schemes have evolved.   All of them, apparently,
were based on flexible patterns of stocking with stock being
purchased or brought in from elsewhere (usually northern
areas) during the times of flood and forage abundance,
fattened or finished over a 6-12 month period, and then moved
out to slaughter, usually towards the southern cities.   In
dry times the individual properties were either completely
destocked or carried only a small dispersed herd.   The
larger pastoral companies held chains of properties, usually
in a North-South axis, to facilitate the movement of stock
from one area to another where the feed conditions were
better (Allen, 1972;   Durack, 1969;   Idriess, 1938).

The movement of stock to evade localized unfavourable forage conditions is a form of nomadism. In Australia, where there is little common land other than the travelling stock routes and where all land is leased or privately owned, nomadic stock moves either on agistment or to separate holdings kept specifically for this purpose. This management strategy is termed 'spatial diversification' (Anderson, 1974).

Whilst a flexible stocking policy with its attendant purchases, sales and transport of stock generally characterize pastoral management in The Channel Country, it appears that management is still unsufficiently flexible (Dawson and Boyland 1974; Dawson *et al*, 1975). On one hand, the peaks of forage production are inadequately utilized, and at the other extreme, stock, particularly breeding stock, are left too long when there has been no flood and forage becomes limited. Three decades ago the lack of access roads for effective motorized transport was regarded as contributing to poor management and overgrazing (Skerman, 1947). Though this has now been dramatically changed we find that in the last decade cattle numbers have been allowed to reach extremely high levels because of external economic constraints, e.g. downturns in the export beef markets.

Given that a pastoral industry based on extensive grazing management principales is the *only* primary production suited to these lands, we can conclude that *The Channel Country* is a unique rangeland type used by an industry that in the future is expected to experience economic and market fluctuations almost as significant as the fluctuations in the carrying capacity of the land. Within these external economic constraints, improved management of the land and financial status of the operator would flow from a better synchrony of use with the potential of the land. Considering the nature of the land and pastures, synchrony of use with forage availability requires timely information on the size of the floods, their extent and duration, so that a carrying capacity of any area of land could be determined for 6-12 months into the future.

The central premise of this chapter is that the provision of this information is now feasible because detailed surveys of the land and its potential, flow records for the rivers concerned, a network of rainfall recording stations and access to Landsat imagery are all now available. The following analysis and integration of these data evaluates their use in contributing to conservative land resource management in *The Channel Country*.

IV.   ANALYSIS

A.   *Catchment Hydrology*

The framework used to incorporate and analyse all relevant data sets is set out as Figure 3.   That is, to delineate the areas of interest;   appraise the available rainfall data for stations;   appraise the available flow records to form general conclusions about the size and frequency of the flows in these catchments and formulate a simple rainfall/run-off model.

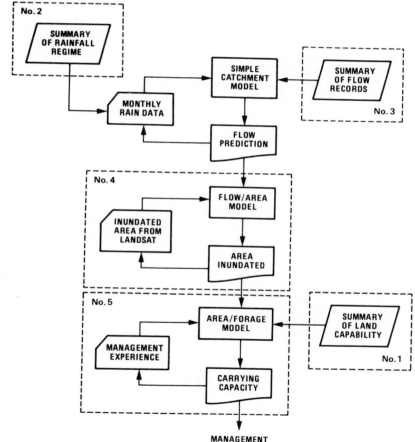

FIGURE 3.   *A schematic diagram of the various stages in the analysis of all available data sources that are relevant to land resource management in The Channel Country.*

The final steps are the formulation of a volume of flow/
area of inundation model and the summary of available information on the carrying capacity of flooded lands.

A previous analysis of steps 1-3 (Graetz, 1980) found
that the rainfall/run-off relationships for two of the three
catchments can be treated at a very simple level using
seasonal rainfalls and annual flow totals only. The derived
relationships were sensible and could be used as rough guides
but more useful predictive models will only be derived by
using event-based data and not using data aggregated over
long time scales.

For land resource management in *The Channel Country* the
most important relationship is between the flood volume or
discharge and stock carrying capacity. This is determined
by two intermediate relationships, the first between flood
discharge and area of flood plain inundated and the second
between the inundated area and the carrying capacity (see
Figure 3, No. 4 and No. 5).

The only cost effective way of determining both of these
parameters for the large areas of land involved is by using
Landsat imagery of a flood; e.g. see Deutsch and Ruggles
(1974, 1978); Rango and Salmonson (1974); and Schwertz *et
al*, (1977).

B.   *Applicability of Landsat*

Landsat multispectral scanner images enable the extent
of flooding and turbidity of floodwater to be determined.
Robinove has analysed Landsat data for the Cooper Creek in
relation to both flooding and rangeland type mapping (1978,
1979). Also standard false colour composites of wavebands
No. 4, No. 5 and No. 7 show ephemeral herbage with a
characteristic blood red signature which declines in intensity
and darkens as the forage hays off. Excellent examples of
the capability of Landsat to discriminate the shallow, turbid
floodwaters of *The Channel Country* from land and to display
the subsequent herbage growth are presented by Robinove
1979).

An obvious method to determine the relationship between
flood volume and area of land inundated would be the retrospective use of Landsat imagery back to launch date of
July 1972. However, until October 1979 acquisitions for this
area were irregular and infrequent and no complete time
sequence of any one flood has been captured.

Therefore, one cannot develop at this stage any predictive relationships between flood volumes (inadequately guaged as they are), area of inundation (from Landsat) and finally forage production and carrying capacity (from property stock records).

However, if one makes the reasonable assumption that the amount of forage production resulting from a flood is dependent only on the area of inundated floodplain and is largely independent of the depth or duration of flooding then parts No. 1, No. 4 and No. 5 of Figure 3 can be condensed and all the flood data can be supplied by Landsat alone with the exception of the 'management experience' loop.

C.   A Landsat-Based Resource Information System

From the above considerations a functional resource information system can be simply constructed as in Figure 4. Here a 'user', either a state agency, pastoral company or individual pastoralist, receives advance warning of the likelihood of extensive flooding either by rainfall or flood warning data, both of which are available from the Commonwealth Bureau of Meteorology. This triggers a request for repetitive image acquisition by the Australian Landsat Station. The present satellite availability means that imagery can be acquired only every 16 days.

After receipt of imagery the user can then interpret and, by using other data sources such as tenure and land system boundaries as overlays, provide an assessment of the pastoral importance of that flood sequence. This assessment could be made available at various aggregated levels; the smallest units being the individual swamps, then up to indivual properties, local government regions and ultimatdly the whole of The Channel Country itself.

Such an assessment could be transmitted in an appropriate form to all of the various users; the individual pastoral managers, the pastoral organizations and stock firms, and government agencies concerned with the administration of matters like drought relief or animal health.

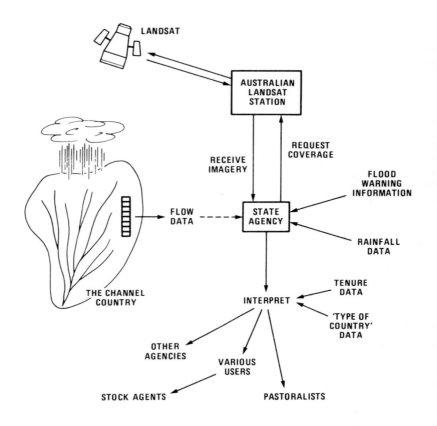

FIGURE 4.   A schematic diagram of the information flow in a Landsat-based monitoring system for The Channel Country.

V.   FEASIBILITY STUDY

A.   *Background*

The most obvious test for the Landsat-based flood information system is operational usage. Unfortunately, to the time of writing, this has not been possible for technical and logistic reasons.

Regular acquisitions of Landsat image data over Australia commenced only in October 1979 with the operation of the Ground Receiving Station at Alice Springs in the centre of the continent. The problem of delay in the processing and provision of Landsat data products was further exacerbated by the line-start-anomaly affecting Landsat-3 at this time. The net result was that the small floods of 1979/80 season could not be monitored with Landsat.

For the 1980/81 flood season some of the images acquired during the time of actual flooding were clouded. Also, the delay in obtaining products had by this time reached 20+ weeks which precluded their use by management in any useful time frame. Post-flood images were used as part of this feasibility study.

However, for the 1981/82 and subsequent seasons the supply of imagery is promised to be sufficiently rapid (2 weeks) to permit a thorough test of the utility of Landsat and, more importantly, to identify all of the 'customers' for the information.

B.   *Digital Processing*

The value of Landsat photographic products for qualitative assessment of flood water and subsequent forage growth has been demonstrated for *The Channel Country* by Robinove. However, digital processing is required to provide the all-important quantitative information (Figure 4) so we have created a demonstration Landsat Image-Based Resource System (LIBRIS) for the Cooper Creek floodplain only. The information system is similar to that developed for rangelands elsewhere (Graetz *et al*, 1982).

Processing in LIBRIS is based on two comprehensive packages developed and supported by the Image Processing Systems Section of CSIRO Division of Computing Research. The first, CSIRO-ORSER, processes Landsat digital image data while the second, SISDIM, contains the routines for integrating spatial data sets.

Two Landsat images, WRS 103-078 and 103-079, acquired 4-3-81, were joined using the SPANY program in CSIRO-ORSER. This required identification of a tie point common to both images. Using 100+ tie points, the joined image was rectified to the Australian National grid and a new image created with resampled 10 acre (4 ha) pixels. This resolution was more than adequate for the purposes of this study. The final image, or map, bounded by the co-ordinates (25°15'S, 141°E) /(28°15'S, 143°E), was 320 x 192 km or 61,440 km^2 in area.

C.   *Forage Indices*

To be of value to pastoralists, the four-band MSS data
must be converted to pasture-related indices. On the basis
of field measurement and mathematical modelling of the
reflectance of landscape components, 'cover' and 'greenness'
indices can be derived using MSS No. 5 and No. 7 (see Graetz
et al, 1982). Cover is defined as the percentage of area
covered by vegetation of any type while greenness is a measure
of the vigour of vegetation cover present.

Briefly, in No. 5 and No. 7 all rangeland reflectance
data lies within a triangular envelope, the base of which is
termed the cover line. All pixels can be projected onto this
line and their cover score determined on a linear scale
ranging from good cover to bare eroded soil. The accuracy of
the cover index depends on these extreme points, which are
located using training sites of known condition. Some areas
like shadowed valleys and badly wind-eroded bare soil will
lie beyond the defined limits of the index.

The distance a pixel lies above the base or cover line
of the data triangle is a measure of the greenness of vegetation present in the pixel. It can range from dry, or not
actively growing, to lush.

Normally, because of the diet selection behaviour of
livestock, pastoral operations can continue long after the
'green' response is gone, so the cover index is more important
for livestock management and landscape stability. However,
as the present study is concerned with a short term pasture
growth response to flooding the 'greenness' index is the
most appropriate.

A 'greenness' index image was created from the rectified
Cooper Creek Landsat MSS data. The index values were then
sliced into six equal classes of forage vigour ranging from
dry to lush.

D.   *Integration of Tenure*

The tenure (property) boundaries were digitized from
1:250,000 scale map sheets and referenced to the Australian
National Grid. The boundaries could be displayed as a graphic
overlay or converted to image format and used to interrogate
the rectified Landsat image or greenness map via the SISDIM
package.

TABLE 3.   Summary of Forage Vigour Classes for
Northern Cooper Creek Pastoral Properties (4.3.81)

	Area of property (Hectares) %	Dry %	Forage vigour 1 %	Forage vigour 2 %	Forage vigour 3 %	Forage vigour 4 %	Lush %
MORNEY PLAINS	58801	3	43	23	14	10	8
WAVERNY	114858	0	69	21	7	2	1
STH. GALWAY	224884	0	40	28	17	9	6
CARRANYA	3053	2	13	22	22	21	20
DIANA	6309	2	32	42	18	4	2
WINDIJO	15041	3	49	19	14	11	4
YERPA	9108	0	43	28	20	6	2
CUDDAPAN	144674	4	75	14	4	1	1
ARRABURY	366832	3	81	11	2	1	1
TANBAR	772258	2	77	16	4	1	1
KEEROONGOOLOO	282572	1	68	17	9	4	2
WINDORAH	31298	0	60	25	10	5	1
HAMMOND	20776	1	48	33	13	4	1
MIOWERI	8351	0	26	26	19	16	13
NANGWARRY	50373	0	50	32	14	4	1
ADMOND	4604	0	26	63	11	0	0
MARAMA	267113	0	91	7	1	0	0
MALAGARGA	71450	0	82	13	4	1	0
MT. HOWITT	249763	3	72	14	6	2	3
ARIMA	70891	2	90	6	2	1	1
DURHAM DOWNS	463064	2	88	8	1	0	0
LAKE PURE	129284	1	82	15	1	0	0
BELLALIE	90051	1	98	1	0	0	0
COOMA	21367	0	92	6	0	0	2
BOONDOOK	20679	0	93	5	1	0	0
MT. MARGARET	33639	1	86	8	2	1	2
QUARTPOT	20258	1	85	9	2	1	1
BODALLA	21525	0	95	2	1	0	2
PARAGUA	31403	1	28	31	13	11	16

Forage vigour reports for 29 properties in the northern
part of the Cooper Creek study area are presented in Table 3.
These reports cover an area of approximately 36,000 km^2.

Unfortunately there cannot be a rigorous assessment of
these results because during this time period no quantitative
ground assessment methods were being applied by CSIRO or any
other land management agency.   However, on the basis of
applications elsewhere in Australian rangelands we regard
both the cover and greenness indices as being consistent and
a reasonably accurate assessment.   Field testing of these
indices is presently in progress.

VI. THE FUTURE

Providing normal catchment rainfall occurs, Landsat imagery for *The Channel Country* will be acquired for the summer seasons to monitor both the floodwaters and subsequent forage production. On the basis of cloud cover figures for this area for the last two years (Table 4) at least one of these events should be suitably imaged on the present 16 day overpass cycle.

TABLE 4. Estimated cloud cover for Landsat image, 103-079 (December 1979 - June 1981)

0%	10%	20%	30%	>40%	Successive dates >30%
20	2	3	2	6	0

(Total of 33 images)

The images will be rectified to the Australian National Grid and reports of area flooded and forage growth will be generated for each property in the study area. It is planned to evaluate these reports with managers and with independent vegetation measurements to complete the all important link between Landsat derived 'greenness' and carrying capacity (see Figure 3 No. 5). It is also planned to rectify existing images of the major flood of 1973/74 to provide a benchmark for flood comparison. This flood was the largest since European occupation of *The Channel Country* and filled Lake Eyre, a usually dry salina in the centre of the continent.

Ancillary data in the form of soil, vegetation or type of country maps have been invaluable in aiding the interpretation of Landsat imagery for rangelands. This also applies to the development of Landsat-based greenness and cover indices. For example, in the Cooper Creek study area the three major types of country - stony tableland, sand dunes and floodplain - lie in different though overlapping regions in Landsat spectral space. Therefore it is intended to incorporate digitized, type of country boundaries into *The Channel Country* LIBRIS so that cover and greenness indices can be tailored to each land type.

Finally, in the future, pastoralists and other land
managers will be assessing the utility of data provided by
The Channel Country LIBRIS.

VII.   CONCLUSION

This chapter has summarized the characteristics that
make *The Channel Country* so unique: the land, the rainfall
and the resultant flooding.   Collectively these inland rivers
constitute a water resource of appreciable size but of great
variability from year to year.

It is now possible to use present day technology and
available data to provide information that is relevant to the
management of these unique lands.   In particular, Landsat
imagery can supply readily interpretable information on the
area of inundations by floodwater and on the *area* of forage
production that results from the flooding.   For this type
of rangeland these two factors may be the two most important
determinants of the carrying capacity over the interflood
period.

Any effective rangeland management process requires at
least two sets of information.   The first must provide quantitative information on the nature and condition of the land
(assessment) and the other must provide information on the
tenure relationships associated with that parcel of land.
In the rangelands of Australia *all* land management must
ultimately be accomplished at the tenure unit or leasehold
level.

The feasibility study described in this paper provides
this information for 29 properties covering approximately
36,000 km^2 centred on the Cooper Creek floodplain.   This could
easily be extended to provide similar reports for the entire
Channel Country.

It is the belief of the authors that a ready market
exists for the information that could be supplied by a government agency monitoring all or part of these lands with
Landsat.   The largest number of 'customers' appear to be the
individual pastoralists but information at regional levels is
also required for the management of the whole industry.

Future research will refine Landsat-derived range
assessment and streamline the information system to suit the
specific requirements of the Australian pastoral industry.

In the global context a Landsat-based information system like LIBRIS is applicable and technologically feasible for monitoring situations of large scale flooding and ephemeral forage growth, for pastorally similar areas, in other lands.

ACKNOWLEDGEMENT

This chapter is based on a paper presented at The International Symposium on Remote Sensing of Environment, First Thematic Conference – *Remote Sensing of Arid and Semi-Arid Lands*, Cairo, Egypt, January, 1982. Proceedings of the Symposium were produced by the *Environmental Research Institute of Michigan*, P.O. Box 8618, Ann Arbor, 48107.

REFERENCES

Allen, A.C.B. 1972. *Interior Queensland: a Geographical Appraisal of Man and the Land in Semi-Arid Australia*, Angus and Robertson, Sydney.

Anderson, J.R. 1974. In *Systems Analysis in Agricultural Management*, (eds.) J.B. Dent and J.R. Anderson, John Wiley and Sons, Sydney.

Australian Water Resources Council 1975. *Review of Australian Water Resources 1975*, Department of National Resources, Australian Government Publishing Service, Canberra.

Australian Water Resources Council 1978. *Stream Gauging Information, Australia*, 4th Edn. Australian Government Publishing Service, Canberra.

Dawson, N.M. and Boyland, D.E. 1974. Western Arid Region Land Use Study Part 1, *Technical Bulletin No. 12 Department of Primary Industries*, Brisbane.

Dawson, N.M., Boyland, D.E. and Ahern, C.R. 1975. *Ecolog. Soc. Aust. 9*, pp.124–41.

Deutsch, M. and Ruggles, F.H. 1974. *Water Resour. Bull. 5*, pp.1023–39.

Deutsch, M. and Ruggles, F.H. 1978. *Water Resour. Bull. 14*, pp.261–74.

Durack, M. 1969. *Kings in Grass Castles*, Hutchinson, Sydney.

Graetz, R.D. 1980. The Potential Application of Landsat Imagery to Land Resource Management in the Channel Country. *Tech. Mem. 80/1, CSIRO Division of Land Resources Management*, Perth.

Graetz, R.D., Carneggie, D.M., Hacker, R., Lendon, C. and
    Wilcox, D.G. 1976. *Aust. Rangel. J. 1*, pp.53-9
Graetz, R.D., Gentle, M.R., O'Callaghan, J.F. and Foran, B.D.
    1980. The Application of Landsat Image Data to Land
    Resource Management in the Arid Lands of Australia.
    *Tech. Mem. 80/6, CSIRO Division of Land Resources
    Management,* Perth.
Graetz, R.D., Gentle, M.R., Pech, R.P. and O'Callaghan, J.F.
    1982. *Proc. Int. Symp. Remote Sensing of Environ.,
    First Thematic Conf. "Remote Sensing of Arid and Semi-
    Arid Lands" Cairo,* pp.257-75.
Hacker, R. 1980. *Trop. Grasslds 14,* pp.288-95.
Idriess, I.L. 1938. *The Cattle King: the Story of Sir
    Sidney Kidman,* Angus and Robertson, Sydney.
Kelley, J.H. 1971. *Beef in Northern Australia,* Australian
    National University Press, Canberra.
Perry, R.A. and Graetz, R.D. 1979. In *Advances in Desert
    and Arid Land Technology and Development,* Vol. 1,
    (eds.) A. Bishay and W.G. McGinnies, Harwood, New York,
    pp.151-82.
Rango, A. and Salomonson, V.V. 1974. *Water Resour. Res. 10,*
    pp.473-84.
Robinove, C.J. 1978. *Remote Sens. Environ. 7,* pp.219-25.
Robinove, C.J. 1979. Integrated Terrain Mapping with
    Digital Landsat Images in Queensland, Australia,
    Professional Paper 1102. U.S. Geological Survey.
Schwertz, E.L., Spicer, B.E. and Svehlak, H.T. 1977. *Water
    Resour. Bull. 13,* pp.107-15.
Skerman, P. 1947. The Channel Country of Southwest Queens-
    land with Special Reference to Cooper's Creek, *Bureau
    of Investigation, Technical Report No. 1. Department
    of Public Lands,* Brisbane.
Yeates, N.T.M. and Schmidt, P.J. 1974. *Beef Cattle Prod-
    uction,* Butterworths, Sydney.

12

LAND ADMINISTRATION, TENURE AND PRESSURE FOR CHANGE IN NORTHERN AUSTRALIA

MICHAEL D. YOUNG

CSIRO Division of Wildlife & Rangelands Research
Deniliquin, New South Wales

I. INTRODUCTION

Some aboriginal, conservationist, mining and urban people are placing pressures on politicians and administrators to alter the nature and extent of land use in northern Australia. This chapter identifies some of the legislative and administrative changes which could arise from these and future pressures.

The chapter sets out to explore the pressures which non-pastoral interests may place on northern Australia's pastoral industry in the 1980's and the 1990's.[1] It describes current land uses and users within the area, then identifies the trends which exist today and finally makes a number of predictive statements about changing situations which may have an adverse effect on pastoralists. Consideration is given to pastoralists' political strength, the growth of the mining and tourist industries, possible changes in the position of aboriginals and possible changes in the grazing rights of pastoralists. It is suggested that in the future more people will attempt to influence the way pastoral lands are used and that increased demands for more conservative land use practices and livestock management techniques will constrain pastoralists' activities.

The purpose of these statements is to identify possible future outcomes so that pastoralists and others may make plans to influence the consequences of possible future events. These statements are not and should not be interpreted as

NORTHERN AUSTRALIA
ISBN 0 12 545080 X

forecasts of the future. The future is always uncertain.
Consideration of the possibilities may lead to ideas and
decisions which will influence these trends.

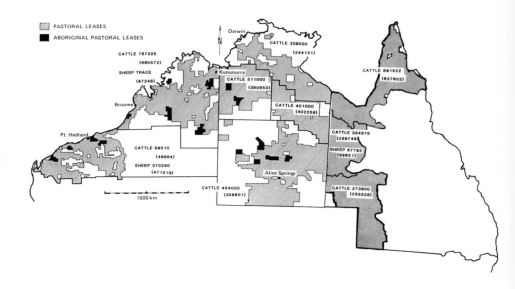

FIGURE 1.   Number of livestock in northern Australia in
March 1981.   Figures in brackets are mean figures for 1963-81
(inclusive) (Source: ABS, Pers. Comm.)

In this chapter the north of Australia is defined to
include the Pilbara and the Kimberleys in Western Australia,
all of the Northern Territory and the western half of Queensland (see Figure 1). While most of western Queensland is
held under pastoral lease, less than half of the Territory
and northern Western Australia is used by pastoralists. In
the 19 years between 1963 and 1981 (inclusive) this region
supported an average of approximately 3.2 million cattle and
0.5 million sheep. During that period these cattle comprised
approximately 16% of Australia's beef herd.

II. CURRENT USES AND USERS OF THE NORTH

In 1977 there were approximately 750 stations in northern
Australia which obtained the majority of their income from
beef cattle production. The value of this production in
1978/79 was approximately $163 million. However, this value
depends upon the prices received and fluctuates widely with
beef prices. In total the north's pastoral industry produces
approximately 0.1% of Australia's Gross Domestic Product and
less than 0.8% of the value of our annual exports.

Mining is the most valuable industry in the north-west.
The mineral discoveries which have occurred over the last
15 years and their associated publicity have probably removed
the image of a pastoral northern Australia from the eyes of
many Australians. Today the majority probably perceive the
north to be the land of minerals, diamonds and beautiful
scenery and not an area of active pastoral land use.

Table 1 indicates the volatile nature of population
growth in the north.[2] Between 1976 and 1981 half of the towns
grew in size and half decreased in size. The reasons for
these changes are complex and vary from town to town. The
1986 census will probably portray a similar picture - both
decline and growth. Development in the north is now heavily
dependent upon Australia's mineral development policies and
the extent of the Federal Government's contribution to the
Northern Territory's budget. Uranium mining policy is of
particular importance as several large developments in the
Territory depend upon approval for the development of a
uranium mining industry. If the development of the Northern
Territory continues, it is likely that more towns will grow
and several new ones will be built. Although of interest,
these changes are unlikely to have a significant effect on
the pastoral industry.

In a spatial sense the two main land users in the north-
west are Aborigines and pastoralists. However, in an economic
sense, the major industry is mining followed distantly by
pastoralism and tourism (see Table 2).

*TABLE 1.   Population of principal towns in northern
Australia with more than 200 people*

Town	Year		Change				
	1976	1981	No.			%	
Western Australia							
Broome	2922	3666	+	744	+	25.5	
Dampier	2727	2471	-	256	-	9.4	
Derby	2411	2933	+	522	+	21.7	
Goldsworthy	989	923	-	66	-	6.7	
Halls Creek	767	966	+	199	+	25.9	
Karratha	4243	8341	+	4098	+	96.6	
Kununurra	1540	2081	+	541	+	35.1	
Marble Bar	262	357	+	95	+	36.3	
Newman	4672	5466	+	794	+	16.0	
Onslow	220	594	+	374	+	170.0	
Paraburdoo	2402	2357	-	45	-	1.9	
Port Hedland	11144	12948	+	1804	+	16.2	
Roebourne	1368	1688	+	320	+	23.4	
Tom Price	3193	3540	+	347	+	10.9	
Wickham	2312	2387	+	75	+	3.2	
Wittenoom	962	247	-	715	-	74.3	
Northern Territory							
Alice Springs	14149	18395	+	4246	+	30.0	
Batchelor	-	308		-		-	
Boroloola	370	420	+	50	+	13.5	
Darwin	41374	56482	+	15108	+	36.5	
Gove (Nhulumbuy)	3553	3879	+	326	+	9.2	
Humpty Doo	-	1265		-		-	
Jabiru	-	1022		-		-	
Katherine	3127	3737	+	610	+	19.5	
Port Keats	882	819	-	63	-	7.1	
Tennant Creek	2236	3118	+	882	+	39.4	
Warrigo Mine	-	991		-		-	
Queensland							
Boulia	272	292	+	20	+	7.4	
Burketown	-	210		-		-	
Camooweal	322	251	-	71	-	22.0	
Cloncurry	2079	1961	-	118	-	5.7	
Cooktown	565	913	+	348	+	61.6	
Julia Creek	650	602	-	48	-	7.4	
Karumba	418	670	+	252	+	60.3	
Mary Kathleen	811	830	+	19	+	2.3	
Mount Isa	25377	23679	-	1698	-	6.7	
Normanton	817	926	+	109	+	13.3	
Total	139136	171735	+	32599	+	23.4	

TABLE 2.   *Gross value of tourism, mining and pastoral products in northern Australia.*

	Mining[a] 1980/81 $'000's	Tourism 1978/79 $'000's	Pastoral Production 1978/79 $'000's
Northern, W.A.	994,646	12,468[b]	14,180
Entire, N.T.	455,218	85,000[c]	106,890[d]
Western, Qld.	713,522	10,000[e]	42,220[f]
Total	2,163,386	117,468	163,190

a   Source: W.A. Dept. of Mines, pers. comm.; N.T. Dept. of Mines and Energy, pers. comm.; and Qld. Dept. of Mines.
b   Value of accommodation estimated by ABS and raised by a multiplier of 2.5 estimated by Dept. of Tourism (1976).
c   Source: N.T. Development Corporation, pers. comm., 1980.
d   Source: ABS, pers. Comm., 1980.
e   Subjective estimate, data for the entire region not available.
f   Source: ABS, pers. comm., 1983.

A.   *Pastoralists and Politics*

The northern pastoral industry comprises 750 stations. Pastoralists and their families account for less than 5% of the area's total population. This suggests that during the next 10 years the industry will find it difficult to influence many of the factors which determine its welfare. By way of comparison approximately 25% of the Northern Territory's population is aboriginal. At a State level the needs of pastoralists may be dominated by the needs of larger sectors of the population. Nationally the northern pastoral industry must expect to be regarded as an appendage to the grazing industries of southern and eastern Queensland and New South Wales which has a dominant voice in agricultural politics. If past trends continue primary producer taxation and other rural policies will be formed in response to demands from wheat, wool and more intensive beef producers. Beef prices will continue to be determined by exchange rates, trade restrictions and international demand; northern pastoralists will continue to be price takers not price makers. The industry must expect to be regarded as relatively

insignificant when Australia's agricultural policies are being
formed. As in the past it is likely that their financial
fortunes will be largely determined by agricultural, national
and international political and market forces.

In the north, markets are distant, an infra-structure is
lacking, rainfall is variable and appropriate technology is
lacking. Nevertheless, the north's rural industries - both
agricultural and pastoral - are of interest to Federal and
State politicians who perceive a need to develop the north.
This desire is not new and can be expected to manifest itself
(Davidson 1966). Several previous attempts, such as those
on the Ord River, at Humpty Doo, Willeroo and Tipperary, have
failed. The new Northern Territory Government is attempting
to establish small agricultural farms which will be run by
individual farmers rather than large corporations. This
change in emphasis from big to small may lead to success. If
the program is successful then some pastoralists may begin to
subdivide their stations. Those that are reluctant to do this
may have some of their land resumed from them. Successful
agricultural development could produce considerable capital
gains for pastoralists who are prepared to subdivide their
stations while failure will not decrease the pastoral value
of the land they lease. The Gulf country of Queensland could
also benefit from successful agricultural development.

B.   *Grazing Rights - Challenges and Re-interpretations*

Throughout the north pastoralists lease from the Crown
the right to graze livestock over an area. These grazing
rights are usually leased for a period of 30-50 years in
Queensland and 50 years in the Northern Territory and Western
Australia. Under current policies the re-issue of these
leases for a further period is not guaranteed. However,
pressures for change are increasing, as pastoralists are
becoming more aware of the fact that others, particularly
aboriginal people, are aspiring to take over these lands. At
each renewal it is possible to change the area leased, modify
conditions of use and change the people entitled to use that
land. When, and probably well before, leases terminate
politicians will be pressured to review these factors and
consider the arguments for and against closer settlement,
attempting to force development and reducing the extent of
foreign ownership.

All Western Australian pastoralists may request to renew
their lease in 1995. In 1964 when this process was last
undertaken conditions were changed:

- from requiring lessees to spend a fixed amount per square mile on improvements to requiring them to submit improvement plans and obtain permission to make them;
- to allow the Minister to determine the manner in which a lease may be stocked (Young 1979).

The next review of pastoral land policy in Western Australia must occur before 1995 and may even occur in the 1980's. It will be interesting to see how many lessees followed their improvement plans and how effective the additional stocking rate controls are. There is no guarantee that all leases will be renewed and it is possible that the conditions attached to the use of any new lease will be radically different.

Following a recent review (1980), the Northern Territory's pastoral land tenure system has been amended to allow pastoralists to apply for a perpetual lease which, subject to compliance with certain covenants and conditions, guarantees the lessee and his heirs and successors a right to use the leased area for grazing purposes for ever. However, before a perpetual lease can be issued the lessee must demonstrate that substantial efforts have been made to eradicate brucellosis and tuberculosis and also, for the Minister to consider the needs of Aborigines to live and hold title to some areas within the lease. This latter provision demonstrates the changing position of Australia's pastoral lands in society. Socio-economic pressures are such that people from outside the pastoral industry are now lobbying for a reduction in the rights of pastoralists. Few of these people consider that perpetual lease, even perpetual lease with periodic covenant review, is the most appropriate way to hold the north's pastoral lands.

Even the existence of a lease does not guarantee that a pastoralist may continue to use that land. Several New South Wales governments have demonstrated that, in response to public pressue for closer settlement, it is possible to withdraw leased grazing land from lease before its lease expires (Young 1980b). The Western Lands Act, which was amended to enable the withdrawal of Western Land Leases issued for grazing purposes, has been described as 'the plaything of governments' and 'one of its biggest nuisances' (King 1957, p.163). However unlikely, it is possible that history may repeat itself and in the north pastoral land laws may become one of the playthings of government.

C.   *Land Management Problems*

In the past, few people have expressed interest in the
resources most pastoralists utilize.   Throughout the world
environmental awareness and the influence of environmental
lobby groups is growing.   A recent study by Payne *et al* (1979)
of the West Kimberleys observed that nearly 30% of the region
(26,700 km^2) has been degraded to poor or very poor condition
and is characterized by erosion and that some 8% (7,000 km^2)
should be removed from grazing to prevent irreversible
degradation.   Similarly, the recent Commonwealth and State
Government Collaborative Soil Conservation Study in Australia
(1978) concluded that 10% of the grazed area of arid Western
Australia and 18% of the grazed area of the Northern
Territory's arid zone supported degraded vegetation and is
substantially eroded.   Opinions formed by examining the
patchwork quilts of differential reflectance, representing
degrees of erosion, which coincide with station boundaries
on Landsat images, suggest that it may be the pastoral
industry and not the mining industry which is mining north-
west Australia.   People associated with the mining industry
are now stressing this point (see, for example, Oliver 1980,
Ewers 1979).

All pastoral land is leased on the understanding that
land degradation will not occur.   Legally pastoralists only
hold the right to graze pastures and are obliged to allow
other people to utilize some of the resources they do not
lease.   Unlike most farmers, pastoralists lease their land
from the Crown on the condition that certain covenants are
complied with.   In the past many of these, including those
which try to prevent overgrazing, have not been enforced.
Perhaps the main reason that land administrators have not
been able to prevent land degradation lies in the way State
Land Acts are written and have been administered.   An
examination of lease records would probably indicate that in
many cases administrators and Ministers of Lands have been
reluctant to enforce compliance.   In the future, concerned
members of the public may demand that all conditions and
covenants relating to land use and land management be
enforced.   This will require governments to identify the
extent to which they should become involved in land use
management.   There is a fine line between the extent of
government which is necessary to protect society's interests
and the extent of government which is paternalistic towards
pastoralists.

Most mining companies are devoting considerable efforts towards identifying ways to lessen their impact on the environment. For example, two have recently commissioned studies to identify ways which will minimize the effect of their exploration activities on pastoral lands (Graetz and Tongway 1979, 1980). The mining industry is generally perceived to be responsible for any damage it causes. Perhaps the approach to the environment which is being taken in mineral exploration is establishing a firm precedent for the pastoral industry.

D.   *Other People's Rights*

Recent experiences in the United States suggest that, as most Australian pastoralists do not officially hold mining, recreation and hunting rights, they should not expect to retain them. Today no American rancher has the right to exclude other people from grazing land he leases from the Federal Government (Public Land Law Review Commission 1970). In the United States the Bureau of Land Management is being influenced by a rapidly growing conservation elite who have successfully reasoned that the leased grazing land is part of the public domain and that it should be simultaneously managed as a grazing, wilderness, wildlife, forest and recreational resource. In 1973 the Natural Resource Defence Committee, with funding from the 115,000 member Sierra Club, successfully filed a civil action against the Bureau for not complying with the provisions of the National Environ- mental Policy Act in their administration of public domain lands in the United States. The ensuing court order requires the Bureau to prepare 212 environmental impact statements to assess the effect of livestock grazing on leased land (Harper 1976). Amongst other things, the Court recognized that wildlife and recreation facilities are valuable public resources.
In 1979 there were 7,349 members of the Australian Conservation Foundation and only 4,793 pastoralists throughout Australia - nearly two members per pastoralist (Australian Bureau of Statistics, pers. comm., Australian Conservation Foundation 1979). It is possible that, in the near future, Australian land resource managers will be persuaded by similar organizations to require pastoralists to maintain wildlife habitats and also manage the wildlife which these habitats support. Those pastoralists whose stations are closest to capital cities such as Darwin and tourist centres such as Alice Springs can expect greater pressure.

E.   *Pressure from Conservation Groups*

There are signs that pressure from conservation groups
is already growing. In 1979 the Conservation Council of
South Australia asked the CSIRO to prepare a background paper
on the arid lands of South Australia that would help the
Council to formulate its policies relating to these lands.
The two ecologists who prepared the paper, Graetz and Foran,
concluded that..."In recent years there have been considerable
changes in the way in which society regards land and land
use. The emerging view is that land has both a commodity
value to its user, and ultimately to the state, as well as
a resource value to society as a whole. This view will
increasingly generate conflict between land users (e.g.
pastoralists, miners) and non-users (urban society) which
will tend to reduce the ability of the users to maximize
private economic gain. The conflicts arise as balances are
sought between (traditional) land users and the ever-changing
public interests. Government as a whole has and will continue
to be very deeply involved in settling conflicts between
competing private interests, separate government departments
and in protecting the public interests."[3] It is possible that
Australia's conservation groups may direct their attention to
the north within the next decade. In fact, the Australian
Conservation Foundation has now held a Conference on
Australia's arid lands and published what they consider to
be an appropriate strategy for them (Messer and Mosley 1983).

F.   *Concern about Wildlife*

In response to thousands of letters from concerned school
children in 1970, the U.S. Congress passed a Bill which
prevents the management of wild horses and burros on leased
grazing land. In passing this Bill Congress noted "that wild
free-roaming horses and burros are living symbols of the
historic and pioneer spirit of the West: that they contribute
to the diversity of life forms within the Nation and enrich
the lives of the American people; and that these horses and
burros are fast disappearing from the American scene.
It is the policy of Congress that wild free-roaming horses
and burros shall be protected from capture, branding,
harassment, or death; and to accomplish this they are to be
considered in the area where presently found, as an integral
part of the natural system of public lands" (Atkins 1977
p.337).

With the appropriate support from television document-
aries it is possible, but perhaps unlikely, that school
children could persuade the Australian Government to take
a similar stance and prevent the control of wild horse, donkey
and camel populations in northern Australia. Alternatively
these children may direct their attention to the further
protection of indigenous species such as dingoes, kangaroos,
wallabies and emus. Regulations introduced in the last
decade such as those which aim to prevent the extinction of
kangaroos are already reducing pastoral income. In 1980 a
society was formed for the protection of dingoes.

Interest in and concern for Australia's wildlife is
growing annually. In the United States the Taylor Grazing
Act, which is used to lease Federal grazing lands, requires
lessees to leave forage for wildlife (Public Land Law Reform
Commission 1970). It is possible that Australian governments
may implement legislation which will require lessees to
allocate forage to wildlife. The forage needs and habitats
of some smaller mammals and reptiles have still to be
identified. It is possible that the needs of some could
require very conservative grazing practices.

G.   *Livestock Management - Pressure from Outside*

The campaign which led to the protection of wild horses
and burros originated in California when a woman discovered
a truckload of bleeding and exhausted horses which had been
mustered by aeroplane (Johnston and Pontrelli 1969). The
first step in this very successful campaign was the banning
of aerial mustering in 1952. Despite evidence that burros
were damaging the environment and recommendations from the
Wild Horse and Burro Advisory Board, the Californian
prohibition of the aerial mustering still remained in 1975
(Cook 1975; Carothers *et al* 1976). However unlikely, it is
possible that within Australia similar legislation to prevent
aerial mustering could be enacted.

Experience in Australia is also indicating that the
public is beginning to demand that farmers pay greater
attention to the way they manage livestock. New South Wales
has recently passed legislation in which all speying and
castration or dehorning of all livestock over six months is to
be performed by veterinarians (Primary Industry Newsletter
1979). Pressure may be brought upon the Western Australian
and Northern Territory Governments to pass similar legislation.
Pressure is also being brought upon governments to regulate
the way livestock is transported. Animal liberation groups

are pressing for legislation to set the maximum distance which livestock may be carried to 80 km per week (Primary Industry Newsletter 1980 a,b).   Although this may seem ridiculous to pastoralists a compromise of 400 km may not seem to be ridiculous to people living in Perth and Canberra.

H.   *Mining - Growth Plus Benefits*

The last decade has witnessed a rapid development of our northern mining industry.   The Northern Territory's Department of Mines and Energy valued mineral production in 1980/81 at $455 million.   This figure should be compared with the 1979 value of $107 million for the Northern Territory's pastoral industry and $85 million for its tourist industry. In Table 2 it is estimated that the annual value of minerals produced in the north of Western Australia is at least $995 million, the pastoral industry $14 million and tourism $12 million.   Similar trends exist in Western Queensland.

I.   *Tourism - Growth and Benefits?*

Reliable information on the size of the tourist industry is difficult to obtain.   The number of people visiting the Northern Territory is increasing at approximately 10% per annum.   Estimates place the gross value of these people's expenditure at $85 million in 1978/79.[4]   By way of comparison the gross value of pastoral production in the Northern Territory was $107m in 1978/79.   The tourist industry is growing rapidly and may soon become the second most important industry in the arid zone of Australia.   At present it comes third after mining and pastoralism (Young 1980b).

The future of tourism depends upon levels of disposable income and the relative costs of different forms of travel. If the relative costs of liquid fuels rise then it is likely that changes in recreation patterns and lifestyles will occur. If the direction of these changes is towards camping treks around Australia and away from overseas travel then, in the future, northern pastoralists may expect more people to pass through and camp upon their stations.   Graetz and Foran[3] point out that the few disadvantages of large numbers of people traversing a pastoral lease may be counterbalanced by improved communication and road services which are provided for these tourists.   This assumes that the costs of the improved communications and roads are borne by society and not the pastoralists alone.   But in some areas these costs are

borne by local governments from rates levied on pastoral land.
If more mines are developed pastoralists must expect more
local visitors on their stations. However they will also
receive greater access to community facilities provided by
mining companies. Some pastoralists may also be able to
profit by selling their products directly to the new markets
which mining towns create and by diversifying into the
provision of recreational facilities for miners.

J.   Aborigines - Conflicts and Gains

During the last decade the position and rights of tribal
aboriginals in pastoral Australia has changed dramatically.
The Aboriginal Land Rights (Northern Territory) Act has been
enacted by the Commonwealth Government and 19 aboriginal
pastoral stations have been established in the north west.   In
the Territory traditional aboriginals have the right to claim
pastoral leases which are held by them and all unalienated
Crown Land (see Figure 2). The restricted freehold title
issued to successful claimants enables them to negotiate

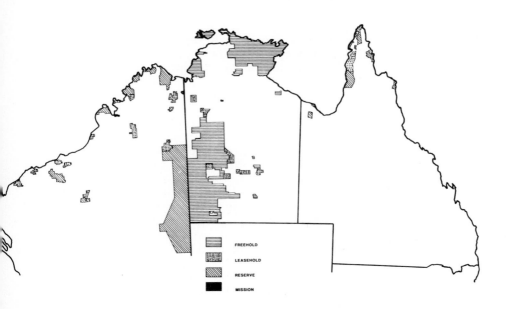

FIGURE 2. Aboriginal land use in northern Australia
(Source: Atlas of Australian Resources, Aboriginal Land and
Population, 1982)

royalties from any minerals found on their land.   It also
enables them to use these lands for pastoral purposes
unfettered by the usual constraints placed on pastoral lessees
via   the Crown Lands Act.

Most of the debate and discussion about aboriginal land
rights is about the disadvantages position of this once
silent minority.   Today aboriginals have a voice which is
heard and is influential.   Their demands are largely for the
replacement of rights which they have specifically lost.   As
Aborigines obtain greater rights to some areas of land it is
likely that pastoralists will also obtain greater rights to
the land they use.   In the Northern Territory pastoralists
are already arguing that it is inequitable to place controls
over them and not place them over Aborigines who use the
land for similar purposes.   In Queensland a different view
is taken of the needs of Aborigines and, under present
policy, no special tenure provisions exist for them.

Conflicts between aboriginal, pastoral and mining
interests are likely to continue.   The resolution of these
rights will be seen by most as a prime responsibility for
governments.   At present Aborigines have the right to enter,
travel over and camp upon pastoral stations and lessees who
interfere with these rights may have their leases forfeited.
In Western Australia legislation may be enacted to enable
administrators to fine pastoralists who do not allow
Aborigines to exercise their rights on pastoral leases.   In
the Northern Territory legislation to enable this was
introduced in 1978.

K.   *National Parks - Growth and Conflict*

In recent years the number of national parks in the
north has doubled (Figure 3).   This will tend to encourage
more people to visit the region.   Some perceive Australia's
pastoral lands as large wilderness areas which have been
relatively untouched by man.   These perceptions coupled with
a general belief that Australia needs more national parks may
lead to the creation of additional ones.   Every park which is
created reduces the area of land available for pastoral land
use.

In 2015 when Western Australian pastoral leases expire
it is possible that some land may be transferred to other
uses.   Between 1963 and 1979, 14 national parks and conservation areas were formed by not renewing term leases when
they expired in the popular box woodlands of western New
South Wales.   Prior to this period of lease termination there

were no conservation areas in the region (Young 1980b).   This
practice has established a precedent for pastoral leases to
be converted to a National Park.

In Western Australia the Conservation through Reserves
Committee has released its report on the Kimberley area
(Western Australia 1978).   This report recommends the
acquisition of parts of one station in the Mitchell River
area.   It also requests the Department of Lands and Surveys
to consider creating a reserve within the Gardiner Range
when any pastoral lease in that area expires.   Present policy
is to attempt to acquire land suitable for conservation and
other related purposes from pastoralists, and when this fails,
to request the Department to reserve the land when the lease
expires.   The lessee would then be paid for his improvements
and in the interim the Department is requested to ensure that
the lessee is not given authority to disturb the land.

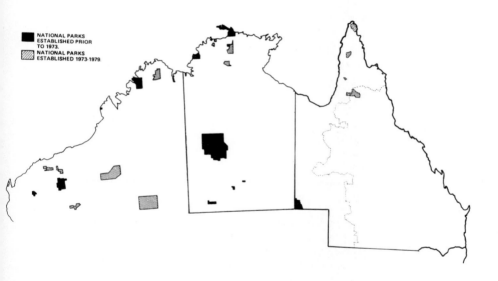

*FIGURE 3.   Growth in the national parks in northern
Australia (Source: Atlas of Australian Resources, Nature
Conservation Reserves, 1978).*

Note: The Tanami Desert Wildlife Sanctuary changed status
to Aboriginal Freehold Land between 1978 and 1982 (see
Aboriginal Land Use, Figure 2).

III.   CONCLUDING COMMENTS

It is difficult to draw together a firm set of conclusions from the scenarios presented. They suggest that in the
next decade northern pastoralists will be affected by
developments in the rights of Aborigines, increased pressures
from an expanding tourist industry, a rapidly developing
mining industry and, above all else, a growing public concern
that our remaining natural vegetation should be carefully
husbanded. More and more people will become interested in
pastoral lands and the way pastoralists use them. As communications improve and research continues, these interested
people can be expected to become more critical of pastoralists
and arid land administrators. It is possible that increasing
numbers of Australians will become concerned about finding
ways which will allow us all to use pastoral lands and
maintain their condition. It is unlikely they will become
very concerned about the well-being of a few pastoralists.

NOTES

1   An earlier version of this chapter which only considers
pressures for competing land uses in the north-west of
Australia was published in Australian Rangeland Journal
3(2):   149-60 (1981).

2   Australian Bureau of Statistics only publish population
data for towns >200 people. The 1976 population for
towns marked - would be between 0 and 200, therefore
the 1976 total, and values for change and % change would
be slightly underestimated. Care should be taken to
note the definition of northern Australia which is used
in this chapter. The eastern part of Queensland is
not included, as is shown in Figure 1;   therefore the
large towns such as Cairns, Townsville, Mackay,
Rockhampton and others are not listed in Table 1.
Reference should be made to the chapter by Parkes (6) on
population distribution and ecological structure.

3   Graetz, R.D. and Foran, B.D.   1979.   The arid land
resources of South Australia: A brief summary.
Reproduced by CSIRO Division of Land Resources Management
as Technical Memorandum No. 79/4

4   Northern Territory Development Corporation.   1980.
     Personal communication.

REFERENCES

Atkins, S.B.   1977.   *Natural Resources Journal*, 17, p.337.
Australia Commonwealth and State Government Soil Conservation
     Study,   1978.   *A basis for soil conservation policy in
     Australia*, 1975-77, Report No. 1, Australian Government
     Publishing Service, Canberra, xii, p.193.
Australian Conservation Foundation,   1979.   *Annual Report*,
     1978-79.   Australian Conservation Foundation, Melbourne,
     p.19.
Carothers, S.W., Stitt, M.E. and Johnson, R.R.   1976.   *Trans.
     North Amer. Wildl & Nat. Resour. Conf.* 41, p.396.
Cook, C.W.   1975.   *Rangeman's J.* 2(1), p.19.
Davidson, B.R.   1966.   *The Northern Myth*, Melbourne University
     Press, Melbourne.
Department of Tourism,   1976.   *Perth Regional Tourism Study*,
     1975-76, Department of Tourism, Perth.
Ewers, W.E.   1979.   In *Mining effects - a perspective*,
     (eds.) R.A. Rummery and K.M.W. Howes, CSIRO Division of
     Land Resources Management, Perth.
Graetz, R.D. and Tongway, D.J.   1979.   Minimizing the impact
     of mining exploration in the rangelands: a case study,
     Management Report No. 4, CSIRO Division of Land
     Resources Management, p.34.
Graetz, R.D. and Tongway, D.J.   1980.   Roxby Downs, South
     Australia: environmental problem of mineral exploration.
     Management Report No. 5, CSIRO Division of Land
     Resources Management, p.30.
Harper, H.A.   1976   *Proc. Western Section, Amer. Soc. Anim.
     Prod*, 27, p.152.
Johnston, V.E. and Pontrelli, M.J.   1969.   *Trans. North Amer.
     Wildl. Conf.* 34, p.240.
King, C.W.   1957.   *Rev. Market & Agric. Econ.* 25(3-4), p.11.
Messer, J. and Mosley, G. (eds.)   1983.   What Future for
     Australia's Arid Lands?   *Proceedings National Arid-Lands
     Conference*, Broken Hill, 21-25 May, 1982.   p.207.
Northern Territory   1980.   *Inquiry into pastoral land tenure
     in the Northern Territory: a report to the Minister
     for Lands and Housing to inquire into the most
     appropriate form of tenure for pastoral land in the
     Northern Territory of Australia*, Department of Lands,
     Darwin.

Oliver, C.O. 1980. In *Northern Australia, Options·and Implications,* (ed.) R. Jones, Australian National University Research School of Pacific Studies, Canberra.

Payne, A.L., Kubicki, A., Wilcox, D.G. and Short, L.C. 1979 A report on erosion and range condition in the West Kimberley area of Western Australia, *Western Australian Department of Agriculture Technical Bulletin No. 42.*

Primary Industry Newsletter, 1979. Traditional farm practices banned or restricted in nation's toughest farm livestock welfare legislation, *Primary Industry Newsletter 691,* pp.1-2.

Primary Industry Newsletter, 1980a. $6,500 million farm production at risk as animal welfare groups zero in on farm 'cruelty' and animal stress, *Primary Industry Newsletter 707,* p.3.

Primary Industry Newsletter, 1980b. Livestock libbers want hauls restricted to 80 km with six-day spells in-between, *Primary Industry Newsletter 711,* p.6.

Public Land Law Review Commission, 1970. *One third of the nation's land: a report to Congress,* Government Printer, Washington, xiii, p.342.

Western Australia, 1978. *Conservation reserves in Western Australia: Report of the Conservation through Reserves Committee on System 7 to the Environmental Protection Authority,* Department of Conservation and Environment, Perth.

Young, M.D. 1979. Differences between States in arid land administrations, *Land Resources Management Series No. 4,* CSIRO Division of Land Resources Management, Melbourne.

Young, M.D. 1980a. *Aust. Rangel. J.* 2(1), p.41.

Young, M.D. 1980b. In *Desert planning: international lessons,* (ed.) G. Golany, Architectural Press, London.

13

NUCLEATED RURAL SETTLEMENT AS A RESPONSE TO ISOLATION: THE LARGE CATTLE STATION

J. H. HOLMES

Department of Geography
University of Queensland
St Lucia, Queensland

I. INTRODUCTION

The assiduous pursuit of closer settlement policies by state governments for a period extending over a century has been decisive in shaping the Australian rural production system, which is now preeminently based on the owner-operated family property. Only in thinly settled areas used for extensive grazing, in the inland and north, have closer settlement programmes faltered. The outcome is a mix of rural production units of widely varying size and internal structure.

Within the extensive grazing lands of northern Australia, family-sized properties were established only in the more accessible lands of inland Queensland, including both productive downlands and mulga and gidgee lands of variable quality, used mainly for wool production. In this zone, shown in Figure 1, an attenuated system of dispersed rural settlement exists, with landholders and their families relying upon a thinly-stretched incomplete fabric of rural services, based on widely spaced small service towns.

In other districts, somewhat larger grazing holdings were established, almost entirely owner-operated but large enough currently to require a labour force equivalent to between three and ten man-years annually to manage herds of 3,000 to 15,000 cattle. These family-run, medium sized properties are located in the Top End (near Darwin) and the Alice Springs District as well as in areas of intemediate accessibility in Queensland, also shown in Figure 1.

NORTHERN AUSTRALIA
ISBN 0 12 545080 X

More recently, the inferior and least accessible cattle grazing lands, formerly unoccupied, have been taken up as "battler's blocks", often occupied and operated only on a seasonal basis and used for cattle gathering rather than husbandry.

This leaves lands of reasonable grazing capacity but low accessibility under the continuing control of large absentee-owned stations. Very large stations still dominate in the main northern cattle districts: Channel Country, southern Gulf Country, Barkly Tableland, Victoria River and southern Kimberley. Commonly these stations carry from 10,000 to 80,000 cattle with annual inputs of between 10 and 30 and up to 50 man-years.

The large cattle station is a seemingly anachronistic survivor from an earlier era and is a continuing challenge to the enshrined national goal of the family-sized property. The large station depends upon a transient population with a truncated social and demographic structure, whose wages, living and working conditions are sub-standard and for whom the problems of isolation are magnified by a regimented, semi-feudal, constricted environment. In most basic attributes, the large station offers a sharp contrast to the smaller owner-operated properties, most notably in labour deployment, working and living conditions, demographic and settlement structure, service demands, dependence on public utilities, urban links and local multiplier effects. These very fundamental contrasts have major implications in relation to the whole spectrum of governmental policies relating to land tenure, land use, labour conditions, service provision and public investment. Yet these contrasts are rarely examined, and the *rationale* for the large cattle station is poorly understood.

A.   *Why Have Large Cattle Stations Survived?*

The various arguments adduced for the survival of the large station deserve close scrutiny. These arguments can be broadly described as: historical-policital; environmental; and locational.

1. The Historical-Political Thesis. It is commonly argued that the survival of particular large station(s) is due solely to the failure of governments to implement closer settlement programmes, with strong hints of behind-the-scenes influence (Kelly 1966, 1971). Such arguments are most cogent where large stations are located unconformably beside small-

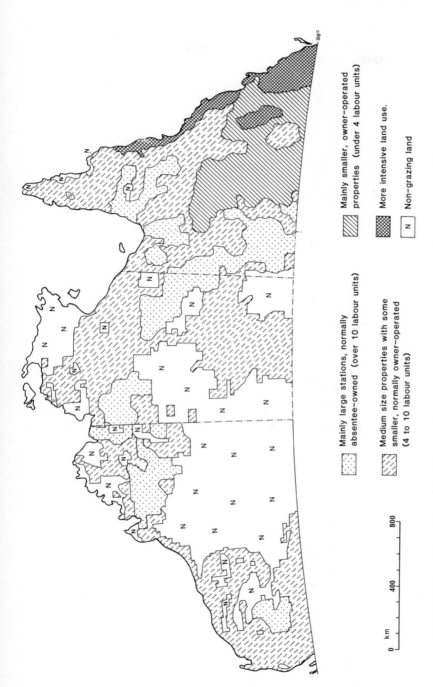

FIGURE 1. The distribution of grazing properties by size of workforce in the extensive grazing zone of northern Australia (Source: Author).

Mainly large stations, normally absentee-owned (over 10 labour units)

Medium size properties with some smaller, normally owner-operated (4 to 10 labour units)

Mainly smaller, owner-operated properties (under 4 labour units)

More intensive land use.

Non-grazing land

holdings, and the cadastral pattern can be described as an
accident of history, whatever the more immediate cause.
However, extensive regions of larger or smaller holdings
cannot be explained solely in terms of political whim,
particularly where these constrasting size patterns are
found within the same state or territorial jurisdiction.   In
some instances the timing of settlement and the origins and
status of the pioneers may have been important.   I have
heard it argued that the early wave of big graziers spreading
westwards from Queensland across the grasslands to the
Kimberleys created the framework for large stations, whereas
the later thrust of South Australian smallholders into
the Alice Springs district precluded a repetition of this
pattern.   The causal effects are too complex to be easily
unravelled.   However, whatever may have been the historical
reasons, it has been difficult for governments to implement
major changes.   It can be argued that in the more remote
areas the initial settlement structure has become firmly
entrenched, not necessarily because of political influence,
but because this settlement system has created the local
conditions for its own survival.   The locational thesis,
discussed below, does incorporate this inertial effect,
presenting arguments why a predominance of large stations
can stifle attempts at closer settlement.

    2.   *The Environmental Thesis*.   This thesis suggests a
correlation between size and environmental constraints.   More
specifically, it is argued that in areas of low carrying
capacity, extreme variability in seasonal conditions and/or
exceptional management problems, viability is associated with
large size.   Such a thesis is readily demolished.   Smallholders have shown an extraordinary capacity to survive
under equally adverse environmental circumstances, though
admittedly aided by governmental support as in drought relief
programmes.   It can be demonstrated that smallholders on
adequately sized holdings can maintain high standards of
husbandry, and investment, while making a reasonable living,
even in areas of higher drought risk and lesser capacity than
some of the major zones of large stations.   For example,
station managers on the Barkly Tableland will readily endorse
favourable assessments of this district, particularly
emphasizing its seasonal reliability and ease of livestock
management.   Only certain exceptional local environmental
circumstances impose constraints on property subdivision
into adquate living areas, the most striking being the
extensive irregularly flooded inland deltas of the Channel
Country where manageable smallholdings cannot readily be
designed.

3.   *The Locational Thesis*.   The main zones of
large cattle stations share one common characteristic, namely
remoteness.   This is most readily observed in the dissociation
between large properties and towns.   For example, in the
Northern Territory, the two districts characterised by very
large stations, Barkly and Victoria River are noteworthy for
their lack of any urban development, while the districts with
viable smaller properties are within reasonable travelling
distance of Darwin, Katherine or Alice Springs.   Similar
relationships exist in Queensland and Western Australia.

   Remoteness, or low accessibility, can be measured
cumulatively in terms of the space-distance, time-distance or
cost-distance to a specified set of services/destinations,
or in terms of receiving services from these destinations.
This concept of cumulative distance is expressed, for example
in a recent unpublished internal document within the Bureau
of Transport Economics, in which a single index of remoteness
is calculated from the cumulated standardised distances to
the nearest town in each of six categories of increasing
population size.   The resultant Australian map confirms the
overall inaccessibility of the zones of large cattle stations.
An additional component, not directly included in the B.T.E.
index is that relating to local accessibility to rural network
services such as all-weather roads, telephones, mail runs
and school bus services, which are poorly developed in the
zone of large stations.

B.   *Size Advantages in Remote Locations*.

   The cost burdens imposed by remoteness are complex,
variable according to individual needs and difficult to assess
in toto.   However, for cattle properties there are two basic
components which have a strong influence on organisational
structure and size.   These two are: the marketing of product
(cattle);   and the delivery of basic services.

   Transport costs do markedly depress the ex-property value
of cattle, both stores and fats, from remote stations.
First-stage costs to meatworks or fattening areas may reduce
values by 10 to 30 percent, compared with more accessible
properties.   Large size can have some value in minimising
marketing disadvantages, by increasing bargaining power in
negotiating contracts for transport and sales, but these
advantages appear to be minor.   Further advantage does accrue
where size is linked to locational diversification, with a
chain of properties used not only to adapt to spatial
variations in seasonal conditions but also to sustain a

regular movement of cattle from breeding to fattening and market.   The most clearly defined movements are from the Barkly Tableland to the Channel Country and from the Gulf Country to the Cloncurry-Townsville corridor.   However, only a few of the major pastoral groups have a regular movement system, and these are only part of a multitude of livestock movements of varying volume, persistence, location, organisational integration and property size, with some smaller remote landholders managing reasonably successfully in the same type of activity.   Large size does not appear to be a conclusive advantage, either in arranging for livestock disposal or in bearing the brunt of lower income arising from remoteness.

However, the locational thesis has much greater cogency when applied to the problem of providing basic services. Large property size provides two related avenues towards minimisation of the burdens of external linkage to distant locations.   The first avenue is self-sufficiency through internalisation in the supply of basic services.   The second avenue is concentration of demand to facilitate the provision of specialised services from distant, non-local origins (See Figure 2).

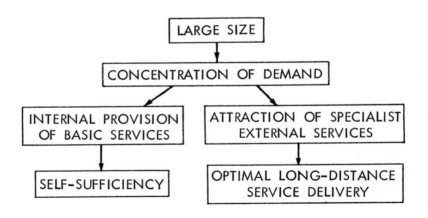

Figure 2.   The relationship between size of holding, concentration of demand, self-sufficiency and optimal delivery of specialist services.

II.   INTERNALISATION OF SERVICES

Remote landholders are required to be very self-
sufficient, since dependency on external services imposes a
heavy contact burden. For the family-sized unit, self-
sufficiency is founded on the resourcefulness of individual
members, who, perforce, are jacks-(and jills-) of-all-trades
in maintenance, repair, minor construction work, livestock
management, emergency response, minor medical, primary
education by correspondence, everyday needs such as energy and
food, and also in leisure activities and all aspects of
child-raising. A wide array of services, provided from
external sources in more accessible locations, are
internalised in less accessible locations.
    This process of internalisation can be accomplished more
effectively on larger holdings, where population numbers and
enterprise size reach the threshold level for entry of
specialist service personnel and facilities. In effect, the
threshold concept in central place theory is transposed into
a within-property rural context. Indeed, some larger stations
are better equipped with urban-style services than are some
small country towns. A full list of all the specialist
personnel and services, possibly to be found on large cattle
stations, must include the following, often with two or
more tasks undertaken jointly by the one person:

(1) Manager, who may also act as book-keeper, storeman,
mechanic, pilot, associate headstockman, supervisor of yarded
cattle or any other task according to need and preference;
(2) Manager's wife, who occupies a pivotal (if poorly
paid or unpaid) role as co-ordinator, facilitator, hostess
and general custodian of the homestead complex and often as
communications supervisor if there is no book-keeper;
(3) Book-keeper/storeman responsible for office,
internal and external radio communication services, well-
stocked store serving station needs and cash sales to staff
and wayfarers, spare-parts depot, hardware, petrol, diesel
and avgas;
(4) Motor mechanic servicing vehicle fleet with garage,
workshop, wrecking yard for additional spares and power-
station for continuous electricity generation;
(5) Boreman, usually with offsider, with well-equipped
truck for maintenance of bores, pumps and windmills;
(6) Grader driver, normally camped out with grader,
caravan, workroom trailer and fuel trailer;
(7) Truck or road-train driver for internal, and some

external, movement of cattle, particular weaners;

(8)   Pilot with light plane and all-weather airstrip;

(9)   Gardener tending ornamentals, lawns, fruit and vegetables;

(10)   Cowboy/butcher for milk, meat and general help;

(11)   Governess or teacher with schoolroom and School of Air radio;

(12)   Station cook with kitchen and staff dining room;

(13)   Domestic help for manager's wife and/or station cook;

(14)   A stock camp comprising headstockman, camp cook and from 5 to 10 stockmen and/or jackaroos, with the largest station having four stock camps;

(15)   Health services, including a clinic well-equipped according to Royal Flying Doctor Service specifications, managed by a partly-trained person (often the manager's wife) and used for scheduled medical and dental clinics with qualified visiting personnel;

(16)   Residential facilities, including: spacious main homestead with gardens and lawns; guest quarters; cottages for married staff (rarely for stockmen); separate bachelors' quarters for some older staff, particularly if permanently located at homestead; quarters for stockmen and jackaroos and other on-property itinerants which may also be used by some visitors; aboriginal huts or "camp";

(17)   Recreation facilities, the most popular and common being a small, briefly crowded canteen when opened for the nightly ration of two to four cans of beer before dinner;

(18)   Caravan area for any contractors temporarily in residence doing fencing, dam construction or bore-drilling;

(19)   The homestead area may also be the location for some other "internalised" services funded by other sources. In one case this includes a part-time government-funded nursing sister. More commonly it includes a school complex of large mobile buildings, used mainly by station aboriginal children, with the few white children having correspondence lessons. The station complex may also include a police station (rarely nowadays), a stock inspector, a contract helicopter pilot and even a contract private veterinarian. These non-company workers are not necessarily fully engaged in servicing the needs of the station.

It can fairly be claimed that almost all the tasks carried out by specialist workers on large stations are also internalised on owner-operated properties, being done by the grazier and his family. However, the large station has more diverse human resources and a more substantial supply of

equipment and spare parts to enable it to carry on with less
reliance upon outside contacts for assistance, equipment or
parts. This larger community is much more self-sufficient
than the family-based rural unit and functions as a self-
contained small town, but with a much reduced level of
external contact than any open town. Basic supplies are
usually received only twice yearly. Social and recreational
activities are poorly developed and the only external social
contacts are at a few annual race meetings, gymkhanas and
rodeos, for most personnel, but married couples may
participate in a few social occasions at other stations. Many
single workers still have an unquenchable urge to "blow their
cheques" at the first opportunity for an extended lost
weekend in the nearest pub. On this issue all managers are
in agreement that remoteness from a pub is an important
asset in labour management.

III.   CONCENTRATION OF DEMAND FOR SPECIALISED EXTERNAL SERVICES

The second avenue towards minimising the costs of
external linkage is to concentrate activity within one large
nucleated rural settlement point which generates sufficient
demand to attract specialist services at reasonable frequency
from distant sources. Most stations have a weekly air service.
Until 1973 this was a highly subsidised scheduled airline
service. Nowadays it is either a private charter plane
charging a standard landing fee or else it is a company
plane, used mainly for delivery of mail, perishables, replacement parts, some personal orders and occasional passenger,
especially for medical treatment. The homestead also acts as
the hub for scheduled radio sessions, internally and
externally. It has been assumed that most stations would
soon obtain telephone services via the domestic satellite,
but this expectation has been dimmed by the October 1983
statement by the Minister for Communications that the user-
pays principle would apply to satellite-relayed telephony.
The delivery of health services is markedly enhanced by rural
settlement nucleation, both for medical emergencies and for
scheduled clinics, and it is recognised that health care
delivery in this zone is superior to that in the zone of
dispersed rural settlement associated with family holdings,
even though population density is lower and distances to town
are much greater.
This differentiation in health care delivery highlights
some important contrasts between nucleated rural settlement

(large stations)   and dispersed rural settlement (family-based properties). The large stations create only limited demands focussed at a few points such that service delivery costs are kept at a reasonable minimum level. Dispersed rural settlement, on the other hand, requires an array of conventional rural network services, such as roads, telephones, and mail services, together with reasonable access to a local service town. However, as described elsewhere (Holmes 1977, 1981) there is a systematic relationship between rural population density, the size and spacing of service towns and the quality and density of rural service networks. In the extensive grazing zone, rural families have been obliged to accept a rudimentary, incomplete array of services. Recent efforts to upgrade these services, most notably with all-weather roads, automatic telephones and itinerant teachers are achieved only at a very high costs per unit of service delivered, and are increasingly dependent upon substantial direct subsidies or cross-subsidies. There is much less demand for upgrading services in the zone of large cattle stations. In any case, substantial subsidies towards services in this zone, particularly business-related services, are politically much less acceptable than in areas of family-based holdings.

IV.   SETTLEMENT INVERSION

The large cattle stations provide a radically distinct solution to the problems of remoteness, pursuing the concept of internalisation and demand concentration to its ultimate conclusion. The outcome is a unique settlement system in which the rural settlements are dominant and the urban, where present, either miniscule entities or else appendages to the rural. This inversion of customary urban/rural relationships is to be seen in terms of the size, functional complexity, centrality and nodality of settlements.

The nucleated rural settlement of the large homestead complex is the residential and working focus for a population of 20 to over 100 people, with a significant proportion pursuing a semi-nomadic existence in the stock camp. "Urban" settlements in the same area are tiny by comparison, comprising only a roadhouse, pub or service station. Some separate urban functions are appendages to the rural settlement. On the Barkly Tableland, these include policemen, stock inspectors, teachers, subsidised nurse, contract veterinarian, contract helicopter pilot and temporary construction contractors.

In functional complexity, many stations can match some
well-established towns in the array of specialist personnel
and facilities they contain. However, whereas urban estab-
lishments strive to attract customers, station facilities
are intended for station purposes and are closed to consumers
in general. Traditionally, stations have felt an obligation
to provide a restricted set of basic services to travellers
generally and particularly to drovers, rations to resident
Aborigines, and a more extensive set of goods and services
to their own staff.

Stations are becoming more reluctant to meet the needs
of travellers, particularly along busy routes such as the
Barkly Highway, and there are growing controversies about
granting access to Aborigines for purchase of petrol, alcohol
and basic rations. The question of service access to
"outsiders" will loom larger in the future with growth in
demand, but with the provision of local open-access services
continuing to be stifled by the self-sufficiency of the
stations. Because of this continuing vacuum effect, it can
be argued that stations will continue to have some obligation
to provide certain basic services which would be locally
available if the settlement system had not been inverted.

Stations also have the attribute of nodality; they are
the focal point for roads which converge on the homestead
complex, such that travellers must pass through the complex,
as if passing through a small town. This is contrary to the
usual rural pattern with homesteads located along private
tracks subsidiary to country roads. The attribute of nodality
is being lost where major through roads on new alignments
have been constructed, and this new through-route alignment
is also leading to the relocation of some independent
services. At Avon Downs the new police station and stock
inspectors housing are located on the Barkly Highway close
to but separate from the homestead complex.

V.   BASIC WORKFORCE STRUCTURE

Internalisation of support services must lead to a high
proportion of station workforce being engaged in service
occupations. According to the census industrial classification, all employees are engaged in primary production
(grazing), but the occupational classification includes a
wide range such as manager, clerk, truck driver, vehicle
repairer, teacher, health worker and domestic service. In
Table I, the basic workforce structure is given for the eight

absentee-owned cattle stations occupying the core grazing
area of the Barkly Tableland. These stations are: Lake Nash,
Austral Downs, Avon Downs, Alexandria, Alroy Downs, Rockhampton Downs, Brunette Downs and Walhallow-Eva Downs-Cresswell
Downs. Anthony Lagoon remained destocked and with only two
resident aboriginal caretakers when this survey was undertaken
in September, 1983.

   Of an average of 30.6 employees on each station, 14.1,
or 47.5 percent were engaged in support services. This
covered all work classifications other than headstockman,
stockman and jackaroo. Individual stations varied between
61.1 and 36.2 percent, with the two largest stations having
the extreme values. While partly reflecting different
management policies and company attitudes, these discrepancies
also arise since some stations are more highly specialised in
breeding, entailing a much heavier burden of mustering and
yardwork for branding, cutting, dehorning, weaning and similar
tasks.

   In terms of weekly man-hours the labour inputs into
stock work are much higher than indicated since stockmen
commonly work at least 75 hours per week whereas support
staff notch up between 50 and 60. These figures are crude
estimates, but are affirmed from a succession of spot
enquiries. However, in terms of man-weeks, stock inputs are
lower since the stock camps normally operate for only 8 to 9
months each year. Further allowance needs to be made for
the stock-work of the manager, especially on smaller stations,
and particularly for aerial stock-work, usually by contract,
which is used increasingly for mustering, particularly while
demands have been increased with the brucellosis - T.B.
eradication campaign. The transfer of mustering work to
aircraft and the demands of the eradication campaign do
combine to place an exceptionally heavy burden of tiring,
dust-laden yard-work on the stockmen, whose working conditions
appear to have deteriorated in recent years.

   This workforce structure is reasonably consistent with
that recorded in 1980 by Holt and Bertram in their survey of
all 25 stations in the Barkly Region. Holt and Bertram found
that 50.4 percent of total labour inputs were provided by the
livestock workforce, and that casual and contract labour
were concentrated on livestock work. Casual and contract
labour accounted for 33.3 percent of all labour inputs and
for well over half of all labour inputs in livestock work.
Of the "permanent" workforce, only 35.8 percent were engaged
in livestock work. The above data were calculated from tables
on pages 114 and 115 of Holt and Bertram (1981), using a
conversion rate of 46 man-weeks of casual labour to one man-
year of permanent labour.

TABLE 1.   Barkly Tableland Stations: Composition of Workforce

Attribute	All Stations	Largest (Any Station)	Smallest (Any Station)	Mean per Station
	Number of Persons			
Livestock Workforce:				
Non-Aboriginal	87	26	7	10.9
Aboriginal	44	18	0	5.5
Total	131	44	7	16.4
Support Workforce:				
Non-Aboriginal	86	16.5	7	10.8
Aboriginal	28	10	0	3.5
Total	114	26.5	7	14.3
Total Workforce:				
Non-Abor. Male	152	40	14	19.0
Non-Abor. Female	21	5.5	0	2.6
Non-Abor. Total	173	43	14	21.6
Aboriginal Male	64	24	0	8.0
Aboriginal Female	8	5	0	1.0
Aboriginal Total	72	27	0	9.0
Grand Total	245	70	14	30.6
Percent of Workers in Support Categories	46.5	61	36	47.5

Obviously the measurement of labour inputs into various station tasks is a very complex undertaking, but one which is of critical interest in any comparison of the internal efficiency of stations vis-a-vis family-sized holdings. My current research into labour inputs of 100 family-based grazing properties in western Queensland, based upon detailed work diaries, will provide a basis for further comparative work. Of course, there are grounds for believing that stock-work is undertaken in a very inefficient manner on large stations, because of youth, inexperience and lack of work motivation of an exceptionally transient workforce.

Supporting opinions on this aspect are readily obtained, from
a variety of sources, but the question merits systematic
study (See Sri-Pathmanathan 1983). My recent research on
large stations, occupying only 25 days provides useful
prelimary information but is an inadequate basis for any
closer interpretation.

Of the stockmen on the Barkly stations, just over one-
third are of aboriginal descent, and the proportion appears
to be increasing, as managers recognise the advantage of
recruiting and maintaining a more experienced component in
stock-work, using Aborigines with a strong local affiliation.
Aborigines are also important in the support workforce,
with males commonly acting as boremen and offsiders and
females as cooks and domestics.

Only 11.8 percent of the workforce is female, being
concentrated mainly as governesses, cooks, and domestics,
with three engaged in stock-work and one pilot.

VI.   STRUCTURE OF SUPPORT WORKFORCE

Of the various occupational categories described earlier,
only on the largest stations would most be represented by a
specialist worker within that category. Various task combinations have evolved in response to the particular needs of
individual stations, the preferences of managers and the
skills of workers. Managers usually resemble owner-operators
in being jacks-of-all-trades. Particularly on smaller
stations with a support workforce of ten or less, managers
become seriously overburdened by an obligation to fill certain
support tasks, most notably bookkeeper/storeman and possibly
mechanic, while also finding it important to engage in some
stock-work in the yards to ensure against inefficiency and
to boost the morale of stockmen currently receiving an
excessive dose of this gruelling, dusty work. Under pressure
from many directions, these managers have an exhausting
workload, normally 14 or 15 hours daily for seven days a week
during the mustering season, which is even heavier than that
of the stockmen. Managers also resemble owner-operators in
being prepared to forego their summer holidays if the wet
season is late to arrive. Companies seem all too willing to
exploit the manager's concern for the proper management of
the stations in their care.

Table 2 shows that the support categories most frequently
present on any station are: manager, grader driver, boreman,
camp cook (all stations), mechanic, cowboy/gardener,

TABLE 2.   Barkly Tableland Stations: Occupational
Classification of Support Workforce

Occupation	Total Workers	Mean Workers per Station	Number of Stations with Worker in Category
Manager	8	1.0	8
Overseer	3	0.4	1
Bookkeeper/storeman	3	0.4	3
Mechanic	9	1.1	7
Grader driver	10	1.3	8
Boreman	23	2.9	8
Driver/general hand	11	1.4	5
Cowboy/gardener	8	1.0	7
Pilot/technician[a]	3	0.4	2
Governess/teacher	9	1.1	7
Domestic	7.5	0.9	3
Station cook	7.5	0.9	8
Camp cook	11	1.4	7

[a]In addition, one manager worked frequently as a pilot for
spotting and mustering.

governess/teacher and station cook (seven stations).
Numerically, the largest category is boreman and boreman's
offsider, followed by cook. Boremen are an exceptionally
large group on the Barkly Tableland because of the urgency
of maintaining water supplies in hot treeless areas during
the dry season, and the absence of natural permanent waters
or artesian bores. Stations do not employ permanent fencing
workers. Fencing repairs are done periodically by various
workers, including stockmen, while fence construction and
bore drilling are done by contractors. Grader drivers are
essential for maintaining a smooth surface on heavy black
cracking clays. Any road not graded each year becomes almost
untrafficable, while graded firebreaks are also important.
     Efficient, safe management of large properties requires
a direct, reliable radio communication system. Transceivers
are installed at the homestead, at each stock camp, in the
manager's Toyota, with each grader driver and usually with
each boreman's outfit. Extended work delays and much
redundant travel have been eliminated by radio contact, and
managers claim to have reduced their annual internal travel
to a current level of between 50,000 and 60,000 km because

of radios. Nevertheless one manager of a large station still
does not wish to instal an internal radio system.

VII.   POPULATION STRUCTURE AND MOBILITY

There are very striking demographic differences between
family properties and large absentee-owned stations. Family
grazing properties have a stable population, with a balanced
age-sex structure save for an under-representation of young
adults and of older people who retire to coastal and near-
coastal locations. Also secondary school children are away
at boarding school for most of the year.

On large absentee-owned stations, the white population
has an unbalanced demographic structure, with very high
masculinity, a predominance of young males, very few married
couples and few children per couple. For all groups,
residential and labour mobility is very high. In our
September 1983 study, no white person was identified who
could be described as a permanent resident. The large
stations have retained their traditional demographic
characteristics, and indeed these may have become further
accentuated, with excessively high mobility among young,
inexperienced stockmen. The large cattle station seems lost
in some time warp, retaining many of the peculiarities of a
frontier society, through its failure to provide working and
living conditions acceptable to a stable workforce in a
modern society. Problems of remoteness are not solely
responsible for the impermanence of the white population.

In September 1983 the white population of the eight
stations comprised 169 males and 55 females, giving a
masculinity ration of 313. An additional 21 males and 15
females resided in station complexes, but were not employed
by the stations, including workers such as schoolteachers,
stock-inspectors and their families. Table 3 shows the very
low proportion of married couples and even lower proportion
of children on stations, compared with national averages,
with a dependency/worker ratio of only 0.29. If non-resident
dependents were included, this ratio would rise slightly.
Large stations still attract some middle-aged men whose
marital status and responsibilities for dependents remain
obscure, to say the least.

For all occupational groups, labour turnover and mobility
are very high. The longest-serving manager has been 13 years,
and the second longest 8 years on the same station. Most have
been there for under 4 years. Holt and Bertram (1981, p.34)

TABLE 3. *Barkly Tableland Stations:* *Composition of Non-Aboriginal Resident Population*[a]

| Attribute | Number of Persons | | | |
	All Stations	Largest (Any Station)	Smallest (Any Station)	Mean per Station
Workers: Male	152	40	12	19.0
Workers: Female	21	5.5	0	2.6
Workers: Total	173	43	14	21.6
Dependent Wives	17	5	0	2.2
Children at Boarding School	2	1	0	0.3
Children: Correspondence Lessons	21	5	0	2.6
Children: Pre-School	7	3	0	0.9
Total Dependents	51	12.5	0	6.4
Dependents per Worker	0.27	0.64	0.0	0.27
Married Couples	25	5	0	3.1
Total Population	224	54	18	28.0

[a]*Some resident workers have dependents permanently living outside the district. These dependents are excluded from the count.*

record that half of the managers in their survey had been on the station for less than one year.

Managers are usually aged in their thirties, originate from Queensland, have served in the same company since jackarooing and have very strong ambitions to buy a small grazing property in east Queensland, but are restrained by lack of finance. All state that they intend leaving the Barkly Tableland in the near future. Headstockmen are even younger with almost all being in their early twenties and only two older than thirty (of the 13 surveyed). Over two-thirds were in their first or second season on the station. Stockmen and jackaroos are even younger, with only a handful aged more than twenty. Turnover is very rapid, with a high proportion staying less than two months, particularly on the larger stations. Smaller stations do succeed in keeping most stockmen for one full mustering season, but less than ten

percent return to the same station for a second mustering
season.   These returners are mainly jackaroos seeking advancement through the company to positions of headstockman and
manager.

The labour turnover is not quite so rapid in the support
workforce.   On average, surveyed married workers have spent
almost two years on the same station, with the longest being
eight years.   Of the 35 single support workers surveyed, only
16 had been on the station for longer than one year.   However,
this group did contain two of the tableland's three "old-
timers", an Englishman employed as a cook with 23 years of
continuous service and a bookkeeper with 13 years.

The aboriginal population, by contrast, has a more
balanced demographic structure with extended family and clan
units in residence.   Mobility rates are much lower and more
localised.   It was not possible to obtain an accurate tally of
aboriginal population, however, because of the fluidity of
certain family ties.   In recent times, there has been a
noticeable   change on the issue of employing Aborigines, and
a growing trend towards selective employment of more reliable
Aborigines, particularly those without an excessive number of
"dependents".   Managers tend therefore to divide Aborigines
into two groups.   One group, more work-orientated, are to be
encouraged to stay on and to affiliate with the station,
while the remainder are regarded as "camp" Aborigines.   In
the near future the latter group will probably be removed to
small settlement reserves, soon to be designated, located
several kilometres from the homestead area and supposedly
independent of the station.

VIII.   ISSUES IN LABOUR RECRUITMENT AND EFFICIENCY

This revival of interest in employing Aborigines is the
outcome of the recruitment problems of large stations.   High
labour turnover is dysfunctional, particularly when stations
must increasingly rely upon inexperienced young stockmen under
the charge of a headstockman whose experience is also very
limited.   Labour recruitment is difficult given the context of
remoteness, harsh environment and limited budgets, with
managers being under constant pressure from head office to
reduce the wages bill.   However, labour problems are partly
self-inflicted, with many stations caught in a vicious circle
of poor working and living conditions, excessively long
working hours and glaring breaches of award requirements.
This lack of care serves only to promote labour inefficiency,

which in turn leads to pressure for longer working hours.

The enforcement of award rates and conditions would create a crisis for large stations, evoking a mix of responses of which the most common would be:

(a) Adoption of more labour-saving methods, including greater use of aerial mustering, trapping, closing off water supplies, and more subdivisional fencing;

(b) Greater attention to recruiting and retaining a smaller number of experienced workers;

(c) Reduced levels of management and reduced mustering frequencies, particularly on land with low grazing capacity and with physical problems in mustering;

(d) Destocking of the most difficult areas, particularly where allied with disease-control programmes;

(e) Pressure towards subdivision into smaller holdings in which labour costs are hidden within the labour unit comprising the owner-operator and his family and where efficiency can be markedly improved through accumulated local experience combined with strong personal incentives towards maximum efficiency in inputs of labour and equipment.

IX.   SOCIAL ISSUES

The large station remains an anomaly within Australian society with its encapsulated, controlled living and working environment, within which neither managers nor workers can escape from each other during the mustering season, save for a few brief periods of time-off. Released from the constricting conditions of station life, station hands work off their frustrations in "lost weekends" which often may extend well beyond the day work is supposed to resume. Managers are generally reluctant to give time-off, other than for recognised annual events such as the Brunette Downs Race Meeting and the Mt. Isa Rodeo, because of problems in resuming station activities after the break.

No matter how effective management practices may be, there is a structural weakness in the need for a closed, controlled society on cattle stations, with problems being reinforced by the sense of extreme isolation and by the pressures on companies to exploit wage-labour in order to remain competitive with owner-operators who are able to forgo award rates and overtime payments in their labour costs. These internal pressures within large stations are a field ripe for sociological inquiry.

X.   ISSUES OF REGIONAL DEVELOPMENT

Large cattle stations have served an important pioneering function in ensuring that remote areas have been used productively. Whether they should continue to do so is another issue. It has been easy for governments to accept the *status quo* in remote areas, since the stations continue to offer a low-cost solution to the maintenance of rural production in such areas. However, the survival of large cattle stations is not merely the result of benign neglect, but also because of their negative impact on regional growth which effectively stifles any potential momentum towards further evolution of the rural and urban settlement system.

This negative effect is not solely the result of extensive land use and sparsity of population, but also because the multiplier effects of the station system are not directed towards local development. As shown earlier in this paper, multiplier effects are either internalised within the station's own structure or are non-local in their impact, creating demands for services from very distant cities. Regions such as the Barkly Tableland and Victoria River District will remain bereft of conventional rural and urban services, and of the workforce to sustain such services, while the station system continues in its present form.

The history of Australian land settlement has shown that significant local multiplier effects, leading to urban and regional growth, can be obtained only through closer settlement into owner-operated holdings. The generally accepted view is that this phase in our settlement history is now behind us and that there are no new rural settlement frontiers of this character. However, the regions currently occupied by large cattle stations do deserve scrutiny in this context. New alternative forms of closer settlement merit close examination, particularly forms which can sustain the high economic and social costs of remoteness, while promoting some regional growth which, in due course, may ameliorate the most severe problems of remoteness. Momentum may now be gathering towards some change in the station system, as the dysfunctional aspects become clearer. Large stations appear to be becoming less attractive for investment. In recent years several well-known stations have been purchased cheaply, with the purchase costs being largely recouped by selling the herd, leaving the station destocked, depopulated and neglected. The Hooker Corporation has had difficulties finding a buyer for Victoria River Downs, and is currently intending to sell the station in five separate lots.

Such indicators suggest that the option of smaller stations, possibly capable of carrying from 3,000 to 6,000 cattle does merit closer examination.

In any transitional phase towards such a system of closer settlement, a transitional rural settlement system would be desirable, capable of coping with the problems of remoteness, while also using existing infrastructure. One obvious design is to retain the existing settlement cluster but have it shared out between possibly three to five individual land-holders, and also including an "open" urban area within which services could evolve in response to demand, including mechanics, contract boremen, grader operators, teachers and others, meeting the needs of the stations and any other local demand. This settlement system would continue to retain the advantages of enhanced external accessibility through nucleation and of "internalisation" through concentration, although it does also retain the disadvantages of low internal accessibility from the homestead to the grazing holdings, which would radiate outwards, in some irregular layout, from the homestead cluster. Although aircraft and road vehicles do enable rapid movement to all points on the property, it might still be necessary to maintain one or two satellite yards with a hut or camping arrangements, to under-take stock work at the more distant parts of the property.

This option does retain many of the advantages of the large cattle station while removing most of the obvious disadvantages. It also provides a context within which urban growth could occur, particularly at those clusters where additional sources of growth were present.

XI.   ISSUES OF PUBLIC INVESTMENT

One of the decisive advantages of the large station has been its minimal demand for public funding in support of essential rural services. In this respect, large stations are in marked contrast with the dispersed settlement system of owner-operated properties in western New South Wales and central Queensland, where rural servicing is sustained only by very substantial direct subsidies and cross-subsidies from public enterprises such as Australia Post and Telecom Australia (Holmes, forthcoming). Any move towards subdivision of the large stations will certainly generate additional costs to the public sector, and this would be yet another component in the very substantial income transfers currently made by the Australian community towards the sparselands generally and the Northern Territory in particular. It might be argued that the cattle lands have been disadvantaged in that they are in the one remote area which is missing out on this munificence. It can also be argued that at present the costs of sustaining the cattle industry are being borne inequitably by the workers engaged in that industry, from manager down to freshly arrived stockmen, all of whom have to accept wages, working and living conditions which most Australians would fine intolerable, and that national goals of equity can best be served by transferring some of this burden from the few onto the community as a whole. Such income transfers would be politically and socially acceptable only if accompanied by changes in land tenure intended to develop a system of family-based properties. At the same time, any restructuring of the rural settlement system should try to avoid past mistakes, which often produce economically non-viable holdings, and also encouraged the evolution of rural settlement pattern which is much too costly to service.

XII.   CONCLUDING COMMENTS

In this chapter, I have highlighted the negative impact of large cattle stations, in relation to internal efficiency, regional development and national goals. It is acknowledged that this presentation is ill-balanced as it fails to consider fully the negative outcomes of alternative rural settlement systems in this locational/environmental context. It is acknowledged that the present system does not place unduly heavy pressure either on the natural environment or on the public purse, and that it does leave governments with options for the future. Certainly, the large stations have not generated the complex array of environmental, social and economic problems arising from misplaced efforts at creating small-holdings in the arid zones of New South Wales and Queensland (Young 1983). Any close scrutiny of these holdings must act as a salutary reminder against strong endorsement of schemes for closer settlement. It is suggested here that a more appropriate tenure system would be based upon properties of intermediate size, with mainly resident ownership and with property size still being positively correlated with remoteness.

ACKNOWLEDGEMENT

This chapter is a revised version of a paper presented at a conference on "People and Economy in the North", North Australia Research Unit, Darwin, December 1983. I am indebted to Michael Young for comments on an earlier draft and to David Murray for information in preparing Figure 1.

REFERENCES

Holmes, J.H. 1977. Population, in *Australia: A Geography*, (ed.) D.N. Jeans, Sydney University Press.

Holmes, J.H. 1981. Sparsely Populated Regions of Australia, in *Settlement Systems in Sparsely Populated Regions: The United States and Australia*, (eds.) R.E. Lonsdale and J.H. Holmes, Pergamon, New York.

Holmes, J.H. forthcoming. Australia: The Dilemma from Sparse Population and High Expectations, in *Providing Essential Services in Rural Areas: International Comparisons*, (eds.) G. Enyedi and R.E. Lonsdale, Westview Press, Boulder.

Holt, R.E. and Bertram, J.D. 1981. The Barkly Tableland Beef Industry 1980, *Technical Bulletin 41*, Department of Primary Production, Darwin.

Kelly, J.H. 1966. *Struggle for the North*, Australasian Book Society, Sydney.

Kelly, J.H. 1971. *Beef in Northern Australia*, Australian National University Press, Canberra.

Sri-Pathmanathan, C. 1983. *Development of the Northern Territory Rural Industries - A Human Resources Perspective*, Northern Territory Vocational Training Commission, Darwin.

Young, M.D. 1983. Resource Management of Australian Arid Lands, in *Design for Arid Regions*, (ed.) G.S. Golany, Van Nostrand Reinhold, New York.

14

ENVIRONMENTAL PROBLEMS OF THE MINING INDUSTRY IN NORTHERN AUSTRALIA

C. D. OLLIER[1]

Department of Geography
University of New England
Armidale, New South Wales

I. INTRODUCTION

In Australia the mining industry, although producing about 35 per cent of the value of total exports, is strangely unpopular, and is severely hampered by unfavourable public opinion. Much of the criticism is aimed at the mining in northern Australia where many large projects happen to be located. More than half Australia's mineral production comes from northern Australia, including most of the iron ore, half the coal and two-thirds of the bauxite. The location of some mines of northern Australia is shown in Figure 1. The 'environment' figures in much of the opposition to mining and some anti-mining people have come to be known as 'environmentalists' or 'conservationists' although, of course, not all people concerned for the environment are totally against mining.

If we look at the impact of mining in detail we can find some genuine environmental problems and some genuine answers from the mining industry. We also find that the standards expected of the mining industry differ from those expected of other activities, that the mining industry is suffering 'selective indignation', for environmental questions affect many more activities besides mining. Furthermore, the

[1]*Present address: Bureau of Mineral Resources, Canberra, A.C.T.*

NORTHERN AUSTRALIA
ISBN 0 12 545080 X

'environment' is often used as an emotive smokescreen for
what are really social and political arguments.

II.   ENVIRONMENT AND PROTECTION AND ENVIRONMENTAL IMPACT STATEMENTS

Before discussing particular problems of the mining
industry in northern Australia, it should be pointed out that
machinery does exist to protect the environment from any
undesirable effect of mining. The image of mining as a
hideous landscape-scarring industry dates from before the
modern concern with environmental issues, and future mining
must conform with modern requirements. This is taken
seriously by the industry and modern mining companies employ
environmental scientists to carry out the work. The Annual
Environmental Workshop run by the *Australian Mining Industry
Council* is regularly attended by over a hundred company
environmental officers.

The most important environmental legislation is the
*Commonwealth Environmental Protection (Impact of Proposals)
Act 1974* and its Administrative Procedures. The main steps
for environmental assessment are as follows:

(1) The Action Minister designates a person or body
as proponent in terms of the Act.

(2) The Department of Home Affairs and Environment
Determines on the basis of preliminary information whether
the Act applies.

(3) Proponent prepares a Notice of Intention.

(4) The nature of the proposal is considered to
determine whether an Environmental Impact Statement (EIS) is
required. If so the Minister directs preparation and
submission of EIS.

(5) Proponent prepares a draft EIS.

(6) The draft EIS is put out for comments from the
public, and Commonwealth Departments.

(7) The proponent is provided with copies of comments
and prepares a final EIS, responding to the comments.

(8) The final EIS is submitted and assessed by the
Department of Home Affairs and Environment.

Within 28 days of receipt of the final EIS the Minister
makes comments, suggestions and recommendations to the Action
Minister. The Action Minister is required to take these into
account in making his decision on the proposal.

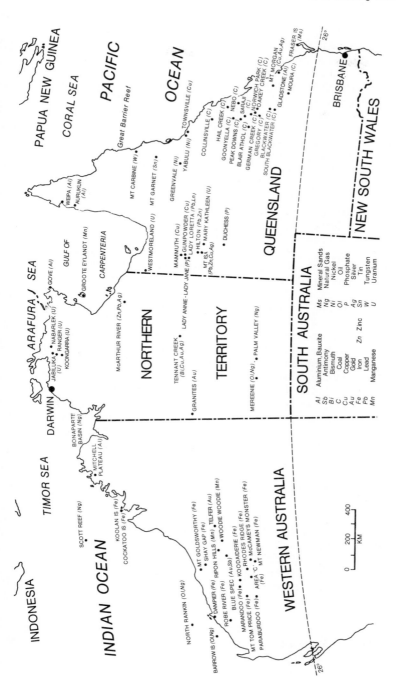

FIGURE 1. Location of some mines of northern Australia

In the early years of its operation thousands of proposals were referred for consideration under the Commonwealth environment protection legislation, but the number soon declined, partly as a result of early screening procedures. In 1982/3 only sixteen proposals concerned mining, of which one led to direction of an EIS.

Many other laws and regulations concern the environmental impact of mining. *The National Parks and Wildlife Conservation Act* came into force in 1975: the Ulura (Ayers Rock - Mount Olga) and the Kakadu National Parks were declared under this legislation. *The Great Barrier Reef Marine Park Act (1975)* is primarily concerned with the preservation of the Great Barrier Reef. The Australian Heritage Commission Act (1975) protects the National Estate. The Queensland Water Quality Council controls discharge of mine waters under the *Clean Water Act 1971-7*. Further environmental protection is provided by State Pollution Control legislation. Even the Customs (Prohibited Exports) Regulations were used in 1976 to prevent environmental disturbance by the mining of mineral sands on Fraser Island. Further information on environmental law in Australia is provided by Bates (1983).

In June 1983 representatives of Federal and State governments and representatives of conservation groups and industry met in Canberra to discuss a proposed National Conservation Strategy for Australia (NCSA). The main plea of the industry group was for avoidance of unrealistic restrictions which strangled development: the next most important appeal was for a reduction in the complexity of environmental legislation and a reduction in delays and duplication of Federal and State government activities.

III.   ASPECTS OF MINING AND ENVIRONMENT

In this section I shall consider some of the main arguments used in discussing environmental problems of the mining industry. Some of them are of a general nature, but some relate especially to northern Australia. I shall look at four problems related to the physical and biological environment:

A.   Despoilment of land
B.   Waste disposal
C.   Rehabilitation
D.   Water management

and three issues of economic, social and political significance that are closely related to environmental arguments.

A.   Despoilment of Land

It is admitted that mining alters the land, and in
parts alteration will be total. But the actual area involved
is very small. It cannot compare with the vast areas where
urbanisation totally alters the environment. Agriculture,
which for some reason is looked on as a more 'natural'
occupation, destroys the natural environment over even
greater areas. Some individual agricultural developments of
hundreds of square kilometres have totally altered the
environment without raising a single complaint, and they did
not require environmental impact statements. A good example
from northern Australia is the Tipperary sorghum scheme where
the natural bush was cleared from a large area, the
aboriginals moved out, and after failure, no attempt was made
to restore the area to its original condition.

The agricultural use of land, with its inevitable
alteration of the environment, is sometimes excused on the
basis that 'man must eat'. It is also perfectly clear that
civilised 'man must mine'. During the lifetime of an average
Australian, he or she will require from the products of
mining 50 tonnes of iron ore, 55 tones of limestone, 4 tonnes
of lead-zinc ore, 200 tonnes of black coal, 175 cubic metres
of crude oil, 12 tonnes of phosphate rock, and quantities of
other earth materials. There is little excuse for the double
standard often applied in assessing the environmental impacts
of agriculture and mining.

Let us stress the areas involved. Agriculture disturbs
about 65% of the continents surface: for Australia as a
whole mining disturbs less than 0.01%. To take an example
from northern Australia, the Ranger mining operation will
disturb 83km^2 which is less than one thousandth of the area
of Arnhem Land, 90,000km^2. Dale and Stock (1983) write,
"While it is true that farming, grazing, forestry and housing
industries disturb very much more land than the mining
industry, it does not follow that this should exempt the
impacts of mining from closer examination". True, but it is
sensible to retain a sense of proportion.

To the effects of mining it is reasonable to add the
disturbance to the environment caused during exploration.
The main feature is the creation of a network of tracks and
roads to drilling sites, trenches and costeans (usually
backfilled), and rubbish sites and abandoned equipment. Such
environmental impacts are generally widespread out of low

intensity. The scars will heal with time, though in arid
environments they may persist for a very long time. Space
does not permit discussion of other relevant topics such as
urban development (see Laverty, 1983), new mining towns
(see Sharma, 1983), or the environmental effects of roads,
railways and pipelines associated with mines.

B.   Waste Disposal, Health and Safety

"Disposal" means to get rid of something, and usually
involves either "dilute and disperse", as when waste is
thrown into a river or emitted into the atmosphere, or
"store", as when fines (slimes, red mud etc.) are stored in
tailings dams. When waste is stored it is important to
ensure that it does not leak and disperse into groundwater.

Waste disposal is a technical problem that can be solved
at any level. The cost of waste disposal has to be built
into the total mine management programme and if impossible
demands are made, the whole project may be scrapped. Various
strategies are possible, depending on the product, but the
miner is aware that this year's waste is often next year's
resource. He often wishes to store waste. The disposal of
radioactive waste poses special technical problems, and this
aspect is commonly confused with the issues of nuclear energy
and nuclear weapons which will be discussed later.

Uranium mines have to dispose of waste rock, tailings
(very fine powder suspended in water), radiation-affected
equipment, and in some places radon gas. The main contributor to the radiation exposure is not uranium itself but the
shortlived radioactive daughter products of the gaseous
emanation radon. The major hazard with radon gas occurs in
underground mining (such as Jabiluka, if it ever opens) and
the situation is much more easily controlled in the open pit
operations envisaged for other northern Australia mines.

The most important radiation hazard to the health of
uranium miners is the induction of lung cancer. A number of
studies have demonstrated that uranium miners who smoke have
about ten times the likelihood of contracting lung cancer
than uranium miners who are non-smokers. The increase may
result from smoking alone and have no connection with radioactivity but it is clear that cigarette smokers 'should
under no circumstances be engaged in mining or milling
operations which increase the likelihood of their inhaling
radon gas or its daughter products' (Barrie and Turner, 1975).

Some of the environmentalists have insisted that if
mining is permitted, Aborigines must be given jobs in the

force. But sociologists have learned that almost all
Aborigines are heavy smokers. It seems that radon levels are
sufficiently low to permit work in the area without ill
effect, for non-smokers, but heavy smokers put up their radon
intake by ten times, and become a health risk.

Other mining operations such as bauxite mining, coal
mining and base metal mining, also have waste disposal problems. These are generally regarded as minor problems and do
not seem to arouse as much excitment as radioactive waste,
although heavy metal pollution is probably a more difficult
technical problem than radiation.

C.   Rahabilitation

Rehabilitation of mined areas can be carried out in many
ways and the cost of reclamation is built into the total mine
management. A company may be required to put up a bond to
ensure reclamation when the mine is no longer generating
income. There are technical problems about how and when to
restore: some operations such as strip mining allow progressive restoration of mined areas, operations such as deep
excavation require restoration as a final operation.
Decisions have to be made on the objectives of reclamation
and there may be advantages in making a new landscape rather
than restoring the old one - a region could benefit by having
a lake rather than filling a hole to the level of the
surroundings. Usually reclamation is only a technical and
financial problem.

Once again the miners seem to get selective treatment in
demands for 'total restoration of the environment'. Farmers
on overgrazed properties, or irrigators in salted-up areas do
not appear to pay for restoration of the original environment
after they have taken their profits. Nor do real estate
developers have to restore the environment if by some
miscalculation their developments are not successful. So far
as forestry is concerned, tax concessions encouraged tree
clearing until the 1983 Budget was brought down, and State
Forestry Departments are still clearing the native forest to
plant exotic pines.

Two environmental changes are associated with mining:
the change from the original situation to mining and its
infrastructure, and when mining is finnished the return of
the land to its original or some other agreed use. In the
old days mining could proceed without attention to the after-
math, and no provision was made for rehabilitation: today no
land is subject to disturbance by mining without provision

being made for subsequent rehabilitation. Ideally the
determination of post-mining land use should be made prior to
the start of mining. In reality few companies are operating
under such pre-determined conditions. Most companies have
programs to restore mined land to something like its original
condition, to prevent erosion, and to contain toxic wastes.
Rehabilitation programs vary from mine to mine, but there are
four common stages:

(1) Determination of the future land use of the area –
objectives of rehabilitation.
(2) Earthworks to provide drainage and prevent erosion.
(3) Establishment of vegetation.
(4) Monitoring to ensure the desired ecosystem is
attained.

An example of rehabilitation that brings out some of the
special problems of northern Australia is provided by Groote
Eylandt. On Groote Eylandt the aim of rehabilitation efforts
was towards the establishment of commercial enterprises on
mined-out and back-filled areas (Farnell, 1979). The main
thrust was the development of beef cattle and forestry
enterprises, to be taken over and utilized by the local
aboriginal community at Angurungu.
Problems with establishing pasture and a beef industry
included:

(1) Fires in the dry season.
(2) Invasion by native grasses and acacias.
(3) Danger that introduced species might spread.
(4) Extremely high levels of nutrients and gypsum
required.
(5) Uncertain economics of beef production.
(6) The small areas (10 ha per year) available for
pasture development.
(7) Hygiene problems associated with slaughtering.
(8) Cost of establishing water point, yards etc.

The problems were sufficient to cause pasture development
to be abandoned.
For forestry, species introduction trails were carried
out on eleven species, of which *Eucalyptus camaldulensis* was
the most successful. Problems included destruction by fire,
destruction by termites, and invasion of plantations by
Acacias. In 1978 a House of Representatives Standing
Committee on Expenditure, which had been investigating
forestry in the Northern Territory concluded that forestry is

plagued by the problems as found on Groote Eylandt, and devastation by cyclones is a further threat; in the past 20 years expenditure on forestry in excess of $30 million brought a return of $150,000: where forestry enterprises had been established for local aboriginal communities great cost had been incurred for negligible return. On the basis of these findings the Committee recommended that the Northern Territory Forestry branch be substantially disbanded. This report combined with the Company's own findings led to abandonment of forestry-based rehabilitation. With the realisation of the severe constraints of the environment – low fertility soils, termites, fires, long dry season, high temperatures, high intensity rainfall, and cyclones – the rehabilitation policy changed. The aim is now simply to attain conservation of the soil and regeneration of native species in the disturbed area.

More favourable conditions for rehabilitation are found in the Central Queensland coalfields (Kelly, 1979). The major development of these mines coincided with an agricultural peak and the government-sponsored Brigalow Lands Development Schemes. Since the government was spending large sums of money to clear land and sow pasture it followed that mined land should be rehabilitated "for purposes connected with grazing", that is pasture.

Mineral sand mining on the coast brings special problems (Lewis and Brooks, 1979), partly because of the delicacy of the original ecosystem of dunes and vegetation, but also because of competitive land use. Between Wollongong and Rockhampton, where most sand mining occurred, 37 per cent of Australia's population lives within 40km of the coastline, so mineral sand mining was a familiar activity. Competing land uses include agriculture, grazing, forestry, recreation, real estate subdivision, and preservation of areas for their "natural" condition. Except for the last use, preservation, all land uses are compatible with mineral sand mining and rehabilitation, and it might seem sensible to organise timing of activities so that the mineral resources can be exploited before the ground is "sterilised". This seldom happens, and a lot of valuable mineral sand is now locked under highways and real estate. The greatest community pressure is from those who wish to retain the coastal area in its natural state, so where sand mining is still permitted rehabilitation programs are generally directed towards re-establishing a self-sustaining ecosystem, similar to the pre-mining condition.

Mount Isa is one of the few large mines in Australia which experiences a monsoonal semi-arid climate. Underground mining produces waste which is split into two fractions: the course fraction is used in back-filling underground excavations, and the fine fraction is pumped into dams and dried out. Rehabilitation of these tailing dams is the ecological problem, and the aim is to cover the slimes with near natural vegetation. Costs in 1979 were about $1000 per hectare. The major problem appears to be the lack of reliability in the rainfall, and the failure of the wet season during experimental rehabilitation has prevented full assessment of the results (Farnell, 1979).

A very unusual exercise in rehabilitation is proposed for Rum Jungle, in the Northern Territory. This is a classical "black spot" which was left in a mess after initial mining of uranium on behalf of the government. The Federal Government now proposes to spend about $16 million on rehabilitation aimed at stabilising the minesite and minimising the escape of pollutants, but no future land use has been established for the area, or any rehabilitation objectives specified.

D.   *Water Management*

The old adage that 'minerals have to be mined where you find them' has meant that mines have had to be developed in the arid zone where groundwater is the only possible sourse, of the large volumes of water needed for town supply, ore treatment, and dust suppression. Commonly therefore the groundwater resources are as carefully assessed in a mining development as the actual ore reserved. The economics of most mineral deposits are such that groundwater sources previously thought unusable in terms of pastoral or town supply because of distance or poor quality or low yield may well be viable for mine supply. Indeed to make some mining projects feasible, the mining companies have had to minimise water consumption and utilise brackish or saline water for processing when possible. Also, in many arid zone situations it is regarded as vald to "mine" stored groundwater within the planned project life.

In the Pilbara Region of Western Australia, development of the iron-ore industry has so far depended entirely upon groundwater for water supply. However new projects are now planned to use the erratic river flows of that region in conjunction with groundwater. Mount Newman, in the eastern Pilbara, has developed an artificial recharge scheme.

In the west Pilbara, towns at the mining centres such as Tom Price and Paraburdoo are supplied with water from local underground sources. Parkes refers to town water consumption

for Paraburdoo in Chapter 18. Suitable local supplies of the
required quantity are not available for the Indian Ocean
Ports, including Dampier, and these are served by the West
Pilbara Water Supply Scheme which pumps water 105km overland
from a borefield in the Millstream calcrete acquifer (see
Parkes, Chapter 18, Figure 4). A conjunctive use project is
planned to supplement the supply. A dam on the Harding River
will be used for Water Supply in good seasons, and the
Millstreams groundwater storage will be used to sustain the
supply through severe droughts.

IV.   OTHER ISSUES RELATED TO ENVIRONMENTAL PROBLEMS

The 'Environment' is frequently taken to mean the physical and biological conditions that influence and affect life
at a place. However, the *Environment Protection (Impact of
Proposals) Act 1975* specifically states that 'environment'
includes all aspects of the surroundings of man, whether
affecting him as an individual or in his social groupings,
and 'environmental' has a corresponding meaning. This makes
it clear that social, economic and political questions can be
considered as part of environmental issues.
    Many of the arguments used by environmentalists are not
concerned with the physical and biological environment but
relate to other issues of which the main ones are:

The economic environment - who finances the mining
operations and who gets the profits? Who are the
"multinationals?" Should the mining industry
finance aboriginal development?
Nuclear energy - some people object to all nuclear
development and oppose uranium mining totally. Their
concern for the environment does not concern mining
specifically, or northern Australia, for they have
total objection to anything to do with radioactivity.
Aboriginal rights - what are aboriginal rights? What
royalties should be paid to whom? Should there be
racial discrimination, with aboriginals getting
different treatment (and royalties) from other
Australians?

Many objections to mining in northern Australia arise because
of its impact on the aboriginal population and because some
of the mining is related to uranium.

E.   *The Economic Environment*

This would not normally be considered as an environmental problem, but with the broadening of the definition of the environment it becomes relevant, and it is important to know how much of the cost of environmental problems can be borne by the mining industry.   It must be admitted that miners are in it for the profit – like farmers, manufacturers, shopkeepers and many others.   It should not be necessary to argue the national interest in an environmental article but it needs to be stressed that mining earns over 35 per cent of the value of Australia's total exports.   In northern Australia, where it seems that agriculture and the pastoral industry will always have severe problems (Mackenzie, 1980; Davidson, 1980) mining appears to be potentially the main source of money.

Because of the high risks involved, the high cost of exploration and the large capitalisation costs, mining is inevitably an international industry (pejoratively termed multinational).   It is not possible to raise sufficient capital within Australia to finance the country's mining operations.   In fact about 40 per cent of Australia's minerals industry is owned by direct foreign investment. Mineral exploration is a high risk enterprise, and risk is particularly great if vast sums of money are spent on exploration without satisfactory security of tenure.   To cite but one example, The Granites gold prospect in the central west of the Northern Territory, is on aboriginal land, and although the company had rights over the area which pre-date the Land Rights Act, Commonwealth legislation required North Flinders Mines Ltd. to reach agreement with traditional owners on development.   A delay arose because the Central Land Council demanded a private royalty payment of 7.5 per cent of gross revenue.   In the end arbitration brought some sort of settlement, the terms of which are not disclosed, but development was held up for *eight* years during which capital costs had doubled.

The receipt of money from mining can also be at political risk.   As Everingham (1983) wrote, "...our biggest mining revenue earner, uranium,...contributed $300 million to a total of $500 million value of production in 1981-82...If the anti-uranium lobby wins the argument we in the Territory can kiss goodbye to a crucial $1 billion worth of investment and hundreds of permanent jobs" (Everingham is Chief Minister of the Northern Territory government).

Australia is generally perceived as a "resource-rich" country, and there is a general desire amongst governments

and other organisations to get a share of the "riches",
especially if it is thought that mining is "booming". It is
sometimes forgotten that mineral deposits only become
"riches" if they are mined and sold economically. Australia
may not be as rich as we think. Some countries are developing better products, like the huge Brazilian iron ore deposits
which are low in phosphorous and aluminium. Some less
developed contries compete because they have massive debts to
some of Australia's customers, who have the choice of being
repaid in minerals, or not being paid at all! Some countries
compete because the buyer has more equity in the mines:
Japanese equity in Canadian coal mines is an example. And
of course some countries can compete because they have lower
environmental control demands than those in Australia.

A pessimistic view of the economic environment is that
world growth of mineral production is declining, there is
substantial excess capacity, and there is likely to be a
buyer's market for the next decade. In facing the tough
overseas competition Australia suffers because of high
production costs, and low reliability as a supplier. Some
existing and proposed environmental projects add to costs and
reduce reliability, possibly to the point where mining is no
longer viable. For instance, at the Western Australian
inquiry into land rights in October 1983, BHP said it found
South Australian negotiating and compensation mechanisms for
dealing with Aborigines unworkable, and as a result "no
exploration had taken place and the potential resources of
the large areas concerned remain unknown".

F.   *Nuclear Energy*

The world needs energy. In the developed world it is
needed to maintain present life styles, and if the developing
nations are to improve their standard of living further, vast
amounts of energy must be expended.

Environmentalists hope to utilise 'alternative' sources
of power, but these seem quantitatively inadequate for future
needs. For example, the US Department of Energy forecast
that solar technology would contribute three per cent of
total US electricity by the turn of the century. It is
desirable that all other sources of energy be developed, but
it seems that nuclear energy is inevitable. If a cleaner,
safer source of energy were available nobody would want to
use the controversial and potentially dangerous nuclear
energy. But this debate is too late - nuclear energy is
already with us.

At the end of 1982 there were 294 nuclear power reactors in operation and 215 were under construction, in a total of 31 countries. Finland was leading in the nuclear share in electricity generation with over 40%.

Nuclear energy is also cheap. Electricity generating costs in 1981 per Kwh are given in Table 1.

TABLE 1. *Electricity Generating Costs*

	Nuclear	*Coal*	*Oil*
UK (pence)	1.65	1.85	2.62
USA (cents)	2.7	3.2	6.9
Japan (yen)	12.6	16.7	21.4

Environmentalists base some objections to nuclear energy on health and safety, but Table 2 indicates its relative safety. UK deaths are 1.35 (coal); 0.23 (oil); 0.1 (nuclear). A very careful analysis is provided by Hamilton (1982), and his general conclusion regarding the relative impacts of coal and nuclear power appear to agree with the figures below.

TABLE 2 . *Occupational Health and Safety in USA Conversion Systems Producing Electricity.*

Fuel	*Deaths**	*Injuries*	*Man-Days*
Coal (underground)	4.0	112	15,280 *(mostly mining, also transport)*
Coal (open cut)	2.64	41	3,091 *(mostly transport)*
Oil	0.35	32	3,609 *(mostly extraction)*
Nuclear	0.15	16.7	271 *(mostly mining)*

*Deaths per year per 1000 MWe at 75% capacity

The most serious objection to nuclear energy is the problem of disposing of high level wastes from reactors, and disposal of the reactors themselves when they are "de-commissioned" after a working life of about thirty years. This remains a worry for many concerned people. Research is providing better options, but dangerous disposal methods are still reported too frequently. However, the disposal of *reactor* waste is very different from disposal of *mine* waste, and it does not seem likely that ceasing to mine in northern Australia would have much impact on nuclear waste disposal or the growth of the nuclear industry.

Other arguments against mining uranium relate to the proliferation of nuclear weapons, and especially the fear that irrational or terrorist elements might get the bomb. There seems little doubt that because of the difficulties and dangers involved in making nuclear weapons, terrorists would rather steal a ready made weapon than steal the materials. Since there are already enough weapons around to kill everyone on earth, the potential terrorist thief has plenty of choice! It also seems too late to curtail the supply of nuclear weapons by limiting the amount of uranium exported from Australia. Ironically the main market for Australian uranium is Japan, the only country to have suffered nuclear attack.

In Australia the 'Uranium Decision' has still to be made: a good account of the background is provided by Saddler and Kelly (1983)

G.   *Aboriginal Rights*

In the Northern Territory the aboriginal population passed the point where it could possibly live by food gathering and hunting in 1971, or about 1963 if people of part aboriginal descent are counted, so there is no going back to a 'foraging mode' (Woodward 2nd Report, 1 74, Sec.249) The same is probably true of the rest of northern Australia.

The aboriginals today are much different from the tribal aboriginals of the past, with different relationships to the environment. Some environmentalists present a pleasant picture of aboriginals living a tribal life in harmony with the environment, but those days have gone. The Aborigines do not live in an anthropological zoo, nor do they wish to. They are, to varying degrees, modern men like all the other Australians, with material needs (cars, guns, tobacco, etc.) and cultural aims (education, political rights). They can have a very adverse effect on the environment, like anybody

else, and indeed the use of high-power guns, four wheel drive
vehicles and packs of dogs has enabled some groups of aboriginals to have a devastating effect on the natural environment
in northern Australia. Even a simple artefact like a jerry-
can may have a huge impact, for a hunter can stay at a dry
waterhole and kill off fauna with an imported water supply,
when in earlier days he would have been forced to move on.
     Although a long way from the old tribal life, many of the
aboriginals of northern Australia prefer a life style
significantly different from that of the white Australians,
but there is no question of their living a simple life as in
the old days. Certainly they have changed, and are changing,
but there is nothing more likely to cause great change than
the sudden accumulation of great wealth. Those environmentalists who affect to admire the qualities of unspoilt
aboriginals should bear in mind, when they argue for greater
royalties for aboriginals from mining, the effects of this
wealth on the aboriginal way of life.
     Because of its very wide interpretation of "the
environment" the Aboriginal Land Rights (N.T.) Act, 1976 has
become probably the greatest environment problem facing the
mining industry in the Northern Territory. Similar
legislation is passed or foreshadowed in other states.
     The Act 'provides for the granting of traditional
aboriginal land in the Northern Territory for the benefit of
aboriginals and for other purposes' and it gives 'inalienable
freehold title to the land on reserves in the Northern
Territory and provides machinery for them to obtain title to
traditional land outside reserves'.
     Aboriginals may make a direct land claim in the case of
unalienated crown land, but even a pastoral lease may be
purchased on behalf of aboriginal interests if they make a
claim to it. This right to claim or purchase land is of
great interest to miners because: 'mineral exploration and
development will be allowed in aboriginal land only with the
consent of the aboriginals'. As is the case for all other
Australian private landowners, the ownership of minerals on
aboriginal land is to be retained by the Crown. But these
rights which require aboriginal agreement to mining, and the
right to prevent it on their land, is a unique privilege not
given to other Australians.
     The situation is that if a mineral deposit is discovered
on aboriginal land, the aboriginals can hold up mining until
they are paid money, and until any other conditions they wish
to stipulate are met (such as all miners to learn the
aboriginal language). They are in a position to thwart any
development of a find, even after much time and money have

have been spent on exploration. But the situation is worse than this. Even if a mineral deposit is found outside an aboriginal reserve, the aboriginals can claim the land (provided the mining title was granted after 1976) and, if successful, impose the same restrictions on development. Thus it becomes impossible to explore anywhere without the threat of a takeover of the land by aboriginals, who can then impose any restrictions they wish, and take large royalties although they had no financial involvement in the exploration.

This situation, where aboriginals can move in wherever a mineral exploration programme is successful and either prevent mining or demand large financial returns, makes northern Australia one of the least attractive prospects for mineral exploration available, and will certainly direct future exploration elsewhere.

Of course some aboriginals see the positive advantages of mining, and times may be changing. According to the *Canberra Times*, November 23, 1983, "the Northern Land Council has been instructed by aboriginal traditional owners to seek from the Federal Government the development of the proposed Jabiluka and Koongarra uranium mines".

A council field representative, Mr. Jacob Nayingul, said that the traditional owners opposed the Government's intention to incorporate the two mines into the World Heritage-listed Kakadu national park, 220 kilometres south east of Darwin.

"They want the mines, not a park", he said.
"The people said if they could not get the mines they would seek the equivalent amount of money that they would have received from them", he added.

It seems to me that the word "environment" has been over-extended and debased in recent years, but using it in its modern form one might say that concern for the economic environment is likely to prevail in the end over concern for the natural environment.

V. CONCLUSIONS

Northern Australia includes large areas with a harsh environment and the mining industry has some environmental problems associated with high temperatures, high winds, floods, great seasonal variation and so on. But these problems are comparatively small compared with the social and political problems that have arisen since the 'Environmental' movement became active and selected the mining industry as

one of its favourite targets. Genuine environmental problems
can be solved and there is legislation and machinery available
to control the effects of mining. The mining industry today
is as concerned as any other group with environmental
problems, but at present the 'environmental' arguments aimed
at the miners are often emotional, political and sometimes
biased. The effect of this could be to discourage mineral
exploration and mineral development in northern Australia,
removing one of the best prospects for economic development
in the region.

ACKNOWLEDGEMENTS

The author and the editor extend their thanks to the
Research School of Pacific Studies, Australian National
University and to Dr. Rhys Jones, Editor of the volume,
Northern Australia: Options and Implications, 1980, for
permission to use a substantial part of a paper, first
published in that volume under the title, Environmental
Problems of the Mining Industry, Especially in Northern
Australia, written by C.D. Ollier.

EDITORIAL NOTE

The chapter by Ollier is not a *reprint*. The chapter has
been prepared specifically for inclusion in this volume and
contains a considerable updating of material from the 1980
chapter, acknowledged above, first prepared for a seminar in
1978, sponsored by the Research School of Pacific Studies,
Australian National University.

REFERENCES

A.M.I.C.   1978.   The 1977 A.M.I.C. Environmental Workshop
    Discussion Papers, *Australian Mining Industry Council*,
    Canberra.
Barrie, J. and Turner, A.C.   1975. *The Background to
    Narbalek*. A summary report on uranium in Australia
    prepared for the Northern Land Council, A.M.I.N.C.O.
    and Associates.
Bates, G.M.   1983. *Environmental Law in Australia*,
    Butterworths, Sydney, chapter 8.

Dale, P. and Stock, E. 1983. The Environment. In *Mining and Australia*, (eds.) W.H. Richmond and P.C. Sharma, University of Queensland Press, St. Lucia, pp.228–256.

Davidson, B. 1980. The Economics of Pastoral and Agricultural Development in Northern Australia. In *Northern Australia: Options and Implications*, (ed.) R. Jones, Research School of Pacific Studies, Australian National University, Canberra, pp.73–84.

Everingham, P.A.E. 1983. Mining Your Own Business, *Bull. Aus. I.M.M.*, 475, pp.23–25.

Farnell, G. 1979. Rehabilitation of Mined Land on Groote Eylandt, N.T. In *Mining Rehabilitation*, (ed.) I. Hore-Lacey, Australian Mining Industry Council, pp.33–40.

Flowers Report 1976. *Nuclear Power Issues and the Environment*, 6th Report of the Royal Commission on Environmental Pollution, Her Majesty's Stationery Office, London.

Ford Report 1977. *Nuclear Power Issues and Choices*, Report of the Nuclear Energy Study Group, sponsored by The Ford Foundation.

Fox, R.W., Kelleher, G.G. and Kerr, C.B. 1976. *Ranger Uranium Environmental Inquiry*, First Report, Australian Government Publishing Services, Canberra.

Fox, R.W., Kelleher, G.G. and Kerr, C.B. 1976. *Ranger Uranium Environmental Inquiry*, Second Report, Australian Government Publishing Services, Canberra.

Hamilton, L.D. 1982. Comparing the Health Impacts of Different Energy Sources. In *Health Impacts of Different Sources of Energy*, International Atomic Energy Commission, Vienna.

Kelly, R. 1979. Rehabilitation of Mined Land in the Central Queensland Coalfields. In *Mining Rehabilitation*, (ed.) I. Hore-Lacey, Australian Mining Industry Council, pp.41–48.

Laverty, J.R. 1983. Urban Development. In *Mining and Australia*, (eds.) W.R. Richmond and P.C. Sharma, University of Queensland Press, St. Lucia, pp.119–149.

Lewis, J. and Brooks, D. 1979. Rehabilitation after Mineral Sand Mining in Eastern Australia. In *Mining Rehabilitation*, (ed.) I. Hore-Lacey, Australian Mining Industry Council, pp.50–68.

MacKenzie, I. 1980. European Incursions and Failures in Northern Australia. In *Northern Australia: Options and Implications*, (ed.) R. Jones, Research School of Pacific Studies, Australian National University, Canberra, pp.43–72.

252   C. D. Ollier

Richmond, W.H. and Sharma, P.C. (eds.) 1983. *Mining and Australia,* University of Queensland Press, St. Lucia.
Saddler, H. and Kelly, J.B. 1983. The Uranium Decision. In *Mining and Australia*, (eds.) W.H. Richmond and P.C. Sharma, University of Queensland Press, St. Lucia, pp.257-283.
Sharma, P.C. 1983. The New Mining Towns - "Outback Suburbias?" In *Mining and Australia*, (eds.) W.H. Richmond and P.C. Sharma, University of Queensland Press, St. Lucia, pp. 150-180.
Woodward, A.E. 1974. *Aboriginal Land Rights Commission,* Australian Government Publishing Services, Canberra.

15

BUILDINGS AND SETTLEMENTS: LIVING IN THE NORTH

BALWANT SINGH SAINI

Department of Architecture
University of Queensland
St Lucia, Queensland

I. INTRODUCTION

This short chapter is one of a selection in which a
number of contributors deal with aspects of building, settlement and life in northern Australia. The climate of this
region, which is uncomfortably hot for the best part of the
year, has been discussed in some detail by Lee and Neal in
chapter 3. In the following chapter (16) Szokolay discusses
the built environment of climatically stressed ecosystems
from an architectual science viewpoint. In chapter 17
Auliciems and Dedear consider, *inter alia*, thermal stress,
the 'cooled' climate of buildings and the human climatic
demands of the outdoors in northern Australia. Subsequent
chapters discuss aspects of life in remote settlements and the
role of infectious disease in the human ecology of northern
Australia, increasingly affected by the growth of settlement
and contact with distant urban processes and places.

It is my aim to offer some general comments on building
and settlements and on living in the north, as a prelude to
the more detailed studies which are presented later.

Approximately three-quarters of the entire Australian
continent is hot and uncomfortable, with temperatures
exceeding 30°C, for more than twenty-five days in a year
(Andrews, 1967). Consequently, most Australians have shied
away from the northern areas of Australia and have preferred
to live on the south-eastern and south-western coastal fringe
where the climate is cooler for much of the year and where

NORTHERN AUSTRALIA
ISBN 0 12 545080 X

rainfall is more reliable. This settlement pattern has been
influenced further by the location of harbours, the exit
points for primary products, grown in or won from the interior
regions. But northern Australia has largely stayed empty
(see Parkes's chapter (6) on population). Relative to area
there have been few jobs created there and together with
difficulties generated by long distances and a sense of
isolation, its climate is perceived as unpleasant by many
coastal dwelling Australians.

So for many years the region saw very little growth,
except for a few single-industry towns (mainly mining) and
small commercial and administrative settlements whose live-
lihood depended upon pastoral and other activities around
them. We all owe a great debt to the first pioneers, who
bravely established a frontier-type economy in which fortunes
fluctuated and social, economic and geographical progress
was irregular. During the early days of white settlement
prejudice against living in hot climates was fairly widespread among most Australians.

Browsing through some old papers recently, I came across
a 1925 publication of the Commonwealth Health Department.
The writer was then an up and coming medical practitioner
named Raphael Cilento, later to become Sir Raphael Cilento.
He wrote:

> *To the great majority the word 'tropical' conjures
> up visions of sweltering mangrove flats, the haunts
> of crocodile; of rank and steaming forests that
> exhale the musky odour of decaying vegetation and
> conceal within their leafy depths 'miasmic' swamps;
> of deadly snakes and of the skulking savage with
> his poisoned spear.*
>
> (Cilento, 1925)

I believe that Raphael Cilento, Grenfell Price and
others did much to dispel the notion that people of European
origin were simply incapable of surviving the harsh environment of the tropics. Their belief has been amply justified
by many third and fourth generation northern Australians who
have successfully adapted to their environment and who are
performing their life work and following their ordinary
vocations, as they would in temperate climates.

II.   MINING AND ITS BUILT ENVIRONMENT

Most pastoral communities had few major problems in adjusting their buildings to the needs of the tropics.   The real difficulties only presented themselves during the late 1950s and 1960s when mining development took off in a big way in northern Australia.   Mining is largely mechanised today and is essentially capital, rather than labour, intensive. Nonetheless, there has been a considerable demand for highly skilled workers, who must be enticed away from the southern cities to settle in remote localities.   This situation has posed a great challenge and expense to mining companies, demanding provision of first class living conditions in new towns so that workers can bring their familites with them, and establish some measure of stability, and of community.

The 1960s and early 1970s saw some very good and innovative plans for new towns, and designs for houses which were fully air-conditioned, acting as an additional bait to attract workers and their families.   Some earlier buildings were lightweight and prefabricated.   Designers took their cue from Australia's early timber and tin tradition which had some merit but also many disadvantages (Saini and Joyce, 1982).

Lightweight structures form an important part of the northern Australia heritage, and they have recently inspired a number of young architects to use traditional timber and tin buildings vigorously.   Such structures have proved effective in warm and humid coastal climates.   The typical house has a generous roof, sitting like a 'digger's hat' over wide verandahs which provide shelter from the sun but which also allow cool breezes to blow across, unhindered. However, these lightweight structures are not so effective in hot, dry, inland climates (see Szokolay, chapter 16). There the sun is very hot and the winds are hot, dusty and dry.   Also, the earliest towns in inland northern Australia were very spread out, and their net living density was far too low.   Today in the Interior rows of small, uninsulated houses sit in a virtual dust bowl and on plots of land which are far too big.   Main roads were usually very wide and unsealed; they became an added, constant source of dust and sand, plaguing 'desert' communities of the Outback.

So it is obvious there is little to be learned of positive design value from these old towns.

Recently a number of talented architects have looked to middle eastern and north African towns for inspiration. They have a similar climate to that of northern Australia, but in a part of the world where labour is cheap, builders

prefer mud houses with thick walls and high mass which keep
the interior cool for the best part of the hot days in summer,
and yet are snugly warm during cold winter nights. Buildings
are piled together in a compact mass, not unlike the local
cacti where cells are structured close to each other. The
buildings thus not only insulate each other but also reduce
the total exposure of surface area of the town to the sun.

In Australia, perhaps the best example of compact planning, though not high mass, can be seen at Shay Gap, in the
north Pilbara region of Western Australia. Here buildings
not only shade one another, but are also connected to each
other in a 'system' to minimise the cost of services. All
the houses are air-conditioned. Community facilities, such
as shops, cafes, clubs and schools, are within walking
distance of homes. But in spite of all these design
facilities, people don't seem to be particularly happy living
there. Their main complaint is that they resent having to
'live under each other's skin' as it were. Housewives would
prefer to need to drive or to 'walk' their children to school,
because this would give them something to do. In fact, they
would do anything to break the monotony of being cooped up
all day in an air-conditioned 'box'(see Parkes, chapter 18).
People also associate lightweight, prefabricated buildings
with temporariness and insecurity this adding to the sense
of isolation which remote settlement inevitably has for
people from an urban experience 'down south'. The mining
town of Nhulunbuy, however, provides a good example of the
new mining town design, both in its physical planning and
in its management. This town looks and feels permanent.
Its buildings are solid and have been designed to last at
least fifty years, in line with the area's mineral potential.
If we can anticipate the growth of other commercial and
industrial activities, this town, like others in the Pilbara
such as Paraburdoo, Tom Price and Newman, could well have an
indefinite life span.

New, small, isolated mining towns inevitably have many
drawbacks, compared to those in the more closely settled parts
of the country, including many parts of North Queensland.
There are limited shopping facilities, and most particularly
a limited 'range' of commodities. Some settlers in
'closed company' towns find the ever present company
oppressive, though they are in fact well aware of the
many fringe benefits provided by the company, usually
far in excess of the facilities and services, including
subsidised rents, to be found in the public domain. Lack of
entertainment and the sameness of social contacts are also

seen as limiting factors, as are educational and health
facilities. School leavers and women have experienced
difficulties in finding work. Many of these problems are
endemic to remote environments, but solutions to some of
these problems have been proposed (Brealey and Newton, 1981)
and others will be found through research and a growth of
experience with these rather new settlement forms.

Economists and political gurus tell us that, in the
foreseeable future, the mining industry in Australia is not
likely to grow as spectacularly as it has done during the
last couple of decades. They may be right. But there is
one industry which does have spectacular potential in the
future, and which is often in close geographical location to
the mining industry, and that is *tourism*. Before I discuss
this, I would like to offer brief comments on some aspects of
multiracial communities associated with location of mining
settlements and housing for Aborigines in the future.

III.   MULTIRACIAL COMMUNITIES

Northern Australia has a number of settlements which
offer unique examples of racial diversity, giving them an
exotic flavour not found elsewhere in Australia. Thursday
Island is a fascinating mixture of Japanese, Australian,
Aborigine, Papuan and other nationalities - descendents of
sailors who happened to pass through the place over the years.
Broome, a pearling town in north-west Western Australia, has
a similar mix and a miniature Chinatown as well. There are
larger urban settlements, such as Darwin, Townsville and
Cairns where in the pubs, the shops, the schools and the
streets, faces of Asia and Europe can be seen to make up the
local communities. All these settlements offer a glimpse
of things to come and they clearly remind us how close we
are to Asia and the Pacific Islands. They are good examples
of settlements which, like Hawaii, for instance, exhibit
racial harmony and the *absence* of conflict, which fails to
make headlines in the news media.

However, a problem which does hit the media fairly
frequently is the troubled relationship between some of our
aboriginal communities and communities of European majority.
These conflicts are sometimes dramatic in areas of mining
activity, though by no means always or exclusively. Conflict
is found particularly when mining settlement occurs close to
aboriginal reserves (Australian Institute of Aboriginal
Studies, 1981). Mining settlements, particularly in their

developmental stages, attract a largely skilled, European workforce, with more men than women. The risk of prostitution and other socially undesirable behaviour is high. A possible way to avoid these particular inter-racial conflicts is to site the new towns away from the reserves and sacred sites where aboriginal people gather. Miners and other workers could then commute to the mine sites from existing towns or from new towns distant from aboriginal lands. The idea is not new. It has been successfully tried in north-western Ontaria, Canada. Brealey and Newton (1981), of CSIRO's former Remote Communities Unit, have made a forceful case for the development of commuter settlements in remote areas, relative to ore bodies in Australia.

As for housing for aboriginal people, research so far has suggested that aboriginal people's priorities and perceptions of housing types and needs are quite different from those of the European Australian. They have a very close physical and spiritual association with their environment and therefore show a very flexible and adaptable response to it. As Parkes has recently argued (1983), movement is an ecological fact of life among people who have traditionally lived in stressed, remote and arid environments. Housing for Aborigines needs to reflect and absorb this flexibility of habitat. Perhaps Aborigines don't give the same importance to houses as Europeans because they tend to see the 'four walls' of conventional houses as barriers between themselves and nature around them.

A further design problem is posed by the close kinship which aboriginal families have within each unit and with one another. Houses designed for a statistically typical or probable European family of four, with its culture-specific behavioural ecology, may not be either large or flexible enough to house aboriginal people who may have to accept a sudden influx of friends and relatives. This phenomenon is no different from many other traditional cultures in Asia and the Pacific, where most houses, including those in larger urban areas, have many secondary semi-sheltered spaces, such as verandahs and courtyards, which provide extra shelter for families and other groupings which may vary in size from four or five to as many as twelve or more. Awareness of these needs is growing, as is evident in the aboriginal housing councils such as Warramunga Pabulu in Tennant Creek and the Tangetjira Council in Alice Springs (see Parkes, chapter 18). There is, however, a long way to go and lack of finance for capital works and management infrastructure is an obvious and important constraint. Parkes has suggested, in correspondence, that access to sources of finance, independent

of government, as through mining royalties from developments
on aboriginal land, could be a most significant factor in
facilitating an indigenous housing form developed for and by
the aboriginal people.

IV.   TOURISM AND THE BUILT ENVIRONMENT

I will now comment on the tourist industry to which I
referred earlier. This industry seems to have a bright future
in northern Australia. There are a lot of things going for
it: climate, scenery, relatively high incomes, mobility,
early retirement schemes and a probable future of more and
more leisure time at our disposal. It seems that northern
Australia is now ready for a major thrust in the development
of tourist industries, which, in turn, will help to diversify
future growth in this region.

Tourism is important for a number of reasons. First, it
is labour intensive. Therefore it *creates* jobs, particularly
in the hotel, motel and other accommodation sectors, and in
transport and recreational services. Second, it provides
purchasing power with a multiplier effect which can be as
high as three times the original amount spent. This, in
turn, helps to generate demand for an infrastructure of roads,
resorts and associated facilities. It is this last set of
items, relating to the built environment response to tourism,
which concerns me and about which I would like to make brief
comments.

It is obvious that many new buildings will have to be
constructed to meet the demands of an expanding tourist
industry. However, there must be serious and coincident
consideration given to the recycling of the older buildings
in the tourist environment. Many of these older buildings,
built by our forefathers with a great deal of love and care,
have great charm and give a special character to northern
Australia's cities, towns and smaller settlements. Such
recycling policies have been followed with considerable
success in Europe and Asia.

In northern Australia there are many outback and coastal
towns with important buildings which are either unused or
under-used and many of them are slowly deteriorating, some
to a point almost beyond redemption. It seems likely that
visitors from within Australia, and also those from overseas,
would welcome the chance to stay a night or two in a
refurbished hotel with vast verandahs and attractive cast-
iron balconies. They would probably enjoy a stay in a

traditional timber and tin homestead, surrounded by a well tended garden with lush, tropical growth, or an expansive desert panorama, more than a night in a nondescript and usually sterile hotel or motel room which is no different from those to be found in cities and resorts in any other part of the world.

We must design new buildings and rehabilitate older buildings in order to give tourists and travellers an Australian experience. Many of the older buildings offer a unique opportunity for developing existing resources to supply this. Such a special Australian experience is the key to open this country to international tourists. Northern Australia has a veriety of colourful environments, where nature is bountiful, *strong* and *dominating*. Reef islands, long stretches of unspoilt sandy beaches and rocky inland desert country are tourist features which are not found easily or in abundance these days in a world which is becoming more and more crowded. They add up to an experience which is unique to northern Australia: its very *vastness* and its *emptiness* are themselves supportive resources, as Parkes has proposed elsewhere (1983). The northern Australian environment, with its diversity of landscape features in the coastal regions and its expanse of unique 'desert' landscape and niches of smaller scale but dramatic features, should not be developed to provide the same kind of experience offered by some of Australia's growing southern resorts. They have their particular place in the tourist system but their future role may lie increasingly in servicing the internal tourist industry and in the supply of permanent and semi-permanent accommodation and recreational facilities for our increasingly aging population. Obviously I am not proposing a policy of separate development, but rather anticipating the need for an awareness of different resource potentials, and the likely development of different market emphases.

What is now needed is a careful survey of northern Australia to identify the locations which are unique, which have lasting qualities and have a significant northern Australian character. Now is the time to set space standards, which means establishing the *ecolocical capacity* of a given area and the nature of the activities which can take place within it. It means carefully setting the limits to which an area may be exploited, and exploring the difficulties involved in restoration. It is essential to do this kind of exercise *before* plans are drawn up for building and other uses, rather than afterwards, when it is often too late to do anything about it. Once our tourist resources are destroyed,

the tourist value of the area will also be damaged, probably
beyond repair.

All we have to do is 'treat our environment, not as
something we have inherited from our forefathers - but as
something which we have borrowed from our children' (source
of quotation unknown).

V. CONCLUDING COMMENTS

As a general rule, architecture has to relate to nature.
If nature is *weak*, then architecture should be *strong*, as in
Australia's Gold Coast. If nature is *strong* and *overwhelming*,
then architecture should be *weak* and *subservient* to its
environment: and this *par excellence* is the condition over
most of northern Australia.

REFERENCES

Andrews, J. 1967. *Australia's Resources and their
    Utilization*, University of Sydney Press, Sydney, p.14.
Australian Institute of Aboriginal Studies 1981. *Social
    Impact of Uranium Mining on the Aborigines of the
    Northern Territory*, Report to the Minister for Aboriginal
    Affairs, Canberra.
Brealey, T.B. and Newton, P.W. 1981. Commuter Mining - An
    Alternative Way, *Proceedings of the Australian Mining
    Industry Council*, Environmental Workshop, Canberra.
Cilento, R.W. 1925. The White Man in the Tropics, *Department
    of Health Services, Publication No. 7*, Government
    Printer, Melbourne, pp.7-8 and 68. Also cited by Lee,
    D.H.K. 1957. *Climate and Economic Development in the
    Tropics*, Harper Bros., New York.
Parkes, D.N. 1983. The Place of Movement and the Movement of
    Place in the Australian Arid Zone, *Proceedings of the
    16th Annual Conference of the Australian Institute of
    Urban Studies*, October, Brisbane.
Saini, B.S. and Joyce, R. 1982. *The Australian House:
    Homes of the Tropical North*, Landsdowne Press, Sydney.

16

THE BUILT ENVIRONMENT OF
CLIMATICALLY STRESSED ECOSYSTEMS

S. V. SZOKOLAY

Architectural Science Unit
University of Queensland
St Lucia, Queensland

I.   INTRODUCTION

The climate of northern Australia has been discussed in Chapter 3 by Lee and Neal and in Chapter 17 Auliciems and Dedear examine the problems of thermal comfort.   These two sets of information are taken as the starting point for the present chapter.   The task of building design can be considered as the creation of a controlled indoor environment, where the outdoor conditions are less than comfortable.   The control task can be defined as

GIVEN CONDITIONS - DESIRED CONDITIONS = REQUIRED CONTROLS

The climate of northern Australia presents a challenge to the designer of human habitat and no generalised solutions can be offered.   At least two distinct climate types must be distinguished, which require totally different responses from the designer of buildings: the hot-dry (inland) areas and the warm-humid coastal zone.   Many locations will have a climate transitional between these two archetypes.

The hot-dry zone is characterised by very high peak temperatures, coupled with large diurnal and annual variations, as well as generally low humidities.   The warm-humid coastal areas may not reach the same peaks, but the temperature is continuously high, with small diurnal and annual variations and regularly high humidities.

NORTHERN AUSTRALIA
ISBN 0 12 545080 X

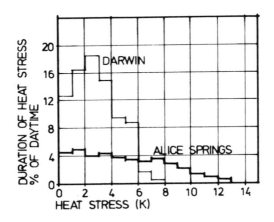

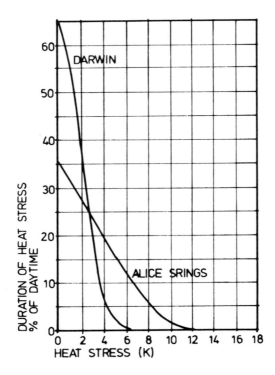

FIGURE 1.   Magnitude and duration of heat stress in two
towns;   top: by stress intervals;   bottom: cumulative.

Kowalczewski (1968) convincingly showed the difference in heat stress (defined as temperatures in excess of 83°F or 28.3°C), both in magnitude and in duration, between Darwin (warm-humid) and Alice Springs (hot dry). His graphs are reproduced in Figure 1, converted to SI units.

The concept of "degree-hours" can be employed to characterise a location from the point of view of thermal control requirements. We prefer the use of the term "Kelvin-hours" (K.h.) to distinguish it from the °F "degree-hours" used in American literature. This is the product of temperature deficit (or surplus) below (or above) an agreed reference level and its duration in hours, summed for a defined period, e.g., a month or a year.

The reference level of 18°C is normally used to assess the heating requirement and we use 26°C as the reference level for the cooling requirement (Szokolay, 1982). The heating and cooling Kelvin-hours are shown in Table 1 for a dozen northern Australian locations, together with the average diurnal range and the annual mean range of temperatures. The latter is defined as the difference between the highest and the lowest monthly mean temperature. The average relative humidity is also included, calculated as the average of 9.00 h and 15.00 h monthly mean values.

As a rule of thumb it can be suggested that a heavyweight building (having a large thermal mass) is beneficial in locations, where the diurnal range exceeds 10 K. (Note: 'K' is used to denote a temperature interval or difference, whilst '°C' refers to a point on the temperature scale.) The thermal inertia of the building fabric tends to even out the diurnal temperature fluctuations and may be able to keep the indoor conditions comfortable whilst the outdoor varies between hot days and cold nights. The building can thus physically control the indoor temperature.

In the warm-humid zones such physical control is very difficult, if not impossible. Under humid overheated conditions physiological mechanisms must be relied on to render the conditions acceptable, i.e., evaporative cooling of the skin surface, perhaps accelerated by air movement. If the indoor thermal environment is to be controlled by "passive" means only, i.e., by the building itself (without mechanical heating or cooling based on some form of imported energy), then the desire for air movement would dictate a fully cross-ventilated building, possibly elevated on stilts to avoid local obstructions to the breeze.

TABLE 1.   *Climatic Indicators for Some Northern Australian Locations.*

	Heating K.h re: 18°C	Cooling K.h re: 26°	Average diurnal range: K	Annual mean range: K	Average R.H. %
NT					
Darwin	–	16 217	8.9	14.5	62
Daly Waters	345	19 607	15.4	27.3	40
Tennant Creek	2 210	19 276	13.2	26.8	30
Alice Springs	14 804	8 901	14.9	32.1	30
Angurugu	45	11 807	12.7	20.8	63
WA					
Halls Creek	675	22 044	13.9	26.2	31
Broome	187	17 679	10.9	20.3	51
Wyndham	–	35 909	11.6	20.5	42
QLD					
Cloncurry	1 893	19 498	13.8	27.2	33
Townsville	597	4 631	9.2	18.2	60
Cairns	–	4 729	8.2	14.8	65
Thursday Island	–	7 906	5.2	8.7	76

II.   PAST EXAMPLES

There are some historical examples of very sound and appropriate solutions to the climatic problem. ("Historical", in the Australian context, is any building more than about fifty years old.) "Adelaide House" in Alice Springs was built in 1925 by the Reverend Flynn, as a 'hospital', or nursing home, to serve as the base for his Flying Doctor Service. It has 450 mm thick stone walls and a verandah all around. Figure 2 shows the building, whilst Figure 3 gives an example of indoor and outdoor temperature readings taken in this house: when the outdoor diurnal variation is over 20 K, the indoor temperature amplitude is less than 5 K (peak-to-peak).

A house type suitable for the warm-humid areas has developed, especially in northern Queensland. Indeed, if there is anything distinctive and different to the rest of the world in the history of Australian architecture, it is the "Queensland House". Saini (1982) produced an excellent

FIGURE 2.   "Adelaide House" in Alice Springs

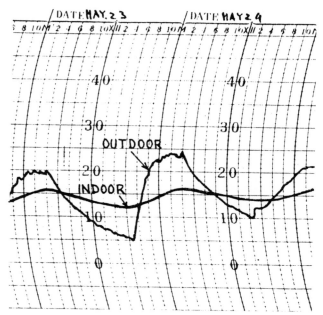

FIGURE 3.   Typical temperature recording in Adelaide House

FIGURE 4.   A "Queensland house" (Rockhampton).

pictorial presentation of this house type.   Figure 4 is just
one example taken from his work.
     In the 19th, and the early part of the 20th century,
building in remote areas was governed by expediency, by the
local availability of materials and skills.   (In this
context the whole of northern Australia can be taken as a
"remote" area.)   It is fortuitous that the available
materials and techniques produced a house type which was
suitable for the warm-humid climates.   The easiest
foundation for hilly terraine is to use stilts (stumps).   The
"high-set" house has obvious advantages in flood-prone areas,
but at the same time it serves the purposes of increased
natural ventilation.   Timber available locally gives the
shortest route from raw material to final product and
carpentry is the most "transportable" building trade, which
was often practised by unskilled, self-taught and do-it-
yourself people.   The only "imported" materials used in
simple buildings were galvanised steel sheets for the roof
and nails for the carpentry.

The same coincidence of what is available and what is needed did not occur in the hot-dry areas. There are only a few examples of heavyweight buildings, such as Adelaide house mentioned above. Most houses in this zone were built with the same materials and same techniques as in the coastal zone, giving very unsatisfactory results, or - at least - missing the opportunity for creating much better indoor conditions.

III. PASSIVE CONTROLS

The thermal control task has been defined as the difference between the existing and the desirable conditions (as in the introduction above). Such controls can be provided by two sets of tools:

(1) passive controls: the thermal control function of the building itself, its positioning, shape and fabric,
(2) active controls: heating, ventilating and air conditioning installations (mechanical equipment), based on some form of external energy supply.

In many instances passive controls can perform the full task, i.e., can ensure the desired indoor conditions without the use of any mechanical equipment, but even if this is not quite possible, such passive controls should be relied on to reduce the (heating or cooling) load on the mechanical equipment as far as practicable. This can be expressed as

REQUIRED CONTROLS - PASSIVE CONTROLS = ACTIVE CONTROLS

This means that the building should be suitable for the climate in which it is located. I suggest that such principles of climatic design should be followed for three reasons:

(1) economic: to employ mechanical equipment imposes a capital cost, but also a recurrent cost, as the energy consumed and the maintenance of the system must be paid for,
(2) ecological/moral: "passive" buildings impose the least load on the ecosystem, consume the least amount of energy, produce the least amount of waste (e.g., waste heat or emissions of fuel-burning appliances)

(3) aesthetic: "passive" buildings, for purely
functional reasons, are more likely to be in sympathy with
their environment, more particular to and characteristic of
their location, less likely to imitate international
examples, thus more likely to increase diversity and interest.
Their design requires more thought than the design of
buildings relying on "brute force" active controls, so the
product is more likely to be of value than the "obvious",
repetitive, often of banal appearance actively controlled
buildings.

The thermal relationship of a building with its
environment can be analysed in "steady state" terms. Heat
flows into or out of a building can be readily calculated.
This is a useful method for actively controlled buildings,
where the required capacity of the cooling (or heating)
plant is to be established on the basis of the "reasonable"
worst conditions. Such design conditions for hot or warm
climates are usually taken as the 85th or 90th percentile
temperatures, both dry- and wet-bulb, and the simultaneous
clear-sky solar radiation.

In reality there is a dynamic thermal interaction between
the building and its environment. The thermal response of
the building over the 24-hour cycle can be considered in
terms of "periodic heat flow", where not only the thermal
conductance of the enclosing elements, but also their thermal
capacity must be taken into account.

It is reasonable to assume, that in the absence of
internal heat generation and direct solar heat input, the
daily mean internal temperature of the building will be the
same as the outdoor mean temperature. As Figure 5 shows, the
sinusoidal outdoor temperature curve is followed by a similar
indoor temperature curve, which is reduced in amplitude and
delayed in phase. The greater the building mass, the greater
the amplitude reduction and the longer the delay or "time-
lag" of the indoor temperature.

If there are significant internal heat gains and solar
radiation entering through windows creates an internal
heating effect, the indoor temperature is likely to rise far
above that outdoors. Ventilation can be relied on to reduce
such internal overheating, but will never produce indoor
temperatures less than the simultaneous outdoor temperature.

Ventilation and solar control are the most important
tools of the building designer. Skilful manipulation of
these can elevate (or depress) the indoor mean above (or
below) the outdoor mean.

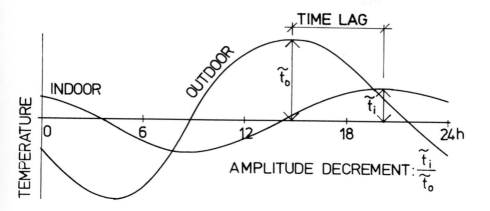

FIGURE 5. Periodic heat flow: diurnal temperature pattern.

It is a mistake to believe that overheating is the only problem in northern Australia. Table 1 showed that, for example, in Alice Springs the annual underheating is much greater than the overheating. However, solar radiation is strong, even in winter, and if a controlled amount is admitted during the day, if the building fabric is sufficiently massive to store this heat for up to 12 hours, and if the night-time heat losses are not excessive, the interior temperatures can be kept comfortable at all times.

Fixed eaves or canopies over windows can provide an automatic seasonal control over solar irradiation for north-facing (and only north-facing) windows. Hence the recommendation that most major windows should face north, i.e., the building as a whole should have a northern orientation. As Figure 6 shows, the zenith angle of the sun at noon on equinox days (March 21 and September 23) is the same as the latitude and the apparent movement of the sun is 23.5° up from this line in summer (December 22) and 23.5° down in winter (June 22).

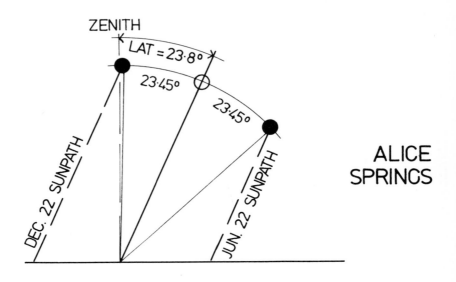

FIGURE 6.   Solar angles at noon: Alice Springs.

As it happens, there is some heating requirement in Alice Springs from March, increasing to July and then decreasing to September, so if the shading device is dimensioned so that the edge of the shadow is at the window sill on equinox days, no radiation will be admitted in the summer half-year, but from March 21, irradiation will increase to June 22 and then decrease to September 23. As the latitude of Alice Springs is 23.8°, the "vertical shadow angle" should be set as (90 - 23.8) = 66.2°.

In locations where there is heating requirement, solar radiation should be excluded at all times. This applies to both warm-humid and hot-dry climates. As adequate shading of east and west windows is very difficult (if it is done, all view would be excluded and lighting diminished to such an extent that the window would become redundant), it is better not to have any windows in the east and west walls.

At lower latitudes (further north) for some part of the year the sun's apparent path will be to the south of the

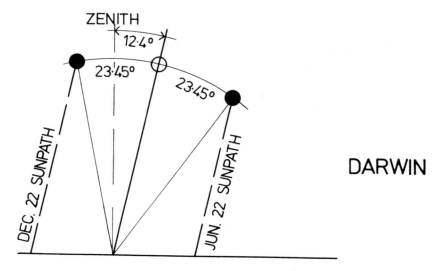

FIGURE 7.   Solar angles at noon: Darwin.

location, so both north and south windows should be shaded.
Figure 7 shows the relevant angles for Darwin.
   In overheated locations the prevention of heat gain is
the first task.   Shading of windows is the first step in this
direction.   The second largest solar gain is that on the
roof.   At the low latitudes of northern Australia the sun's
path is always high, so that the roof is the most exposed
surface.   The first line of defence is the surface itself.
It should be reflective.   The heating effect of solar
radiation on a dark coloured roof can be more than four
times as much as on a light, reflective surface.   If heat
emission at night is desired, a white painted surface will
be better than a bright aluminium.   The two may have the
same absorptance for solar wavelengths, but at normal
terrestrial temperatures (0-50°C) the white paint behaves as
a 'black body':   it has a high emittance ($\cong$0.9), whilst the
emittance of bright aluminium is very low at all wavelengths
($\cong$0.1).
   The second line of defence is the roof fabric.   It
should be well insulated, whether the building is light or
heavyweight, whether it is actively cooled or not.   A poorly
insulated roof, even if the outside surface is white, can
produce a ceiling surface temperature 10 K higher than the
air temperature (with a dark roof this differential can be

as much as 35 K). This would not only increase the inward
heat flow, but also elevate the mean radiant temperature of
the interior, thus directly affecting the thermal sensation of
occupants.

Heat flow through the roof-ceiling combination is partly
radiant, partly conductive. Conduction can be reduced by
insulating (porous, or fibrous materials). These would be
best placed on top of the ceiling, as this is a much smaller
area than the actual roof. If the roof is insulated, then
the gables and eaves (i.e., all parallel heat flow paths)
should receive the same amount of insulation. It is
suggested that 50 mm glass fibre batts or the equivalent
would be the minimum insulation required.

Ventilation of the space between the ceiling and the
roof (or the attic) is advisable. If it is not ventilated,
the temperature of this space can far exceed the outdoor air
temperature. Ventilation will not affect the radiant heat
transfer from the underside of the roof to the top of the
ceiling. This can be drastically reduced by lining the
underside of the roof with a low emittance aluminium foil,
and by placing a similar foil on top of the ceiling
insulation, where its low absorptance will be beneficial.
Unfortunately, as dust settles on this surface, it will
increase its absorptance, so its benefit will diminish. If
only one layer of foil is used, it is better to place it
face downwards, under the roof covering.

Differences between the warm-humid and hot-dry climate
buildings will be obvious in all other respects. Where
there is a large diurnal variation (i.e., in hot-dry
climates), the heavy building with its large thermal
capacity will be beneficial. Concrete slab-on-ground floors
would thermally couple the building with the earth below, so
partly utilise the thermal capacity of this earth.

Such buildings should be operated with windows and doors
closed during the day. By the time the temperature wave
through the heavy walls reaches the inside, the outdoor air
temperature would be reducing, so doors and windows can be
opened to promote the dissipation of the stored heat by
night ventilation.

The ultimate in thermally massive buildings are the
various types of earth-covered or earth-sheltered houses.
The earth temperature at about 3 m depth is practically
constant throughout the year, and the building could make use
of this. There are many such dwellings around the Coober
Pedy area of South Australia and they are becoming
increasingly popular in the U.S.A. They would be ideal for
the Alice Springs region, but there are no examples in

existence known to this writer. (A few have been built in
recent years around Sydney, in the Hunter Valley and around
the Gold Coast area of Queensland. One is known to function
very well at the Atherton Tableland.)

   In warm-humid climates, where temperatures are high even
at night, the delayed release of the stored heat would add to
the discomfort, especially in the late evening hours. For
such climates a hybrid building has been successfully tested
in Darwin, a prototype of which is shown in Figure 8. This
would have the rooms occupied during the day on a concrete
slab-on-ground floor, with massive walls, which would remain
cool during the day. The bedrooms would be in an elevated
lightweight wing, with full cross-ventilation, which may
become overheated during the day (when unoccupied), but would
cool down rapidly, to reach the level of outdoor air
temperature by the time the occupants go to bed. It is quite
a puzzle to this writer why this hybrid building type has not
achieved a greater popularity.

   One possible explanation is that people prefer an "open
living" in such climates, they can't accept the closed-up
environment, even if it is cooler than the outdoors. This is
probably why most designers give up the attempt for keeping
the indoors cooler than the outdoors and concentrate on
keeping the indoors not much warmer than the outdoors. They
can achieve this by full cross-ventilation, total exclusion
of the sun, well insulated roof, either shaded or insulated
east and west walls and the installation of a few ceiling
fans to ensure air movement (and thereby physiological
cooling) at times of little or no wind.

IV.   ACTIVE CONTROLS

   Two methods of active cooling are available and widely
used in northern Australia:

   (1)   evaporative cooling,
   (2)   refrigerated air conditioning.

   Evaporative coolers make use of the latent heat of water
from liquid to vapour. This process absorbs some 2.4 kJ of
heat from the environment per gramme of water evaporated. As
the air stream is passed through such a cooler (Figure 9), its
dry bulb temperature is reduced, whilst its moisture content

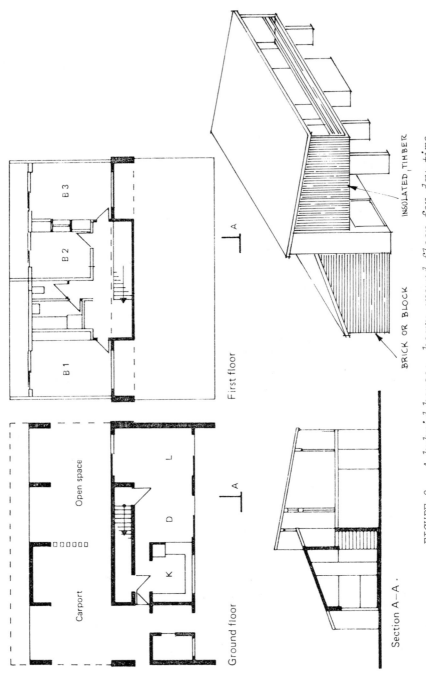

FIGURE 8. A hybrid house: heavy ground floor for day-time use, lightweight upper floor for bedrooms.

is increased.   Its relative humidity is increased by both
effects.   The theoretical limit of cooling for such devices
is the initial wet bulb temperature of the air, but this is
not quite achievable.

FIGURE 9.   An evaporative cooler installed in a house at
Wyndham.

The system does not include recirculation. A large
volume of air is delivered (1.5 - 2.5 m^3/s for an average
house), cooled, allowed to flow through the house and
discharged through some doors or windows kept open. Many
people prefer these systems to the "sealed box" operation of
air conditioning.
The limitation to the use of these systems is two-fold:

(1) if the initial humidity is high, the evaporation
rate, thus the cooling effect will be small
(2) the air supplied is of a high humidity, thus it may
have adverse effects, regardless of its temperature.

Many indirect evaporative coolers have been developed in
the last thirty years - or so. The common feature of such
systems is that one air stream is evaporatively cooled and
passed through a heat exchanger, where it picks up the heat
of another air stream (without mixing), which is then
supplied to the space to be cooled. Due to the less than
100% effectiveness of the heat exchanger, the supply air
temperature will not be quite as low as that of the primary
air stream, but no moisture is added to supply air, thus its
relative humidity is only slightly increased (due to the
reduced temperature). Figure 10 shows the performance of
both systems in terms of the psychrometric chart.
The most successful indirect evaporative cooler has been
developed by the CSIRO (then) Division of Mechanical
Engineering. It is based on an inexpensive plate heat
exchanger and is now commercially produced by a firm in
Adelaide. Pescod (1976) examined the performance of both
direct and indirect evaporative coolers. Figure 11 shows
his results for Alice Springs and Darwin. The energy
consumption of these coolers is only about one-fifth of that
of refrigerated air conditioning.
Refrigerated air conditioning is extensively used in
public and commercial buildings, but its use in residential
buildings is by no means widespread. Small packaged units
("room conditioners") are sometimes installed, especially in
bedrooms, but operated only a few hours a day. The
situation is different in mining towns, where fully air
conditioned housing is usually provided by the employer,
including the supply of energy, at no cost to the occupant,
or at least heavily subsidised. Such a "full house" (ducted)
air conditioning system would use some 20 000 kWh of
electricity in a year in a location such as Nhulunbuy (Gove
Peninsula). From a survey conducted by Saini & Szokolay
(1975) it appears that user behaviour influences such energy

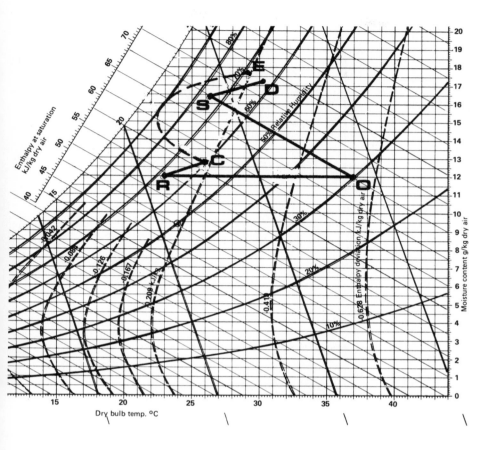

O   outdoor air condition
O-R   indirect evaporative cooler effect
O-S   direct evaporative cooler effect
R & S   room supply condition with the two systems
C & D   room condition
C-E   evaporative effect on exhaust air (indirect)

FIGURE 10.   Psychrometric chart: status changes with
direct and indirect evaporative coolers.

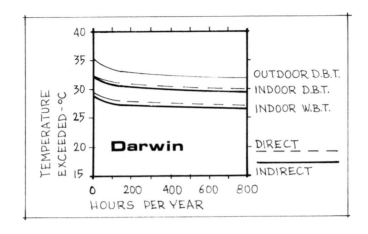

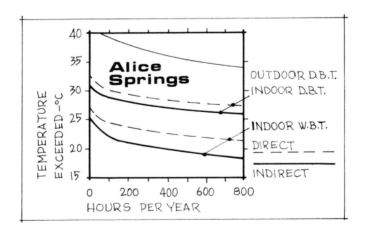

FIGURE 11.   Performance of evaporative coolers: hours
of year when indoor temperatures shown would be exceeded.

use more than the design and construction of the house and
the system. Some users may run the system continuously and
if it is too cool, they open a window.

In a hot-dry climate an air conditioner can be installed
at any time during the 'life' of a building. Design and
constructional requirements are identical for a passively
and an actively controlled building. Not so in the humid
regions. Here the two sets of requirements are quite
different. Here a good 'passive' house may be totally
unsuitable for the subsequent installation of air
conditioning. The passive house may have practically no
north and south walls, only screens or louvres, and if a
room air conditioner is installed, the load due to air
infiltration will be far greater than what the unit could
cope with. The first requirement for an air conditioned
house is that it should be airtight and well insulated.

This however can create its own problems. In some
instances it has been found that, in order to maximise the
cooling effect, the fresh air supply is shut off, the unit
is used in the 100% recirculation mode, resulting in a
rapid deterioration of indoor air quality.

Where such mechanical cooling is employed, condensation
can occur and cause serious problems. Two broad types of
this problem can be distinguished: steady-state and
transient condensation.

(1) Steady state: if the indoors are kept at (say) 25°C
and 50% RH, the vapour pressure will be around 1.6 kPa. The
outdoor vapour pressure will be much higher (e.g., at 38°C
and 70% RH it is 4.8 kPa), thus vapour will permeate the
building fabric. Where such vapour reaches a layer, the
temperature of which is less than the dew-point temperature
corresponding to that vapour pressure, it will condense out.
For the above example this dew-point temperature is 31.5°C.
Thus condensation will occur somewhere within the wall or
roof (interstitial condensation). The condensed water will
saturate the building material, which will thus lose its
insulation properties, so condensation will increase.
Remedy: install a vapour barrier, a continuous uninterrupted
membrane near the outer surface of the enclosing elements.
This is not easy and great attention to detail is necessary.

(2) Transient: if a room is cooled, all internal
surfaces will be close to the room temperature (e.g., 25°C).
If a door or window is opened, the outdoor warm and humid air
will rush in and almost instantaneously profuse surface

condensation will occur, as these surfaces are below the dew-point temperature of the outdoor air. Wall linings, carpets and furnishings will be wet and even if the openings are quickly closed, the space will develop a musty stale smell. The wet conditions will promote the growth of mildew and fungi. Remedy: if air conditioning is used, windows should not be openable and doors should communicate with the outside through air locks.

To air condition a house is a major decision and its successful operation will demand an understanding and a strong behavioural discipline, which is rarely realised by the users.

V.   ASSOCIATED PROBLEMS

Cyclones

All coastal areas of northern Australia are cyclone-prone. Recent major cyclone disasters (Althea, Townsville, 1972, and Tracey, Darwin, 1974) instigated a revision and tightening of building codes and regulations. Cyclone-proofing imposes stringent requirements on the structure and construction of buildings. The occurrence of cyclones may be rare (see Chapter 3), but buildings should be able to withstand even the most infrequently occurring wind loads, or else life may be at risk. Human memory is short: only eight years after the last cyclone the sentiment is rising in Darwin that the cyclone-code requirements are exaggerated and impose an unnecessary additional cost on building.

Dust

A more everyday problem is the keeping out cf dust. Inland areas, particularly after a long draught, can be very dusty. Householders are often engaged in a heroic struggle to keep the immediate surroundings (their gardens) covered with vegetation, as a means of reducing the dust problem. Water is also a scarce resource, so efforts are concentrated on using water most effectively for the watering of gardens. Drip-feed irrigation techniques are often employed. The water supply is usually metered and the consumed quantity must be paid for. In several mining towns householders receive a "free-water allowance", as an incentive to develop gardens. Apart from reducing the dust problems, vegetation can have a beneficial effect on the microclimate.

At times of strong winds dust is transported over long distances and there is no way of combating it at the source. The only protection is the practically air-tight closure of the house.

Insects

Flies and mosquitoes, but also thousands of other flying insects, present a problem: flies during the day, the others mostly at night, gathering to the light. All doors and windows must be fitted with insect screens, all gaps and unprotected openings must be avoided. It is a virtual state of siege, but the inhabitants quite readily adapt to it and consider it as a normal way of life.

Outdoor activities at night are conducted under dim lighting, with some bright light (a flame or an insect-trap light) operating at some distance away. To the visitor the most disturbing are the ubiquitous cockroaches, against which the inhabitants are fighting a losing battle.

Termites and white ants present a different problem. If they get access to any wooden component of the building, they will consume it from the inside, which may ultimately lead to structural collapse. It has been customary to place timber buildings on short brick or concrete piers and laying the floor bearers on galvanised steel "ant caps". With the increased use of concrete slab-on-ground floors, normal practice is to spray the ground with a long-life insecticide before the concrete is laid. With brick or masonry buildings often metal door and window frames are used and even with timber framing, the bottom plate is often replaced by a pressed steel sheet channel. There are also timber preservative treatments available which would reduce the likelihood of insect attack.

Energy

In a 'passive' house about half of the annual energy consumption serves the purposes of producing domestic hot water. Solar water heaters are rapidly increasing in popularity and a well-designed system in northern Australia will require little or no auxiliary energy input.

Electricity is very expensive, in spite of heavy government
subsidies. In remote areas, not connected to the grid, the
supply cost varies between $0.30 and $2.40/kWh, whilst the
consumer would rarely pay more than $0.20/kWh.
       Table 2 shows the cost breakdown for an average small
diesel power station. One of the reasons for such high
costs is the very low load factor (often less than 0.3 in
annual terms) with a ratio of maximum-to-minimum demand
around 5:1. The other main reason is the very high
maintenance cost which includes the long distance travelling
cost of maintenance personnel.
       Bottled gas and kerosene are still important fuels,
especially for cooking purposes. In recent years significant
advances have been made in solar and wind energy applications.
       Solar photovoltaic systems are now competitive with any
alternative for small (<1 kW) power requirements in remote
areas, not so much because of the 'free' source of energy,
but for their high reliability and low maintenance costs.
With sophisticated telemonitoring equipment they can operate
unattended and without personal maintenance calls for
extended periods. Many such systems are now operated by
Telecom Australia, of which an example is shown in Figure 12.
       It is expected that with the reduction in the cost of
'solar cells', the economical size of such systems will
increase, that these will prove to be more successful than
the solar-thermal-mechanical generating systems and in a few
years' time they would be competitive with diesel generators
up to the 10 kW range. Hybrid wind-solar systems (Figure 13)
seem to be especially promising.

*TABLE 2.   Cost of Electricity: A Small Diesel Power
Station.*

Generation:	fuel	12.0 ¢/kWh	
	operation + maintenance	44.5	
	capital charges	9.9	66.4 ¢/kWh
Distribution:	operation + maintenance	1.3 ¢/kWh	
	capital charges	4.8	6.1
			72.5 ¢/kWh

(Sawyer & Amadio, 1981)

FIGURE 12.   *Telecom Australia, solar powered microwave repeater station, near Alice Springs.*

FIGURE 13. Wind (5 kW)/solar (1.8 kW) hybrid power
system: an experiment of the Solar Energy Research
Institute of Western Australia.

VI.   INHIBITING FACTORS

If the earth-covered building consumes the least amount
of energy and ensures the greatest thermal comfort in inland
areas of northern Australia, why is it not widely used?   The
answer is probably two-fold:
- lack of awareness of this form, lack of information,
- social and psychological unacceptability.
It is not 'normal', it is not what people expect.
Australian public taste seems to be far more conservative
and far less adventurous than - for example - the North
American.   Even when a person would like such a house, he or
she would be reluctant to invest in it, worried about future
resale value.

If the next best thing for hot-dry climates is the heavy
masonry building, why are so few examples around?   The
reason is probably economic: transportation of heavy
building materials over long distances and often poor roads
is very costly.   Even Darwin, one of the largest towns in the
north, had to import its bricks (until the recent opening of
a brickyard) and it was often cheaper to bring in bricks from
England, than from Queensland.

Mud-bricks or stabilised earth blocks would be
appropriate in many cases for inland areas, certainly for
owner-builders, but in the view of the general public (and
often of the building inspectors) this is not quite a proper
and respectable building material.

In some locations stone would be readily available, but
not the skills to use it, or the labour cost would be
prohibitive.

In many ways this is the proverbial 'chicken-and-egg'
situation, arising from the lack of organisation of the
building industry.   Most builders and clients operate within
a very limited framework: they need a building, they want to
get the most economical solution, which is the most readily
procurable.   This leads to mundane, immediately expedient and
obvious products, which themselves will reinforce and
perpetuate existing trends.   To produce anything new or
unusual is very expensive, builders are not 'geared up' for
it and the market is so small in remote locations, that the
differentials are even more exaggerated - but such solutions
are also unexpected and frowned upon by the users.

A rather sad example is the reconstruction of Darwin,
after 1975.   The new stringent cyclone-codes forced the
introduction of new and unusual building types, but the
preoccupation with structural requirements was so dominant,

that many other factors, such as the thermal performance of
buildings was neglected.   An opportunity seems to have been
missed.   Many inhabitants are now quite bitter and develop
an exaggerated nostalgia for the 'good old' lightweight,
breezy houses.
     It is suspected that one of the reasons for this is a
lack of understanding of how these new building types work,
consequently the lack of behavioural adaptation.   This
reinforces my repeated suggestion that for every house the
designer should provide a set of 'instructions for use', as
normal with the purchase of a motor car or any appliance.

VII.   THE BUILT ENVIRONMENT

     The immediate environment of the house, to a large
extent, depends on the occupants' attitude: whether they
feel 'at home', i.e., consider themselves as permanent
residents, or they are transient, short-term occupants only.
The former would take pride in developing a pleasant garden,
which takes some effort, but has its rewards in both
physical and emotional terms: it creates a sense of
belonging - a self-reinforcing process.   The latter would
consider any effort in this direction a waste of time, would
live in an air conditioned box, with total disregard for the
environment.   Figures 14 and 15, juxtapositioned, illustrate
the contrasting results.
     At the broader, community level this distinction fades,
but is still noticeable.   There are towns which grew up in a
spontaneous way, without 'articifial' support, such as
Cairns or Broome, where there is a complex economic base,
where many individual efforts add up to an often haphazard,
but organic whole.   At the other end of the scale are the
fully planned 'townships', the almost overnight products of
mining companies: the single-industry company towns.   Some
are quite successful, but even the best show the 'from the
top down' management system, they lack the complexities
produced by an organic growth.   In a typical mining town,
such as Nhulunbuy, the average residence period for families
is one to two years and for single persons some eight months.
They have no time to get acclimatised and they do not even
have the desire to establish any emotional 'anchors'.
Obviously, if a townscape is to be formed, it must be done
by the management.

FIGURE 14. A "transportable" house in Wyndham: the occupants consider it as a temporary abode.

FIGURE 15. House in Kununurra: the well kept garden testifies to the owners' attitude.

Between the two extremes are the towns (such as Darwin or Alice Springs), which owe their existence, or at least their size and state of development to determined central government action and many years of subsidy. In these towns deliberate planning is apparent and the suburbs show the town-planning vogue of the date of their origin. Some of the older, more mature mining towns (e.g., Mount Isa) are similar in character. The single-minded original planning has been softened by a multitude of individual contributions to their development.

In no case does settlement layout seem to have been influenced by climatic considerations, whereas a significant body of knowledge exists which should govern planning and subdivision layout practices. In all cases the best orientation is north-south (i.e., the house elongated in the east-west direction, with the long sides facing north and south). Streets should therefore run east-west and the individual plots should be wide enough to accommodate the length of the houses. If, for reasons of topography or for a desire to avoid monotony, the streets run in different directions, the plot should be wider, to allow the diagonal dimensions of correctly orientated houses between the boundaries.

In coastal (humid) climates the capture of the breeze is one of the main requirements. In a typical suburban situation the houses are so close together that they block the breeze. Figure 16 shows some possibilities for layouts which would allow the maximum of breeze exposure to all houses.

In the dry, inland areas there is no need for such wide spacing. In fact, the wide roads and large open spaces often become deserted no-man's-land, unusuable dust-bowls. These climates would allow, and indeed, require a much denser settlement pattern.

In the individual houses the provision of sheltered (perhaps covered) open spaces is an important necessity. Recently some towns made a conscious effort to create sheltered public open spaces. Several such shopping malls are quite successful.

Some of these towns already have a distinct "central business district", with multistorey office blocks, department stores, shops, restaurants, cinemas, etc. In such buildings air conditioning is a must. Whilst - from the thermal point of view, domestic scale buildings are usually "envelope-dominated" - these larger scale buildings are generally "internal load dominated", meaning that the heat output of occupants, lighting and other equipment

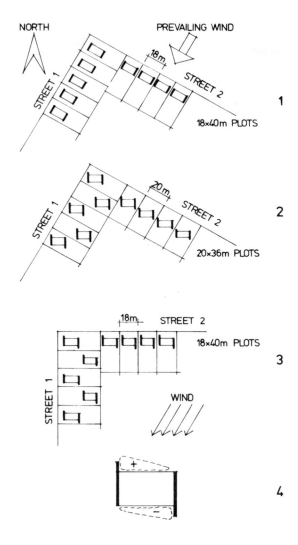

FIGURE 16.
(1) Houses in street 1 are deprived of wind. Orientation
for shading incorrect.
(2) Correct orientation requires wider plots in street 2.
(3) If streets run N-S and E-W correct solar orientation is
possible.
(4) Wing-walls can ensure adequate ventilation with oblique
wind incidence, by creating positive and negative pressure
zones.

constitutes a greater thermal load than the heat entering
through the envelope. The principles of heat gain reduction
outlined for houses, still apply, but the channels available
for heat dissipation are insufficient.

There may also be secondary reasons why work-environments
may need air conditioning. In the home one can dress quite
casually, whilst in the office a "minimum standard" is
expected: shorts, knee-high socks, short-sleeved shirts. In
the home one can have ample air movement - in the office
papers would be blown about. Paper stretched on a drawing
board, or a whole series of other work processes may demand
humidity control. In many instances however, air
conditioned buildings are overcooled - a subject which has
been discussed in the next chapter, by Auliciems and
Dedear.

Humans seem to be highly adaptable - if they want to be.
If their attitude is positive, if they develop a sense of
belonging, a commitment, then satisfaction is relatively easy
to achieve. In some cases this may come from a pioneering
spirit, in other cases from an appreciation of the easy-
going life.

In much of northern Australia the built environment is a
result of improvisations, either because of the inherent
difficulties, or because of this easy-going attitude of the
people, summed up by the succinct expression: "she'll be
'right, mate!"

There are however some reasons for a cautious optimism,
that this attitude is changing. People are becoming
conscious of quality and of the engineering motto: "do it
once - do it well". More and more people are becoming aware
of energy problems and of the possibilities offered by sound
climatic design and passive principles. Prejudices and
preconceived ideas can only be overcome by better information.
It is a slow process, but it is the only way. Only a better
informed public will create a better built environment.

REFERENCES

Kowalczewski, J.J. 1968. A method of evaluating and
     comparing human thermal stresses in different climates,
     Mech. & Chem. Eng. Trans., Inst. Eng. Aus. MC4(1):55.
Pescod, D. 1976. Energy savings and performance
     limitations with evaporative cooling in Australia,
     CSIRO Div. Mech. Eng. Techn. Report 5.

Saini, B.S. & Szokolay, S.V.   1975.   Evaluation of housing
    standards in tropical Australia, *Report to CCPSO*,
    University of Queensland, Department of Architecture.
Sawyer, J.R. and Amadio, J.O.   1981.   Problems with remote
    area power generation, *Solar Energy for the Outback
    Conference*, Alice Springs, September 28-30.
Szokolay, S.V.   1982.   Climatic data and its use in design,
    RAIA Education Division, Canberra.

17

THERMAL COMFORT: COOLING NEEDS
AND CRITERIA FOR DARWIN AND THE NORTH

ANDRIS AULICIEMS
RICHARD DEDEAR

Department of Geography
University of Queensland
St Lucia, Queensland

I. INTRODUCTION

A. *Climatic Determinism and the North*

Equatorwards of 20° latitude, the coastal climates of
Australia have been likened to those of Sudan, Central India
and Bengal. According to Griffith Taylor (1920) these homo-
climes are normally settled by non-white races. Comparative
statistics, as shown in Table 1, were used earlier to demonstrate that the northern lands of Australia simply were not
suitable for close European type settlement. Despite eloquent
argument by Cilento (1925), white man was believed to be
incapable of adaptation to the hot temperature-humidity
milieu.

> *An open-air active occupation such as stock riding*
> *has little to fear, but strenuous field labour,*
> *sedentary indoor life and especially domestic work*
> *and the care of young children cannot be carried*
> *on under favourable conditions at present with*
> *high wet-bulb temperature of this order. (Taylor*
> *1920, p.292)*

The sweeping observations of human progress by historian
Toynbee (1945) and geographer Huntington (1924, 1926) tended
to lend broad empirical observations to support this viewpoint. The natural climate of the North, at least over
several generations as argued by Markham (1947), could do no
other than lead to an inevitable degeneration of settlers and
their initial achievements.

NORTHERN AUSTRALIA
ISBN 0 12 545080 X

TABLE 1.   *North Australian Locality and its Foreign Homoclime*[α]

	Average Temperatures °C			Average Precipitation mm	
	Annual	Hottest month	Coldest month	Annual Average	Wettest month
Broome	26.6	29.9	21.3	23	6
Banana (Congo)	25.5	27.5	22.5	29	6
Darwin	28.3	28.8	25.0	62	15
Cuttack (N.E. India)	26.6	30.5	21.1	55	12
Daly Waters	26.6	30.5	20.6	27	6
Quixeramobim (Brazil)	27.2	28.3	26.1	27	6
Townsville	25.6	27.8	18.9	49	11
Calcutta (India)	25.6	27.8	18.3	60	12

[α]*Adapted from Taylor, 1920*

B.   The Energy Balance

That the residents of Tropical Australia are homeothermic organisms requiring to maintain constancy of core temperatures there can be little doubt. The basic energy balance applies equally to humans here as elsewhere:

$$M \pm R \pm C - E = \pm S \qquad\qquad (Eq.1)$$

where M is metabolic rate, R and C are radiation and convection, E is evaporation and S storage of heat within the body. Over time homeothermy demands that S = O.

It has been demonstrated experimentally that the above rates of energy exchanges to and from the environment are determined by four atmospheric parameters in particular; air temperature, humidity, air movement and radiation fluxes. The two other main physical determinants are the amounts of clothing insulation worn, and of course, the rate of heat production by work done. A variety of coefficients are available nowadays from laboratory and field observations for each variable. These may be combined to produce mathematical estimations of strain within the body, or to enable quantitative assessment of the thermal stress of particular environments.

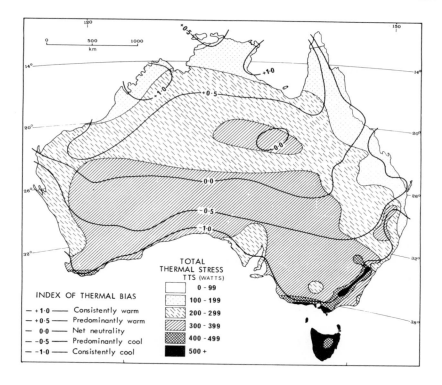

(Auliciems and Kalma 1981)

FIGURE 1.   Daytime thermal stress on outdoor man[α]

[α]Total thermal stress = cold stress + heat stress (calculations represent average energy flux for naked persons facing sun at 0900 and 1500 local standard time in mid-July and mid-January).

Application of the energy exchange equation on a continental scale using an adaptation of Givoni's (1969) Index of Thermal Stress certainly does show mostly positive S values in the North. Except for parts of the northwest, however, some subjectively perceived cool weather is experienced during the low sun season (i.e. in Figure 1 the isopleths of the Index of Thermal Bias show values <+1.0). Indeed to outdoor man with minimal clothing, thermal stress (i.e. cold stress + heat stress in absolute terms) in the North is considerably lower than in the southern part of the continent.

Thus, although the exposed individual heat stress obviously does occur during the height of summer in all areas, in the quantitative terms of energy flux the climates of the

North are not inevitably as forbidding as believed by some
determinists. In large part adequate compensation should be
possible by microclimate modifications often in terms of
relatively simple rationalization in orientation, shading,
insulation, ventilation and only minimal active energy inputs
for cooling (see previous chapter, *The Built Environment of
Climatically Stressed Ecosystems*, by Szokolay).

C.   The Thermal Constancy Hypothesis

Naturally the quest for quantification of optimum thermal
environments also has been applied to building design and
specifications of performance by heating ventilating and
cooling devices. The notion of human constancy is a convenient one, and, as in the above outdoor classification, it has
become widely used by heating and ventilating engineers
as well as climatologists.

Thus for example Fanger (1970) in discussions of his
Predicted Mean Vote (P.M.V.) technique[1] has argued that preferred temperature as well as energy budgets regress to a
universal optimum value once the six variables of heat
exchanges as defined earlier are accounted for. Neutrality[2]
it is claimed can be predicted by mechanistic calculations
alone, Table 2. Repeated claims from laboratory studies have
been made that age, sex, race and acclimatization, i.e. the
general thermal experiences of individuals, do not alter the
temperature of thermal neutrality nor the preference of
particular body states or sensations as shown in Table 3 (e.g.
Fanger, 1973, Fanger, Højberre and Thomsen, 1977, McNall *et
al*, 1968, Nevins *et al*, 1966).

In general, the influential American Society of Heating,
Refrigerating and Air Conditioning Engineers (ASHRAE) has
agreed with this, and for a long time had advocated constant
thermal levels indoors. Recent concessions that a general
reduction of 2°-4°C in winter could be maintained (ASHRAE
1981) is related more to perceived economic advantages of
reduced indoor : outdoor temperature gradients and seemingly
does not represent a shift in philosophy.

The constant temperature hypothesis has led also to a
disregard of spatial differences. In the Carrier System
Design Manual, for example, summer and 21°C with 20% in the
heating season are recommended across the whole of the
Australian continent.

Such simple approaches to definition of optimum environments also appear to have suited earlier European "pioneers"
to the North. The necessity of air cooling to typically
British concepts of comfort was advocated by Griffith Taylor
(1920) and Markham (1947) who otherwise saw no prospect for

TABLE 2.   *ASHRAE and Bedford Sensation Scales*

ASHRAE scale	Bedford scale	Common numerical coding
Hot	Much too warm	+3
Warm	Too warm	+2
Slightly warm	Comfortably warm	+1
Neutral	Comfortable	0
Slightly cool	Comfortably cool	-1
Cool	Too cool	-2
Cold	Much too cold	-3

TABLE 3.   *Assumed relationships in traditional thermal comfort research[a]*

Environmental warmth	Thermoregulatory response	Thermal sensation	Assumed comfort level
hotter than neutral	sweating	warm-hot	unacceptable
nearly neutral	vaso-dilatation	slightly warm	acceptable
neutral	minimal	none	maximum
nearly neutral	vaso-constriction	slightly cool	acceptable
colder than neutral	thermogenesis	cool-cold	unacceptable

[a]*From Nishi and Gagge (1977) and Auliciems (1981).*

their settlement, in the long term at least. The seeming cornucopia of abundant fossil fuels of the middle Twentieth Century certainly did not hinder the spread of air cooling equipment, with recent trends indicating a market penetration up to 80% over the next decade (Auliciems and Kalma, 1981).

Yet surprisingly, neither the benefits of air conditioning nor its temporal and spatial usage patterns, have ever been evaluated in terms of human impacts.

D.   *Variable Human Response*

As discussed by Cilento (1925) and more recently by Sargent (1963) amongst others, homeothermic man obviously does not merely passively accept adversity of thermal stress.   His tendency is to reduce its impacts by dynamic responses which may be conveniently classified into two main groups of thermoregulatory mechanisms:

physiological and	*behavioural*
Vasomotor activity	voluntary postural adjustment
cardiovascular change	alteration of activity
thermogenesis	relocation
sweating	interposition of clothing
involuntary postural	change of diet
adjustment	construction of housing
	heating and cooling

Air cooling is only one of several options and man is capable of choosing any of the alternatives, either singly or more often as observed in sets of responses.   Adjustment clearly will be related to a host of factors including individual and cultural preference, physical fitness, types of activities envisaged, economic circumstances, resource availability and general technological levels.

E.   *Prediction of Acclimatisation to Prevalent Warmth*

That these adaptive mechanisms may actually play a role as important as that of the physical energy budget parameters has emerged repeatedly in field surveys of subjective responses of larger samples of people in their everyday environments as opposed to artificially simplified climate chambers. The results of such field surveys, conducted around the world during the last 50 years and consisting of some quarter-of-a-million individual responses, particularly on the Bedford (1936) and other subjective scales of warmth as shown in Table 2, have been summarized by Humphreys (1975, 1978) and subsequently updated and reanalysed by Auliciems (1981, 1983).

These results are graphically shown in Figure 2 where each point represents a neutral temperature for a single field survey. A considerable diversity of neutralities is observable ranging between 18° and 32°C and closely related to the prevailing warmth of a place as estimated either by indoor or outdoor air temperatures.   It is particularly pertinent to note that the available information from Australia conforms well to the established relationships with indoor and outdoor temperatures as predicted by least square estimates of regressions.

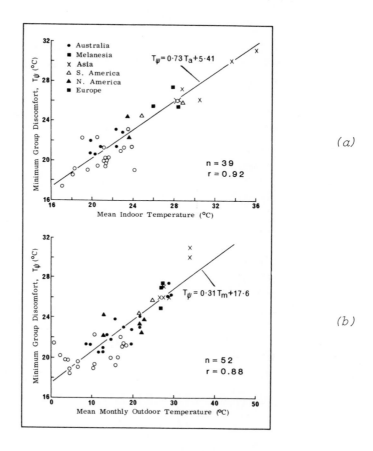

(a)

(b)

FIGURE 2.   *Group neutralities as functions of prevailing atmospheric warmth (details in Auliciems 1983, Humphreys 1975, 1978).*

The remarkably high correlation coefficient of 0.92
between neutrality and mean indoor dry-bulb temperature exceeds even those values obtained in controlled laboratory
experiments with far more homogenous samples than those upon
which field studies have been based. Man quite evidently
adjusts well to his environment: calculations of energy
fluxes fail to explain the 14°C observed differences. The
best single predictor of thermal neutrality is:

$$T\psi_{ii} = 0.73T_a + 5.41 \qquad\qquad (Eq.2)$$

where $T\psi_{ii}$ is group neutrality as predicted by mean
indoor dry-bulb temperature t_a as measured over periods
of several weeks.

The association of neutrality with environmental warmth
is even further enhanced when allowance is made for mean outdoor temperatures. In multiple correlation R = 0.94 and all
terms in the equation below again are higher significance
than P<.0001.

$$T\psi_{iii} = 0.48T_a + 9.22 \qquad\qquad (Eq.3)$$

where $T\psi_{iii}$ is temperature of group neutrality as predicted by T_a and T_m and where the latter is monthly mean
temperature as calculated in the usual manner from
maximum and minimum observations.

Alternatively, although in multiple regression T_m only
served to explain an additional 5-10% variability, its
predictive value of neutrality ($T\psi_{iv}$) as a single independent
variable is also remarkably high (r = 0.88):

$$T\psi_{iv} = 0.31T_m + 17.6 \qquad\qquad (Eq.4)$$

II.   DESIGN AND THE DARWIN STUDY

A.   *Scope and Objectives*

With particular reference to air conditioning in
northern Australia, assessment of thermal environments and
their management requires answers to several questions
including:

(1)   What thermal conditions may be defined as optimum
for facilitating homeothermy?
(2)   What conditions are subjectively defined as the
most desirable?
(3)   What are the perceived and measurable costs and
benefits of air conditioning and especially air cooling?
(4)   To what extent are such conditions achieved?

(5)   What are the economic and energy implications of
current practices and possible alternatives?

These broad questions are currently being investigated
by the authors in several locations including Darwin,
Melbourne and Brisbane. The present report is focussed on
the former as representative of northern Australia,
especially those parts experiencing a monsoonal buildup of
hot-moist atmospheres. These discussions should also be
considered in relation to the earlier physical climate
description by Lee and Neal and the previous chapter by
Szokolay on house design. Obviously extrapolation of the
present information to other workplaces, homes and locations
needs to consider the diversity of physical and biological
characteristics of climates, dwellings and inhabitants.

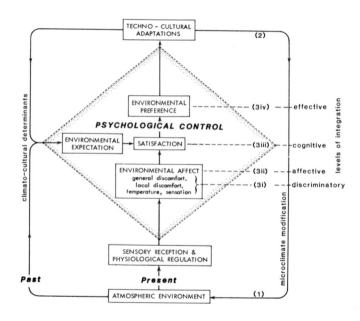

Based on Auliciems (1981, 83)

*FIGURE 3.   Proposed general model of human factors in
atmospheric modification.*

The study was designed to yield information on components of the model shown in Figure 3. This represents a proposed general model of human adjustments involved in atmospheric modification and one based on the previously discussed concept of variable-neutrality psychological control in thermoregulation (see 1F above). The brackets in Figure 3 indicate methods employed as summarized in Table 4.

TABLE 4. Conceptual Framework for Methods Employed in Darwin Surveys

Component in General Model (Fig. 3)	Method of Assessment
(1) Atmospheric environment [Thermal environment]	(i) microclimatic measurement (ii) weather data
(2) Techno-cultural adaptation	(i) questionnaire items on residence times, types of housing, use of air conditioning (ii) clothing insulation inventory (iii) metabolic inventory

(3) Psychological control
    discriminatory level
 	
    affective level
 	
    cognitive level
 	
    effective level
	(i-ii) sensation scales ASHRAE, Bedford
	(ii-iii) Equal interval attitude scales
	(iii) overall assessment of thermal impressions, open ended attitude questions
	(iv) preference scale, directed questions

Increasing integration (vertical label between the two columns in component 3)

Although specific levels of integration are differentiated in the scheme, obviously there is considerable overlap both in the psychological processes involved and in the methods employed in the study. Thus Figure 3 and Table 4, and subsequent references within the text by bracketed numbers, should be regarded more as a general guide to research procedures and less as a definition of discrete concepts.

Obviously also, since the primary functions and the energy costs of air conditioning are related to lowering of air temperatures, discussion in large part is devoted to thermal considerations. Unavoidably, however, the non-thermal considerations need come into play when people are asked to express attitudes towards microclimatic controls in general.

B. *Survey Design*

Two field trips to Darwin were timed to coincide with the least thermally stressful season ("the Dry") in July and August and most trying season called "the Buildup" or more colourfully "the suicide season". The latter usually falls in late October through November until December by which time the monsoonal rain or "the Wet" has set in. The typical course of the seasons in Darwin already has been defined by Lee and Neal in Chapter 3.

Air conditioned buildings were selected from those owned or leased by the Northern Territory Public Service. Permission to interview employees at their customary workplace was granted by the N.T. Public Service Commissioner and relevant Government Department Heads. The final choice was made upon advice of officers in the Department of Transport and Works, resulting in 15 buildings, two of which were libraries and the remainder office spaces.

Interviewers attempted to see each of these respondents on three separate days and on each contact the full "thermal" questionnaire was completed, which took on average about five minutes.

C. *Physical Measurements*

The microclimate data (1i in Table 4) collected were aimed at defining the convective, radiant and evaporative components of the heat balance as shown in Equation 1. Ambient air and wet bulb temperatures were recorded with an aspirated Assman Psychrometer in the usual manner. Radiant temperatures were assessed by a custom-built miniature (38 mm) globe thermometer which yielded comparable results at much quicker reaction times than is normally the case with the traditional 15 cm diameter Vernon globe (Dedear, in press). These globe readings were converted to the more familiar mean radiant temperature with an equation which made the appropriate adjustment for the increased ratio between convective losses and radiant gains[3]. Air velocity was measured by taking the average of three separate readings using a Kurz hot-wire anemometer.

Clothing insulation and metabolic heat production related to the subjects, not the thermal environment. Both were assessed by means of questionnaire, the former consisting of two clothing item checklists, one for each sex. Each item in the lists had a clothing insulating value in clo units as observed in laboratory studies (Sprague and Munson, 1974, Nishi and Gagge, 1977). Metabolic heat production was also assessed in the field by using a check list, in this case consisting of various indoor and outdoor activities typical in an office worker's day and covering the period one hour immediately preceding the interview. Heat production in watts per square metre of body surface area was assessed by reference to several empirically observed rates according to the nature of activity involved (e.g. Astrand and Rodahl, 1977, Durnin and Passmore, 1967).

Outdoor physical parameters (1ii) consisted of air temperature, wet-bulb temperature, wind speed and cloud cover recorded at Darwin Airport by the Bureau of Meteorology. Since no building was further removed from this site than 4 km, it is believed that the latter's observations closely approximate those at sample locations.

D.   *The Questionnaire*

Both Darwin surveys used essentially the same question-naire as based on pilot work in Brisbane and Darwin. The first interview with each respondent took on average 20 minutes to complete. During the two follow-up interviews only a subsection of the original form was administered and took five minutes to finish. Respondents were generally enthus-iastic about participation to the extent that less than ten of the total 380 selected individuals refused to co-operate.

The questionnaire consisted of four main parts described below in rough order of presentation.

1.   Subjective Assessment of Thermal Conditions (3i) at the time of interviewing which could be related to simultane-ous observations of physical conditions. The extensively used ASHRAE and Bedford Scales as already shown in Table 2 were used with reference to microclimatic conditions at that time. Both thermal scales were presented in a linear or continuous format and respondents were informed that they could vote either directly on the labelled points of the scales or in between them. This allowed a greater degree of sensory discrimination for respondents who wanted it, and also came closer to fulfilling the assumptions necessary for parametric statistical analysis.

In addition the respondents were asked to assess upon similarly structured scales their perceived humidity, indoor air motion and impressions of stuffiness/freshness.

2.   *An Inventory of Clothing Insulation* (2ii) according to predetermined values for particular garments and of *activity levels* (2iii) in three ten-minute intervals in the previous half hour.

These latter responses were subsequently employed to estimate metabolic heat production per unit surface area and together with insulation values were used to calculate average P.M.V.[1]

3.   *Classification Questions* including age, sex, length of residence in Darwin and various other factors relating to the recent *thermal experience* of respondents including use of airconditioning at home, house type and building materials (2i).

4.   *Attitudes to Air-Conditioning*[4] both at the work place and the home were assessed with a variety of techniques which included the following:

a.   *Free-answer* questions on the perceived cost of benefits of air conditioning (3iii). The questions were given as "What are the (disadvantages/advantages), in your opinion, of air conditioning in (your workplace/the home)?" The respondent was prompted for three disadvantages and three advantages for both home and workplace, and the answers were transcribed verbatim. A sample of 100 completed questionnaires was used to establish a coding frame for these date (Oppenheim, 1980).

b.   *Specific* questions relating to the months of the year in which air conditioning was preferred (3iv).

c.   *Assessment* of overall impressions of warmth on the ASHRAE scale (3iii).

d.   *Equal-interval* attitude scales on air conditioning (3ii). Here a pair of scales was used, one specifically dealing with office air conditioning and the other with home air conditioning. In the case of the latter, no attempt was made to differentiate between various degrees of home air conditioning; room air conditioners through to ducted systems were included in the subject of the scale. The same generality was sought in the design of the office air conditioning scale, so that the instruments could be broadly interpreted as measuring the affective components of global attitudes towards both applications of air conditioning.

The office version of the scale was introduced to the respondent with: "The following list of items is designed to

explore some of your feelings and opinions about office air
conditioning in Darwin at the present time of year".

In contrast to earlier sections of the questionnaire,
the equal interval scales were not restricted to the space-
cooling aspect of air conditioning. Instead the instruments
related to the total concept of air conditioning, which
incorporate issues such as air quality, indoor air pollution,
improvements to the "paper worker's" environment, implications
for floor-space arrangements and perhaps employee productivity.

Overall the "attitudes" section of the questionnaire took
about ten minutes to complete and progressed in from very
general and unstructured feelings about air conditioning to
more specific quantification of the intensity of attitudes as
recommended by Gallup (Oppenheim, 1966). Further particulars
relating to these questions appear in the respective sections
of *Results*.

III.   DARWIN RESULTS AND DISCUSSION

Results of the Darwin part of the study are presented
here in terms of the conceptual framework established in
Figure 3 and Table 4. In parts it is again more convenient
to use alternative headings to those in Table 4, although
for clarity cross-reference should be again made to particular numbers therein. Some assessments such as outdoor
weather, metabolic activity and personal details are not
reported separately, but are incorporated incidentally in
discussions of main results.

Some details of subjects in the two samples are shown in
Table 5. It is believed that the samples although unrelated
were closely matched and well represent the office workforce
in Darwin.

TABLE 5.   Samples in the Darwin Study

	"The Dry" Sample	"The Build-up" Sample
Number of females	99	105
Number of males	75	92
Number of individuals	174	197
Median age (years)	31	32
Mean length of residence in Darwin (years)	9	10
Total number of thermal questionnaires	493	555

A.   Indoor Climates (1$_i$)

A summary of means and variability in physical thermal
measurements is shown in Table 6.

In the 15 air conditioned buildings studied during the
two Darwin surveys, the indoor climate came remarkably close
to the Carrier Manual's Australian summer recommendations of
24°C air temperature with 50% relative humidity. During "the
Buildup" surveys the mean indoor air temperature was 23.7°C
with 56% R.H. while "the Dry" survey recorded averages of
23.3°C and 47% R.H. Strictly applied, since there is no
heating season in Darwin, temperatures were below the Carrier
recommendations in the latter survey, but the degree of over-
cooling is reduced when the average indoor mean radiant
temperature of 23.7°C is taken into account. When this
convergence between recommended standards and air conditioning
practice in Darwin is the result of trial-and-error on behalf
of local building service engineers, or a rigorous application
of the only available guidelines is open to debate. The few
consultants who could be contacted in the course of the two
field trips did give the latter impression, claiming that
locally generated data were needed.

The indoor mean radiant temperatures recorded during both
surveys were on average higher than the ambient temperatures.
The difference was more pronounced in "the Buildup" survey
when it amounted to a full degree. This finding reflects the
increased solar loads placed on buildings during "the
Buildup", but also the effectiveness of external shading
devices characteristic of Darwin's high-rise office buildings.
Without them the difference between ambient and radiant
temperatures would have been considerably greater.

The recorded air velocities indoors were generally below
0.15 metres per second which is fairly typical for air
conditioned buildings. Of interest was the increase in air
velocities during "the Buildup", probably to offset the
slightly increased humidity and radiant loads.

B.   The Clothing Response to Cool Temperatures (2$_{ii}$)

Insofar as *clothing* adjustments represent one of the most
important behavioural responses to the thermal environment,
the survey results warrant further discussion. As shown in
Tables 7 and 8, observations in Darwin's air conditioned
offices indicate that the average clothing insulation values
being worn were higher than the 0.3 - 0.4 clo units[5] which
Fanger (1970) nominated as typical for tropical clothing
ensembles. "The Dry" survey recorded a mean insulation value

TABLE 6. Summary of data from the Darwin Surveys

Variables	The Dry Survey		The Build-up Survey		t statistic between survey means	significance of t at .05 and d.f. 1046
	Mean	Standard Error	Mean	Standard Error		
Mean monthly temperature (°C Met. Bureau)	25.0	N.A.[1]	28.8	N.A.	N.A.	
Outdoor dry bulb temperature during office hours (°C)	27.9	0.105	30.6	0.078	20.6	Yes
Outdoor wet bulb temperature during office hours (°C)	18.1	0.10	25.1	0.066	58.4	Yes
Outdoor wind speed during office hours (knots)	10.1	0.19	7.6	0.16	10.1	Yes
Cloud cover during office hours (eights)	1.7	0.079	3.1	0.071	13.2	Yes
Outdoor Mean Radiant Temperature during office hours (°C)	44.3	N.A.	45.5	N.A.	N.A.	
Ambient indoor air temperature (°C)	23.3	0.064	23.7	0.065	4.4	Yes
Indoor wet bulb temperature (°C)	16.1	0.054	17.8	0.067	19.7	Yes
Indoor globe temperature (°C)	23.5	0.069	24.1	0.068	6.2	Yes
Indoor air velocity (m.sec^{-1})	0.07	0.006	0.14	0.005	9.9	Yes
Intrinsic clothing insulation (clo)	0.49	0.006	0.43	0.004	8.3	Yes
Metabolic rates in hour preceding interview (W.m^{-2})	70.6	0.780	68.7	0.37	2.2	Yes
Indoor mean radiant temperature (°C)	23.7	0.082	24.6	0.088	7.5	Yes
Indoor relative humidity (%)	47	N.A.	56	N.A.	N.A.	
Indoor partial vapour pressure (mmHg)	10.1	N.A.	12.3	N.A.	N.A.	

[1]N.A. = Not applicable or not available.

TABLE 7.   *Typical Clothing Ensembles Worn in Darwin's Air Conditioned Offices*[1]

Men's Ensemble		Women's Ensemble	
Item	clo	Item	clo
Briefs	0.05	Bra and Panties	0.03
Light short-sleeve shirt	0.14	Half slip	0.13
Light long trousers	0.25	Light knee-length skirt	0.16
Shoes	0.04	Light short-sleeve blouse	0.18
Ankle socks	0.04	Panty hose	0.01
		Sandals	0.02
	$\Sigma = 0.52$		$\Sigma = 0.55$

Intrinsic clo = 0.113 + 0.727
(Σ individual items)
= 0.49 clo

Intrinsic clo = 0.05 + 0.77
(Σ individual items)
= 0.47 clo

[1]*Clo values compiled from ASHRAE (1981), Sprague and Munson (1974)*

of 0.49 clo but this reduced to 0.43 clo during "the Buildup". The seasonal decrease was statistically significant at the .05 level of confidence (Table 6). Disaggregating these averages according to sex of respondent men in both surveys were consistently more heavily dressed than women.

Another sex difference in clothing behaviour observed in Darwin is seen in the standard deviations of the insulation values (Table 8). It appears as though women's clothing is more variable than men's, suggesting that women are capable of closer behavioural adaptation to thermal variations.

Common to both sexes was the relatively low level of clothing variability. Standard deviations for the clo values were 0.17 and .14. Whether this clothing uniformity is simply the result of limited clothing options as the population approaches the culturally acceptable minimum, or rather the general lack of variability in indoor climates, is inconclusive on the current data.

TABLE 8.   Mean Clothing Insulation in Darwin

		"THE DRY"	"THE BUILD-UP"	
SEX	WOMEN	Mean clo = 0.47 St. dev. = 0.14 St. error = 0.008 n = 268	mean clo = 0.41 st. dev. = 0.113 st. error = 0.007 n = 301	t statistic = 5.64 significant at .05
	MEN	Mean clo = 0.52 St. dev. = 0.098 St. error = 0.007 n = 211	mean clo = 0.46 st. dev. = 0.069 st. error = 0.004 n = 254	t statistic = 7.44 significant at .05
		t statistic = 4.7 SIGNIFICANT AT .05	t statistic = 6.2 SIGNIFICANT AT .05	

C. Subjective Assessment of Warmth (3_i)

ASHRAE votes observed in the surveys are summarized in Tables 9 and 10.

Overall the Darwin samples found conditions at work below observed neutrality ($T\psi_0$) as calculated by the Probit technique described by Finney (1971) and applied to comfort scales by Chrenko (1955) and Ballantyne *et al* (1977).

As shown in Table 11 there was general agreement between $T\psi_0$ as measured on either sensation scale. "The Dry" survey's neutrality of 24.2°C may be seen to be approximately one degree warmer than mean conditions (see Table 6). "The Build-up" neutrality of 24°C appeared to exceed means by nearly 0.5K and on the average conditions in the offices were interpreted as being "slightly cool" or "comfortably cool" on the average with significant numbers of individuals recording sensations below those usually considered as acceptable (10% in "the Buildup" and 19% in "the Dry") (see Tables 9 and 10).

This increased exposure to temperatures below neutrality may also be related to the considerably increased widths of the -1 probit category in comparison to the +1 category also shown in Table 12. Presumably this may be explained by the earlier noted and seemingly inappropriate increases in insulation levels of clothing for tropical climates.

TABLE 9. *Frequency Distribution of ASHRAE Thermal Sensation Votes in the "Dry"*

Temp.	Mean Vote	-3	-2	-1	0	1	2	3	Total
18.6-19.5	-2.50	3	0	1	0	0	0	0	4
19.6-20.5	-1.57	4	2	4	3	0	0	0	13
20.6-21.5	-1.21	4	9	8	3	5	0	0	29
21.6-22.5	-1.00	10	17	30	27	6	0	·0	90
22.6-23.5	- .75	3	23	43	33	14	1	0	117
23.6-24.5	- .14	1	16	39	35	42	10	0	143
24.6-35.5	.49	0	1	5	29	22	9	0	66
25.6-26.5	.64	0	0	4	7	12	4	1	28
26.6-27.5	2.50	0	0	0	0	0	0	1	1
	-0.43	25	68	134	137	101	24	2	491

TABLE 10. *Frequency Distribution of ASHRAE Thermal Sensation Votes in "the Build-up"*

Temp.	Mean Vote	-3	-2	-1	0	1	2	3	Total
19.6-20.5	.00	0	0	0	1	0	0	0	1
20.6-21.5	-.95	1	2	4	6	0	0	0	13
21.6-22.5	-.86	5	22	41	30	11	1	0	110
22.6-23.5	-.41	3	9	53	54	28	5	0	152
23.6-24.5	-.26	5	6	28	51	30	3	0	123
24.6-25.5	.31	0	0	11	24	22	7	1	65
25.6-26.5	.85	0	1	1	18	26	14	2	62
26.6-27.5	1.30	0	1	1	0	11	7	3	23
27.6-28.5	2.00	0	0	0	0	2	1	3	6
	-0.16	14	41	139	184	130	38	9	555

D.   *Perceived Advantages and Disadvantages to Air Conditioning in the Workplace (3$_{iii}$)*

As explained in Section 11D4, responses to these open ended questions were coded into broad subject areas on the basis of a sample of completed questionnaires. The frequency breakdowns of perceived advantages of air conditioning in the workplace for both the "Dry" and "Buildup" surveys have been

TABLE 11. *Observed Thermal Neutralities* (T_4)[1] *and Probit Model Slope Terms for Darwin Surveys*

		"The Dry"	"The Build-up"
ASHRAE SCALE	NEUTRALITY TEMPERATURE $T_4(^{\circ}C)$	24.2	24.1
	(95% confidence interval estimate)	(23.9-24.5)	(23.8-24.4)
	B PARAMETER OF PROBIT MODEL	-0.420	-0.413
	(standard error of B)	(0.051)	(0.041)
BEDFORD SCALE	NEUTRALITY TEMPERATURE $T_4(^{\circ}C)$	24.2	23.9
	(95% confidence interval estimate)	(23.8-24.6)	(23.6-24.2)
	B PARAMETER OF PROBIT MODEL	-0.362	-0.360
	(standard error of B)	(0.048)	(0.040)

[1]T_4 *is commonly used to denote the temperature which coincides with the mid point of the 4th category on the 7 point sensation or comfort scales.*

TABLE 12. *Thermal Scale Category Widths (K)*[1]

Scale	Survey	Sensation Category		
		-1	0	+1
ASHRAE	"The Dry"	2.4	2.1	2.0
	"The Build-up"	3.2	2.3	2.4
BEDFORD	"The Dry"	2.2	2.7	1.6
	"The Build-up"	2.3	2.9	1.5

[1]*Widths were derived from successive Probit transition temperatures for the three central categories on each scale.*

collapsed into Table 13. In that table, the raw tallies are presented in three columns according to whether the response was given as the first, second or third disadvantage of air conditioning. The weighted scores in the table were simply derived by multiplying first responses by three, second responses by two, third responses by unity and then adding

TABLE 13. *Perceived Disadvantages of Air Conditioning in Darwin Offices*[1]

| | RESPONSE TALLY | | | | | |
RESPONSE CATEGORY	1st	2nd	3rd	TOTAL 1+2+3	WEIGHTED SCORE	WEIGHTED RANK
No response	40	128	244	412	620	1st
Too cold	62	37	6	105	266	4th
Thermal gradients indoor/outdoors	41	37	23	101	220	5th
General health problems	65	64	38	167	361	2nd
No control of micro climate	47	31	7	85	210	6th
Lack of fresh air odors, stuffi- ness, smoke	63	43	17	123	292	3rd
Breakdowns of air- conditioning plant	29	17	9	55	130	7th
Deprivation of acclimatisation	13	5	11	29	60	9th
Other/miscellaneous	11	9	16	36	67	8th

[1]*"What are the <u>disadvantages,</u> in your opinion, of air conditioning in your <u>workplace?</u>"*

the three products. These weightings were based on the assumption that the first reply given was the most accutely felt disadvantage and so on down to the third response to each question.

General Health Problems emerged as the salient disadvantage of air conditioning in Darwin offices although the prominence given to this grievance could have been biased by the fact that a short health questionnaire was completed earlier in the interview session. Nevertheless, there would appear to be a substantial concern about indoor air quality and this was borne out by discussions with representatives from various employee unions and also articles in the popular press (Brennan, 1982). The typical response in this category hypothesised that the lower levels of fresh air ventilation associated with air conditioning gave rise to a more rapid spread of minor infections, particularly those of the upper respiratory tract. Many respondents went further

to say that once an infection had been contracted, shaking it
off was a very slow process in cool air conditioned environments.   If nothing else, the high frequency of responses in
this General Health Category indicate the need for a carefully
designed epidemiological research project.

Second in the ranking of disadvantages was the "Lack of
Fresh Air, Odors, Stuffiness and Smoke" category. This
response was probably interrelated with the General Health
Problem answer, but clearly has an emphasis on the aesthetics
of air quality. Third in the ranked array was "Too Cold".
Also included in this category were complaints of having
to wear temperate zone clothes in the tropics. The fourth
rank of response was the large thermal gradients between
indoors and outdoors. Typical responses in this category
referred to the 'thermal shock' experienced upon entering
or leaving the building and numerous mentions were made of
the suspected health hazards posed by having to enter a cool
environment with considerable wetness of skin, thereby
increasing susceptibility to "chills" and "sniffles".

Considerable numbers of respondents also listed inability
to control microclimates as a significant disadvantage.
Complaints were made about who was allowed to adjust thermo-
stats and anecdotes were told relating to section supervisors.
In the category about air conditioning plant breakdown the
complaints of inability of individuals to effect changes again
were predominant. In this case fixed and sealed windows may
be seen as the main problem.

Although attracting only 4.1% of total responses, the
identification of deprivation of acclimatization is quite
unexpected. This constituted one of the hypotheses postulated
for the study, but such volunteered opinion came as a
considerable surprise.

Raw tallies and weighted rankings of responses to the
open-ended question on advantages of air conditioning appear
in Table 14. Fewer advantages (540) were offered than disadvantages (701). The rank orders in Table 14 confirm the
notion that beyond thermal comfort and productivity, there is
not a lot that one can say about the advantages of air
conditioning in offices. Nevertheless it would be wrong to
conclude from this substantial difference that the people
surveyed were anti-air conditioning in the workplace because
the quantitative Thurstone or Equal Interval scaling devices
to be discussed later suggest the opposite.

TABLE 14.   *Perceived advantages of air conditioning in Darwin Offices*[1]

	RESPONSE TALLY					
RESPONSE CATEGORY	1st	2nd	3rd	TOTAL 1+2+3	WEIGHTED SCORE	WEIGHTED RANK
No response	20	208	345	573	821	1st
Productivity/ concentration/ ease of paper- work	87	89	8	184	447	3rd
Coolness/comfort	235	42	3	280	792	2nd
Clean environment	5	12	6	23	45	6th
Quietness with shut windows	3	2	5	10	18	7th
Constant temperature	11	6	1	18	46	5th
Others	10	12	3	25	57	4th

[1] *"What are the advantages, in your opinion, of air conditioning in your workplace?"*

E.   *Perceived Advantages and Disadvantages to Air Conditioning at Home (3$_{iii}$)*

The format was identical to that of the workplace, but the pattern of responses that emerged was sufficiently different to justify the separate treatment of the two forms of air conditioning in the current research project.   The definition of home air conditioning given to the respondents was similar to that used in the workplace section in that the refrigeration process was emphasized.   However, the respondents were also instructed to include all types of domestic air conditioning, ranging from full household ducted systems down to the more familiar room air conditioners (RAC).

The results from analysis of these questions are
contained in Tables 15 and 16. The salient feature of Table
15 is the overwhelming priority accorded the "Running Cost"
grievance. It would seem that the high domestic electricity
tariff applied to Darwin residents is keenly noticed in the
running of inefficient appliances like air conditioners.
Many respondents commented on the fact that they were unaware
of just how expensive air conditioning was to run until it
was installed in their home and that, had they known before
the event, more attention might have been applied to the
alternatives. In assessing the statistical significance of
this difference in perception of domestic air conditioning
disadvantages, it should be noted that the rankings of
disadvantages in Table 15 were broadly similar for both owners
and non-owners of home air conditioning, but that there was
a greater concentration of responses in the "Running Costs"
category for owners than was the case for non-owners.

Table 15 indicates that after "Running Costs", "Lack of
fresh air" and "Health issues" were the most frequently cited
concerns. This time, however, especially in the case of
non-owners, deprivation of acclimatization to the tropical
climate was seen as a major drawback.

TABLE 15. *Perceived Disadvantages of Home Air
Conditioning in Darwin*[a]

| | RESPONSE TALLY | | | | | |
RESPONSE CATEGORY	1st	2nd	3rd	TOTAL 1+2+3	WEIGHTED SCORE	WEIGHTED RANK
No response	74	198	308	580	926	1st
Running costs	157	26	14	197	537	2nd
Health issues	26	23	10	59	134	4th
No fresh air/ odors/stuffy	22	35	10	67	146	3rd
Have to seal up house	20	19	6	45	104	7th
Thermal gradients indoors/outdoors	24	22	7	53	123	6th
Noise of machine	2	6	0	8	18	9th
Deprives acclimatisation	29	19	7	55	132	5th
Too cold	4	3	0	7	18	10th
Others	13	20	9	42	88	8th

[a] *"What are the disadvantages, in your opinion, of air
conditioning in the house?"*

TABLE 16. *Perceived Advantages of Home Air
Conditioning in Darwin*[1]

RESPONSE CATEGORY	RESPONSE TALLY			TOTAL 1+2+3	WEIGHTED TOTAL	WEIGHTED RANK
	1st	2nd	3rd			
No response	97	276	352	725	1195	1st
Better night's sleep	121	33	3	157	432	2nd
Clean/dust free house	10	8	4	22	50	4th
Cool, comfort, relief	120	25	1	146	411	3rd
Kills outdoor noise	4	11	2	17	36	7th
More microclimate control	12	6	2	20	50	5th
Better moods, temper	1	5	1	7	14	8th
Others	6	7	6	19	38	6th

[1] *"What are the advantages, in your opinion, of air
conditioning in the house?"*

Respondents also frequently asserted that having air conditioning at home tended to interfere with outdoor lifestyles. Indeed several respondents suggested that their dependence on the machine was causing them to spend disproportionate amounts of time in the bedroom where the devise was installed. Alternatively readers may also appreciate the assertion of a few respondents who blamed air conditioning for marital disruption. It seems that partners with substantially different nocturnal temperature preferences were forced to have separate bedrooms, one with and the other without air conditioning.

Concluding the discussion of free-answer questions in the attitude questionnaire, the tabulated answers on the advantages of air conditioning in the home are presented in Table 16. As with office air conditioning, the respondents found it more difficult to enumerate the advantages (388) of home air conditioning than the disadvantages (533).

F.   *Overall Assessments of Warmth* (3_{iii})

Distribution of the Darwin samples' overall assessments
of workplace thermal conditions are presented in Figure 4.
Both surveys gave negative mean overall assessments (cooler
than neutral) and these results are broadly consistent with
the instantaneous assessments on ASHRAE and Bedford scales
(see Section 111C). However the overall impressions were more
displaced from the centre of the scale in Figure 4 than one
would expect on the basis of the mean temperatures 23° – 24°C
(Table 6) and the observed neutralities shown in Table 11.
This may suggest that either the field measurements taken on
an average of eight occasions per building were systematically
biased towards warm temperatures, or as is more likely, there
was a tendency for residents to "overshoot" in their overall
impressions of thermal discomfort over time.
    Not surprisingly, given the mild conditions during
Darwin's low sun period, people reported home conditions to be
very close to neutral at -.0.2. Equally predictable given
only 40% market penetration of air conditioning in the present
study, a generally poor climatic design of many buildings as
discussed under 111G(2i), and the lack of heat acclimatization
in the early Buildup period, heat stress is often experienced
at home.

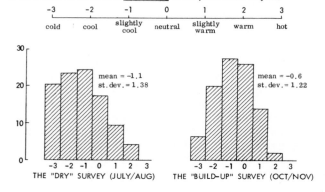

*FIGURE 4.   Overall assessment of thermal conditions
for Darwin office buildings.*

G.   House Types (2_i)

Assuming that heavy construction concrete slab-on-ground designs are in general not desirable in the tropics, it would seem a reasonable hypothesis that occupants of such homes would have greater need for active climate controls such as air conditioning. The classification questions asked of the respondents in the current surveys included items on their home's building materials. Aggregating the two surveys there were 185 respondents who had moderately heavyweight homes. This classification covered brick, brick veneer, concrete, concrete block and stone constructions. The remaining 156 respondents had homes classified as lightweight constructions making use of timber, fibro-cement sheets and enamelled metal-cladding material.

The χ^2 statistic for the hypothesis that there would be proportionally more air conditioners in heavyweight houses than their lightweight counterparts was 0.143 which was insignificant at the 0.05 level of confidence. A possible intervening variable in the hypothesized relationship would be household income, which was not measured in the questionnaire. Since much of the Northern Territory Housing Commission's post-cyclone Tracey housing stock has been of a heavyweight type of construction, the occupants of such could possibly be on lower incomes and thereby less capable of affording air conditioning. That is, household income could be a much more important determinant of air conditioning ownership than simply the warmth or climatic unsuitability of houses.

TABLE 17.   Combined Darwin Surveys:

H_O: There is no difference in frequency of air conditioners between heavily and lightly constructed houses.

	Homes with Heavy Construction	Homes with Lightweight Construction	
Houses Air Conditioned to some extent	fo 71 fe 72.7	fo 63 fe 61.3	134
Homes UNair conditioned	fo 114 fe 112.3	fo 93 fe 94.7	207
	185	156	$\Sigma = 341$

Since no greater incidence of air conditioning could be observed in the heavier constructions, a comparison of the overall warmth impressions of those two home classifications could be expected to show the lightweight constructions to be more comfortable (3iii). The mean impression of warmth for occupants of lightweight houses was 1.0 (n = 80, St.dev. = 1.13) which was below the whole survey's mean of 1.26 which was in turn below the mean impression of "heavy" home occupants (n = 109, $\bar{x}$ = 1.4, St.dev. = 0.96). The difference between "heavy" and "light" warmth impressions had a t value of 2.56 with d.f. 183. This reached significance at the .05 level (See chapter following by Szokolay for theoretical aspects).

H.   *Seasonal Preference for Air Conditioning (3_{iv})*

Clearly once asked during which months the residents would prefer air conditioning, thermal considerations assume a dominant role during the wet season and particularly during "the Buildup".

Preferences peak during "the Buildup" month of November when seasonal heat acclimatization is not yet completed with 90% preference at the workplace and over 60% in the home. But particularly interesting is the difference of preference for "the Dry" period. Here only in isolated cases was air conditioning preferred in the home while some 15 - 20% still wanted it in the workplace. No doubt amongst the latter appear those individuals who actually felt the workplace to be warm during "the Dry" (see Figure 4 and Tables 9 and 10). Probably also evaluation of non thermal attributes of the atmosphere as subsequently discussed (3ii) would have influenced opinion.

I.   *Equal-Interval Attitude Scales (3_{ii})*

The notable difference between "the Dry" and "the Buildup" is a positive shift in distributions. Obviously this reflects people with favourable dispositions towards air cooling both in the home and workplace during the stressful period.

Unlike the findings on monthly preferences as reported in the preceding Section H, the Equal-Interval scales show a majority of respondents actually favouring air conditioning in the workplace even during "the Dry". Here the median scale score was +0.52, which, assuming the true zero on the psychological continuum does coincide with 0 on the scale as established, appears to differ from the earlier results of 19% preference for air conditioning in June–August. The reasons for this apparent discrepancy are probably those of the

increased scope of the Equal-Interval scale as discussed
before. Here elements of affect extend far beyond simple
thermal preference.

In house air conditioning, the response distribution
shown in Figure 8 more closely resembles that in other
attitude questions. There very clearly is a majority of
feeling against it in both "the Dry" and even "the Buildup"
with medians of -0.91 and -0.56 respectively.

IV.   SUMMARY AND CONCLUSIONS

Air conditioning was generally regarded with favour in
office buildings. However it is perceived as having two
related dimensions, one thermal and one non-thermal.
Specifically as regards the latter, perhaps at this time it
is not necessary to do more than note that, within buildings
with sealed windows, mechanical ventilation is obviously
needed. No doubt, the elevated preference is related to this
particular problem.

A.   *Thermal Neutralities and Atmospheric Conditions*

Thermally it is possible to assess environments in Darwin
using five methods based upon

empirical observation of subjective responses
energy balance calculation
mean indoor temperature recordings
mean indoor plus outdoor temperature recordings
mean outdoor temperature recordings.

A comparison of neutral values using all the above is
shown in Table 18. Two observations in particular may be
made. Firstly, observed subjective neutralities may differ
significantly from those predicted by energy balance calculations even using such sophisticated approaches as the P.M.V.
Secondly, persistent exposure to cool temperatures tends to
suppress neutrality below levels customarily associated with
warm environments.

B.   *Argument for Rationalization of Indoor Climates*

This reinforces recently developed models of thermoregulation which recognize the role of prior experience in
addition to the immediate responses of physiological
mechanisms. This is not inconsistent with the processes of
heat acclimatization (e.g. see Edholm 1978, Kerslake 1972), or
with common observations of a seeming "insensitivity" to heat
by longer term residents of the North. It is likely that over

TABLE 18. *Comparison of Neutralities (°C) from Darwin using Five Methods of Assessment*

Method	Input	Equation	"The Dry"	"The Build-up"
1. $T_{\psi0}$ observed (Bedford) (95% confidence limits)	observed sensation votes	probit	24.2 (23.8-24.6)	23.9 (23.6-24.2)
2. $T_{\psi i}$ calculation by heat balance	PMV^1	(1)	24.7	25.6
3. $T_{\psi ii}$ indoor mean temperature	T_a	(2)	22.4	22.7
4. $T_{\psi iii}$ indoor and outdoor mean temperatures	T_a, T_m	(3)	23.9	24.6
5. $T_{\psi iv}$ outdoor mean temperature	T_m	(4)	25.4	26.5

[1]See footnote 2

time physiological and psychological adjustments become enhanced. Presumably such changes are further facilitated by alterations to behavioural responses in rescheduling of activities and other adaptations to life styles more appropriate to the climate of a place. Thus, not surprisingly, even in air conditioned environments, as shown in Figure 2 and Table 18 the best predictor of group neutralities is a composite function consisting of both indoor and outdoor temperatures.

Mean clearly adjusts to the totality of his thermal experiences and the maintenance of indoor conditions simply cannot refer to some idealized condition unrelated to the climate of a location. In Darwin, at least, both house construction and maintenance of sensible indoor temperatures at times fall short of being optimum in terms of human comfort and energy expenditures. What is needed here, as no doubt in other parts of the North, is a rationalization of house design and management approaches as appropriate to climatic and human needs.

C. *Potential for Energy Conservation*

These considerations raise the real possibility of significant energy savings. If indoor conditions, whether achieved by passive or active energy systems, are altered in response to outdoor temperatures, over time neutralities may be expected to drift with acclimatization and thermal expectation according to Equation 4 and the regression line in Figure 2B.

These variable neutralities would in effect reduce the

outdoor-indoor temperature gradient and thus also decrease proportionally, since heat flow is linear to temperature differences, the amount of thermal adjustment necessary indoors. It is possible by such adjustments to conserve up to 40% of energy inputs. Speculating upon Szokolay's (chapter 16) figures for energy consumption in air condition-ing at Pilbara, these savings could represent an overall reduction of some 10% of energies used in the North. It is considered worthwhile, therefore, to provide appropriate design temperatures for the North according to the above criteria in Figure 5.

The potential energy savings with variable neutrality according to Equation 4 are also illustrated in Figure 6 for five northern locations. Comparison is made to a constant indoor temperature of 24°C as found in Darwin with savings defined as that part between the constant and predicted T_ψ curve tending to reduce the indoor and outdoor gradient. In Darwin of course there is no heating season and any savings are those of air cooling, but in colder inland locations the indications are also of substantial reductions in the need for heat inputs in winter. An impression of the magnitude of energy savings may be gained for any month by comparison of areas between curves, i.e. energy savings : energy required.

TABLE 19.   *Potential Energy Savings with Elevated Neutralities*

Indoor/outdoor temperature gradients	"The Dry" °C	"The Build-up" °C
1. $T_o - T_a$	4.6	6.9
2. $T_o - T_{\psi iv}$	2.6	4.2
Gradient 1 minus 2	2.0 (43.5%)	2.7 (39%)

Values from Table 6.

T_o = *average outdoor dry bulb temperature during office hours,* 27.9/30.6

T_a = *average indoor dry bulb temperature during office hours* 23.3/23.7

$T_{\psi iv}$ = *neutralities from equation 4 using mean monthly temperatures,* t_m *at* 25.0/28.8

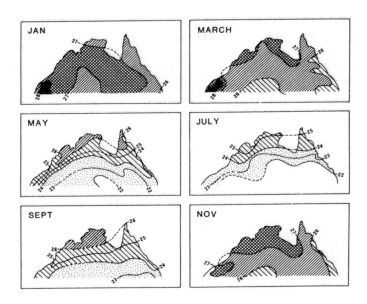

FIGURE 5.   *Variable indoor temperature requirements in Australia based on mean activity outdoor temperatures*

D.   *Air Conditioning and Its Control*

Within the home, in either season, air conditioning does not appear to be greatly desired nor does it appear to be essential to the achievement of preferred conditions. Climatically sound house construction is indicated as the appropriate solution.

In the office buildings during "the Buildup" and probably the remaining wet period some air conditioning is necessary. The amount of cooling, however, needs to be considered in terms of thermal neutrality as achievable under variable atmospheres in accord to seasonal changes.

The management of such variable atmospheres clearly requires the abandonment of the concept of thermo*static* control in favour of what should be termed the*rmobile* control (Auliciems, in press).

In any case the present practice in office buildings of maintaining temperatures considerably below neutrality is unnecessary and wasteful. It certainly does not increase indoor comfort, nor facilitate seasonal outdoor clothing adjustment and heat acclimatization, nor reduce temperature gradients.

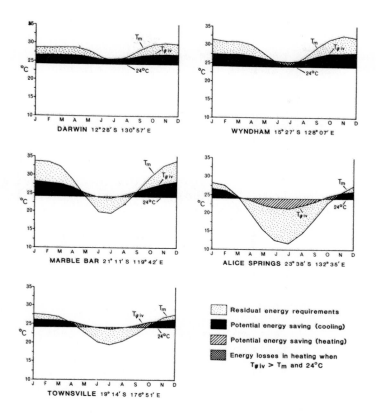

FIGURE 6.   Potential energy savings with thermobile controls

NOTES

$$^1PMV = [0.352.c^{(-0.042\frac{M}{A_{Du}})} + 0.032]\{\frac{M}{A_{Du}}(1-\eta) - 0.35\ [43-$$

$$0.061\frac{M}{A_{Du}}(1-\eta) - P_a] - 0.42\ [\frac{M}{A_{Du}}(1-\eta) - 50] - 0.0023\frac{M}{A_{Du}}$$

$$(44-P_a) - 0.0014\frac{M}{A_{Du}}(34-T_a) - 3.4 \times 10^{-8}f_{cl}\ [(T_{cl}+273)^4$$

$$- (T_{mrt}+273)^4] - f_{cl}.h_c(T_{cl}-T_a)\}$$

where $T_{cl} = 35.7 - 0.032.\frac{M}{A_{Du}}(1-\eta) - 0.18I_{cl}\{3.4 \times 10^{-8}f_{cl}$

$$[(T_{cl}+273)^4 - (T_{mrt}+273)^4] + f_{cl}.h_c(T_{cl}-T_a)\}$$

where $h_c = \{ {}^{2.05(T_cl-T_a)^{0.25}}_{10.4\sqrt{V}} \}$ whichever is larger

A_{Du} is Du Bois body surface area (M^2)

f_{cl} is clothing area factor; ration of clothed body surface area to nude body surface

h_c is convective heat transfer coefficient ($k\ cal.hr^{-1}/M^2.°C$)

I_{cl} is intrinsic clothing thermal insulation (clo)

M is metabolic rate ($K\ cal.hr^{-1}$)

P_a is partial pressure of water vapour in air ($mmHg$)

T_a is ambient dry bulb air temperature ($°C$)

T_{cl} is mean surface temperature of the outer surface of clothed body ($°C$), (solved iteratively with h_c)

T_{mrt} is mean radiant temperature ($°C$)

V is relative air velocity ($m.sec^{-1}$)

η is external mechanical efficiency for body at given metabolic rate

[2]Neutrality for the present is taken to be temperatures at which (with constant air motion, radiation and humidity) maximum responses are observable upon sensation scales as shown in Table 2.

Several designations are subsequently recorded in the text relating to their method of prediction:

$T_{\psi o}$ is subjective neutrality calculated from empirical observation for a particular sample

$T_{\psi i}$ is calculated from measures of parameters involved in homeothermic models

$T_{\psi ii}$ is estimated from mean indoor air temperatures

$T_{\psi iii}$ is estimated from mean indoor plus mean outdoor air temperatures

$T_{\psi iv}$ is estimated from mean outdoor air temperature

[3]The equation used for mean radiant temperature (derived from Hey 1968) was

$$T_{mrt} = \frac{T_g - T_a\ [1-(\frac{1}{1+1.13v^{0.6}\ d^{0.4}})]}{(\frac{1}{1+1.13v^{0.6}\ d^{0.4}})}$$

where T_g is globe thermometer temperature ($^\circ C$)
   d is diameter of globe (m)
   v is relative air velocity ($m.sec^{-1}$)
   T_a is ambient dry bulb temperature ($^\circ C$)

[4]The term air conditioning was operationally defined for the respondents as the refrigeration process which controlled temperature, humidity and circulation speed of air in a room or building.

[5]One clo is defined as the amount of insulation offered by a business suit with corresponding underwear, or an average heat transfer coefficient of 6.46 $Wm^{-2}K^{-1}$.

ACKNOWLEDGEMENTS

Project supported by University of Queensland Special Project Grant 1982, Social Sciences Research Grants 1982 and 1983 and Australian National University North Australia Research Unit Field Grant in 1982.
We are grateful to the following institutions and individuals: Northern Territory Government for permission to use various buildings and occupants in surveys, J.W. Spencer of the Division of Building Research C.S.I.R.O., A. Barnes of the Science Faculty and S.V. Szokolay of the Department of Architecture of the University of Queensland for statistical and technical advise, P. Loveday and the North Australia Research Unit for hospitality and accommodation during the field work, also to E. Savage of the Department of Geography, University of Queensland, for graphical work.

REFERENCES

ASHRAE, 1981. *Handbook of Fundamentals*, American Society of
   Heating, Refrigerating, and Air Conditioning Engineers,
   Atlanta.
Astrand, P.O. and Rodahl, K. 1977. *Textbook of Work
   Physiology*, 2nd ed., McGraw-Hill, New York.
Auliciems, A. 1981. *Int. J. Biometeor*, 25, p.109.
Auliciems, A. 1983. *Proceedings of the Ninth International
   Society of Biometeorology Conference*, Stuttgart-
   Hohenheim, September-October, 1981.
Auliciems, A. (forthcoming). *Thermobile controls for human
   comfort*.
Auliciems, A. and Kalma, J.D. 1981. *Australian Geographical
   Studies*, 19, p.3.

Ballantyne, E.R., Hill, R.H. and Spencer, J.W.   1977.   *Int. J. Biometeor,* 21, p.29.

Bedford, T.   1936.   *Rep. Ind. Hlth Res. Bd.* No. 76, London.

Brennan, P.   1982.   *National Times,* July 11-17, p.13.

"Carrier System Design Manual",   (undated)   Part 1, Load Estimating and Psychrometrics, Australia.

Chrenko, F.A.   1955.   *J. Instn. Heat. Vent. Engrs.* 23, p.281.

Cilento, R.W.   1925.   The White Man in the Tropics, *Department of Health Service Publication (Tropical Division) No. 7,* Commonwealth of Australia.

Dedear, R.   (forthcoming)   *Ping-Pong Globe Thermometers for Mean Radiant Temperature.*

Durnin, J.V.G.A. and Passmore, R.   1967.   *Energy, Work and Leisure,* Heinemann, London.

Edholm, O.G.   1978.   *Man - Hot and Cold,* Edward Arnold, London

Edwards, A.L.   1974.   In *Scaling: A Sourcebook for Behavioural Scientists,* (ed.) G.M. Maranell, Aldine, Chicago, p.113.

Fanger, P.O.   1970.   *Thermal Comfort,* McGraw-Hill, New York.

Fanger, P.O.   1973.   Annexe au Bulletin de l'Institut International du Froid, *Proceedings of Meeting IIR (Commission E1), Vienna.*

Fanger, P.O., Højbjerre, J. and Thomsen, J.O.B.   1977.   *Int. J. Biometeor,* 21, p.44.

Finney, D.J.   1971.   *Probit Analysis,* 2nd Ed, Cambridge University Press, Cambridge.

Givoni, B.   1969.   *Man, Climate and Architecture,* Elsevier Publishing Company, Amsterdam.

Guilford, J.P.   1954.   *Psychometric Methods,* 2nd Ed, Cambridge University Press, Cambridge.

Hey, E.N.   1968.   *J. Sci. Inst. (J. Physics E.),* Series 2, 1, p.955.

Humphreys, M.A.   1975.   *Building Research Establishment Current Paper CP76/75,* U.K. Department of Environment, Garston.

Humphreys, M.A.   1978.   *Building Research and Practice,* V.G (2) p.92.

Huntington, E.   1924.   *Civilization and Climate,* Yale University Press, New Haven.

Huntington, E.   1926.   *The Pulse of Progress,* Vimpson, New York.

Kerslake, D. McK.   1972.   *The Stress of Hot Environments,* Cambridge University Press, Cambridge.

Markham, S.F.   1947.   *Climate and the Energy of Nations,* Oxford University Press, London.

McNall, P.E., Ryan, P.W., Rohles, F.H., Nevins, R.G. and Springer, W.E.   1968.   *ASHRAE Trans,* 82, p.219.

Nevins, R.G., Rohles, F.H., Springer, W.E. and Feyerherm, A.M. (1966. *ASHRAE Trans*, 72, Pt. 1, p.283.

Nishi, Y. and Gagge, A.P. 1977. *Aviation, Space and Environmental Medicine*, V.48, p.97.

N.T.E.C. 1982. *Northern Territory Electricity Commission Annual Report 1981-1982*, Northern Territory Government, Darwin.

Oppenheim, A.N. 1966. *Questionnaire Design and Attitude Measurement*, Heinemann, London.

Sargent, F.I. 1963. *Proceedings of the Lucknow Symposium*, UNESCO, Paris.

Sprague, C.H. and Munson, D.M. 1974. *ASHRAE Trans*, 80 (1), p.120.

Taylor, Griffith 1920. *Australian Meteorology*, Oxford University Press, London.

Taylor, Griffith 1959. *Australia*, 7th ed. Methuen, London.

Toynbee, A.J. 1945. *A Study of History*, Vol. 2, 3rd Ed. Oxford University Press, London.

EDITORIAL NOTE

Following photoreduction of tables it was noted that foot-notes to some tables were indicated by a numeral and not an α symbol as in all other tables. Please accept my apologies on behalf of the authors.

18
ASPECTS OF THE HUMAN ECOLOGY
OF REMOTE SETTLEMENTS

DON PARKES

Department of Geography
University of Newcastle
New South Wales

I. INTRODUCTION

Through zone allowances, the Australian Taxation Office
recognises that certain disadvantages are experienced by
those who live in the remote regions of Australia. Figure 1
is based on a map produced by the Taxation Office. It
indicates the areas which attract higher basic rates of
rebate on income tax. The hatched areas are more than 250
kilometres by straight line distance from a settlement of
2,500 or more people. For northern Australia, Zone A therefore defines the official *remote regions*. A claim may be
made for a remote area tax allowance by any person who
"resides or spends time" at a place which is more than 250
kilometres from the nearest population centre of 2,500 or
more people, measured along the "shortest practicable surface
route". About 70% of northern Australia falls within Zone A.

This chapter discusses aspects of the remote settlement
as a distinct geographical and ecological *habitat*, and in
particular reference is made to two places, Paraburdoo in the
Pilbara region of Western Australia and Tennant Creek in the
central zone of the Northern Territory (Figure 2).

Discussion centres on the *pre-perceptual* environment of
the habitat. This pre-perceptual or *ecological* environment as
Barker (1979) has described it, "consists of the regions [and
the objects, and people] surrounding a person as they exist
independently of his perception of them". The *location,* the
area of the space occupied, the *time,* the *duration* of an
event, the *numbers of persons* present and the *numbers and type*

NORTHERN AUSTRALIA
ISBN 0 12 545080 X

of material objects which are necessary to behaviour are all
examples of elements of the ecological environment which
have properties common to all individuals, the observed and
the observers.   This ecological environment is distinct from
the *psychological* environment, sometimes called the perceived
environment which is *constructed* from questionnaire surveys
and other experimental designs.   It is an approximation to
the ecological environment.   In the following chapter by Neil
et al it is the psychological environment which is discussed.
Parkes and Burnley (1981) and Parkes (1981, 1983) have also
reported on aspects of the psychological environment of Alice
Springs.
    Barker has proposed that the "psychological environment
[includes] the regions and objects outside the person as
they are perceived by him" (Barker, 1979).   He argues that
human *molar* behaviour is associated with two environments,
the ecological and the psychological.   To date almost all of

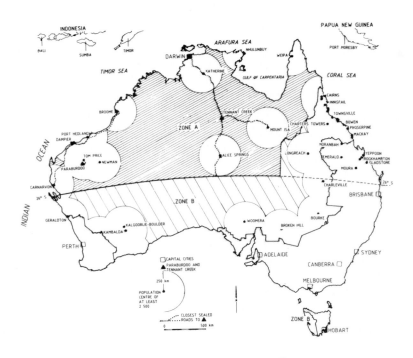

FIGURE 1.   *Remote Zones of Australia*[a]

[a]*Derived from a map prepared by the Australian Taxation
Office.*

the published research on remote settlements in Australia has been derived from attitude surveys which contribute to the construction of the psychological environment. The encompassing ecological environment, which is independent of the individuals within it, is the prior environment. The psychological environment relates to the ecological environment. Molar behaviour occurs in the ecological environment and is adapted from time to time by the adjustment of elements in the ecological environment as the result of perceived advantages; for instance a relocation in space or time. It involves the presence of the whole person within a measurable, spatially and temporally bounded zone. Material objects are also always present and each has dimensions which are measurable independently of any *particular* observer. Similar properties apply to the people who occupy the ecological environment within what are known as *behaviour settings* (Barker 1968). The behaviour which occurs within each setting is *regulated* by rules and observable conventions. Individuals bring a set of variable personal, psychological properties into the ecological environment and also develop new psychological properties and attitudes, from contact with it.

The behaviour setting is an integral part of the ecological environment. "[It is] a bounded, stable, objective entity whose properties are independent of the perceptions of the inhabitants; nonetheless they coerce the inhabitants" (Barker 1979).

Associated with the concept of behaviour settings is the notion of *manning* levels. In remote settlements *undermanning* is possibly a positive factor in the ecological environment, in remote places. Favourable comments about social relationships, friendliness, "the wide range of things to be involved in" are typical responses in field questionnaire studies which have been undertaken in Australia's remote communities (Brealey, 1972; Brealey and Newton, 1978; Parkes, 1981). Undermanning may well be one important explanatory factor for these findings.

In the space available here it is not possible to present an analysis of the ecological environment of remote habitats. A forthcoming book (Parkes, forthcoming) on the human ecology of remote settlements in the Australian arid and semi-arid regions will include a detailed study of the ecological environment of a remote mining settlement. The observations presented here are derived from field work in Paraburdoo and Tennant Creek during July, August and September 1983. Earlier periods of field work in northern Western Australia and in central Australia in 1980, 1981 and 1982 also make their contribution.

The objective of the chapter is to create for the reader a reasonably focussed image of remote settlements in northern Australia; hopefully one which will be acceptable to those who have first hand experience of living in such places. The chapter does not aim to provide solutions to any particular problems. So-called social pathologies, conjectured or real which are so often the focus of comment in academic research and in the media are not addressed here. There is no claim that the image which is to be built, is complete. There is no advocacy for larger or smaller settlements. There is no advocacy for more or fewer settlements.

Evidence from the recent past suggests that in the decades ahead many new ecologically complex but possibly quite small settlements will be built in Australia, Outback. Their built forms and the activities which take place within them will be as different from those of the present as the recently developed remote settlements are from those of the past. Parkes (1983C) for instance discusses the possibility of settlements which are mobile, integrated complexes; they are designed for change in both *location* and *ecology*. Better understanding of the components and mechanisms of the ecological environment of existing remote settlements will aid design and management of future habitats for a 'life after death', so to speak. The really difficult task associated with the building of a new settlement is not its physical appearance or its mechanical impact on the natural environment, but the preparation of a stage for flexible behaviour settings.

II.   CONTEXTUAL SETTINGS FOR TWO REMOTE SETTLEMENTS

1. Physical Settings. Across northern Australia, human settlements are found in a variety of climatic and physiographic settings. In chapter 3 Lee and Neal have discussed the climate of northern Australia and Auliciems and Dedear have discussed aspects of the relation between climate, building and human behaviour in chapter 17.

Laut *et al* (1980) have prepared a provisional classification of Australia into environmental regions, combining a common set of elements relating to climate, terrain form and features, lithology, soils, natural and cultural vegetation and land cover. Figure 2 shows the 33 provisional environmental regions of northern Australia, derived from their analyses.

The hatched regions lie within a 250 kilometre radius of Paraburdoo and Tennant Creek. Plate 1 illustrates some typical landscape in the vicinity of each place.

With few exceptions, notably in the south-east of northern Queensland, these environmental regions each extend over a considerable area. This homogeneity of environment is itself

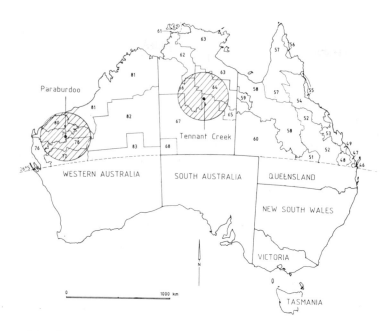

FIGURE 2. *Provisional Environmental Regions of Northern Australia*[a] *and the Situation of Paraburdoo and Tennant Creek.* [a]*Based on Laut et al (1980).*

Pilbara Landscape near Paraburdoo, W.A.

Barkly Tableland Landscape near Tennant Creek, N.T.

(Photos: Author)

PLATE 1. *Typical Landscapes near Paraburdoo and Tennant Creek.*

considerable area. This homogeneity of environment is itself
an ecological factor which influences behaviour of residents
in the remote settlements. "You have simply got to go a hell
of a long way to see anything different, and that costs
heaps", is a typical comment.

Paraburdoo is situated in region 79 and extends over
75,000 km^2, "inland along the Ashburton River into the
headwaters of the Fortescue River. Its landscapes vary from
alluvial and erosional plains to rugged hill lands. On the
plains the low euclaypt and acacia woodlands are used for
extensive livestock grazing. Much of the rugged hill lands
are not used and there is a large national park between the
two major iron mining centres of Mt. Tom Price and Mt. Newman
(Laut *et al*, 1980, p.160).

As Tennant Creek is situated in the northern-most portion
of region 67, its ecological setting is also related to
regions 64 and 66. Region 64 is the Barkly Tableland. This
is a "region of plainslands and overmixed sedimentary rocks
with extensive areas of alluvial plains. Its natural vegetation cover of astrebla grassland and low eucalypt woodland
are used for extensive livestock grazing". The region extends
for nearly 140,000 km^2, eastwards to the Queensland border.
Region 66, an area of 204,110 km^2, includes "the catchment of
the Victoria River and is predominantly plainsland but has
some hill lands in the south west. In the south east [near
Tennant Creek] the plains are frequently overlain by sand
ridges ... It has a natural vegetative cover predominantly of
open eucalypt with some acacia shrubland and tussock grasslands. In the north it is used for extensive livestock
grazing, elsewhere the vegetation is unmodified" (Laut *et al*,
1980, p.158).

2. *Human Settings*. The most pervasive features of the
environment of northern Australia which affect human settlement are the great distances which separate places and the
small sparsely distributed population. Away from the
coastal towns of eastern Queensland, urban settlements are
either *service towns* to the pastoral industry, such as
Katherine, Camooweal, Wyndham, and to an increasing extent,
Tennant Creek, though its recently established abbatoirs are
fraught with difficulties associated with industrial disputes,
government and *tourist* industry centres such as Alice Springs,
and the many *new mining* towns which have developed during the
past two decades. The most impressive swarm of new mining
towns has developed in the Pilbara. Until recently all of
these new mining towns were *closed* towns, managed by mining
companies independently of a Local Government Council.
Newman, Tom Price, Pannawonica, Paraburdoo, Dampier; all are

examples of mining settlements built between 1962 and 1972 and
now operated as *open* towns following what has become known as
a *normalisation* procedure. This process of normalisation has
recently occurred in Paraburdoo and Tom Price with agreements
being signed between Hamersley Iron Pty. Limited, and the
West Pilbara Shire, in November 1982. Effective transfer of
town management to the Shire occurred in June 1983. This
change from a closed to an open town may induce a major
change in the ecological environment of the settlement as the
authority for environmental management shifts from the company
to an elected council. For instance, in July and August 1983
residents of Paraburdoo expressed concern that the 'new
management' would not be able to maintain the high standard
of material quality in the townscape which 'The Company' had
been able to undertake. The change from closed to open
settlements alters the control mechanisms by which the
ecological environment is related. For instance, so-called
cottage industries are a substantial part of the economic
base (the non-basic or town serving part) of the settlement's
economy. At present, the regulatory mechanisms which applied
when Paraburdoo was a closed town still persist; these
activities are located on residential sites, within the
residence itself. A wide range of goods and services are
provided, among others including sales and servicing of
bicycles, toys, clothes, motor car spares, photography,
hairdressing, home computer games and software, taxation and
investment advice. About 60 cottage industry enterprises
exist. Normalisation could mean the end of such an informal
economic sector of the town if local government zoned areas
have to be used in future, and rents and rates have to be
paid for the facility. About 10% of households in Paraburdoo
are associated in one way or another with cottage industries.
As a closed town the operation of cottage industries was not
subject to local government authority. The mining town
behaved as a distinct human eco-system and appropriate
mechanisms developed which suited that system. Under the
authority of a local government shire, one which is already
responsible for other settlements in its domain, it is
possibly difficult to allow distinctions of the sort which
previously prevailed to continue. A new ecological environment will evolve as regulations relating to the location of
activities, their timing, as in open and closing times, the
material structure and scale of the built space in which
behavioural settings might operate, each replace the existing
self-regulatory mechanisms. The remote mining settlement
which was established as a closed town has a number of
distinct ecological structures which may not be adaptable to
regulations drawn up for more traditional settlements.

Tennant Creek is an open town with its own Local
Government Council. However the administration of the town
through a Local Government Council only dates back to 1978
with the establishment of self-government for the Northern
Territory. Previously it was administered from Canberra.
There was been mining-related settlement in Tennant Creek
since 1925 by fossikers and explorers, but persistently since
1933, first associated with gold and more recently (since
1952) with bismuth and copper at Warrego. Some silver is also
mined. However uncertainty about the future of the township
has been a recurring ecological condition. For instance in
1948 the Northern Territory Administrator (equivalent in most
ways to the Governor of other Australian States) "appointed
an Advisory Committee ... to enquire into the likely life of
the Tennant Creek mining field, the degree of permanency
which the township can be regarded as having [and] the
suitability of the existing township site and other matters"
(Tuxworth 1980).

Recent substantial funding by the Northern Territory
and Commonwealth Governments has been committed to roadworks,
public service buildings, schools and a developing association
with the tourist industry. There is a shift in Tennant
Creek's economic base, away from mining towards tertiary and
quaternary activity. However any future mining developments
will be enhanced by this broadening of the economic base, and
the consequent growth of diversity in the human ecology of the
habitat. At present road transport *dominates* the ecology of
Tennant Creek!

Uncertainty about the future is a common feature of all
mining settlements. High rates of population turnover are
also a typical feature of the ecology of these places. A
sense of isolation is often expressed by many residents. A
lack of political significance and effectiveness appear as
elements of the ecological and psychological environments.
They are an element of the ecological environment because of
small population size and youthful population structure. They
are a part of the psychological environment because a sense of
frustration is experienced. In ecological terms small population and youthful structure mean a lack of *energy* to influence
Federal policy to provide appropriate remote-area-compensating
mechanisms, especially in relation to communications, transport costs and materials.

Youthfulness of the population and the common aspirations
of a large proportion of the population, which were in turn
the initial motivating drive for migrating to the settlement,
are too easily overlooked as positive features of the remote
community. Most migrants to remote mining towns have come
from larger metropolitan areas and bring with them a rich

range of skills and diverse experiences. High population turnover is an ecological resource, a source of renewable energy
to the settlement, acting also as a basis for the development
of new adaptive mechanisms which generate a stability and
continuity to the behaviour settings which compose the living
ecological structures of the town.   There is continuity, a
persistence to behaviour settings, in part at least, *because*
of high rates of population turnover. High population turnover
may be an economic cost to employers but it unduly narrows our
understanding of the way remote mining settlements work to
equate this ecological mechanism with solely negative
products. What may otherwise appear to be a weakly-linked
social system because of frequently fractured networks is also
possibly a human ecosystem which satisfies its members more
than the settlements from which they have moved, usually
voluntarily. Proper evaluation of this relation can only be
established if the time horizon of *intended length of stay* is
taken into account and then addressed in terms of the actual,
day-to-day behaviour of the residents. When the ecological
domain of behaviour is better understood, when we have some
datum from which to undertake further study of the actual
structure of daily ecological behaviour, the psychological
environment, derived from attitude questionnaires or other
experimental designs, which paint the perceived environment in
which behaviour occurs, will have greater value.

The remote mining settlement's human ecology is a product
of a different set of inputs of people and the energies which
they supply and demand, to that found in more densely settled
areas.   It is geared to *change* of population members yet
maintains an essentially steady total population size from
year to year, or even growth of that population so long as
the *limit* to the mineral resource itself is not too closely
approached.   It is geared to *movement* in ecological and
geographical domains through the very nature of its population
dynamics and its remote location.

3. *Living in Remote Settlement.*   Brealey (1972)
presented the first detailed results of studies of residents'
attitudes to living in remote communities in northern
Australia.   That study was to become the forerunner of many
more field-based attitude surveys, by the CSIRO Remote
Communities Environment Unit.

Among the towns in northern Australia which have been
studied since 1972 are Newman, Dampier, Mount Isa, Katherine
and recently, as reported in the next chapter, most of the
mining settlements in the Pilbara region of Western Australia
(Neil *et al*).   Parkes and Burnley (1981) and Parkes (1981 and
1983a) report on results of field surveys in Alice Springs in

1980 and 1981 in which the CSIRO design was replicated.

Referring to the 1972 study by Brealey, Newton (1981) summarises some of the results from that work as follows: "It is evident that the majority of residents' comment is unfavourable, with most negative responses centering on high temperatures, isolation, transportation problems, aspects of administrative policy at work, township and regional level deficiencies in entertainment and recreational facilities, as well as shopping and commercial services, and the cost of living". He continues that there are also "unfavourable responses generally related to high levels of population turnover, excessive emphasis on money-making, boredom, depression and loneliness; while the favourable comments normally stressed the relaxed way of life and friendly atmosphere of the towns" (Newton, 1981).

Althouth Newton was writing in 1981 his references were to work undertaken about ten years earlier at a time when many of these towns were in their nascent phase of development. New mining towns such as Tom Price and Paraburdoo had only recently been completed and the dust of the construction stage had hardly had time to settle. Nor had there been time for a good 'wet' to bring life to the newly planted lawns, shrubs and trees. The routines of daily life had not been established. There was no reference points from the past. Now in 1983, the established townscape and landscaped environment of these towns (Plate 2), the improved communications and transport, especially national network colour television and local commercial video based television, which were no part of the ecological environment to which reference was being made in the late 1960's and early 1970's, all present a markedly different physical and activity setting for daily life.

(Photos: Author)

PLATE 2.   Paraburdoo Townscapes in 1983[a]
[a]Relate back to Plate 1.

A substantial core of long-term residents and their
growing experience of living in remote, climatically stressed
environments becomes a store of ecological capital which
newcomers can now draw upon. There is a growing fund of
experience which can now be passed on to the new arrival.
This presents an entirely different ecological environment to
that which existed over a decade and a half ago.

Residents of remote settlements such as Paraburdoo and
Tennant Creek do express a sense of isolation in their
responses to questionnaire surveys. They will also express it
during unstructured interviews and conversations. To date,
however, we have rather little understanding of the meaning
of isolation. There is no published research that I am aware
of which proposes the dimensionality of isolation: what
precisely is it? What is its structure? Trying to evaluate
the impact of isolation on behaviour therefore becomes an
impossible task at present.

Remoteness, which is a geometric and geo-economic
condition can be overcome by material improvements to the
vehicles and media of communication, and by cost-price adjustment. *Isolation* which is an element of the psychological
environment rather than the ecological environment, though
possibly related in some instances, is not open to effective
treatment until we know what it is. It may be that isolation
is a many-dimensioned sensation, periodically varying in
intensity. It may also have positive properties in relation to
behaviour and ecological environment structuring.

Among the positive features of life in remote mining
towns (and in some other settlements, such as Alice Springs),
which have been reported in published studies referred to
above, friendliness, mateship, easy-going way of life, no
'rat race' all frequently appear. These expressions of
positive community are usually also supported by favourable
comments on the advantages of a small population, and a
convenient physical scale to the settlement, which 'makes it
easy to get around and meet people'.

The growing maturity of many remote settlements which
have recently been established as mining towns and the
changing ecological base of other settlements such as Tennant
Creek and Alice Springs are leading to a richer sense of
community, to an *extension* of the initially set or intended
period of residence. External factors such as the recent
economic recession have made the choice to leave more difficult but it seems to be oversimplifying the position to argue
that the increasing modal period of residence is above all a
function of the difficulty of finding alternative employment
in other larger and usually metropolitan regions. It is also

possibly an oversimplification to relate high turnover of population to dissatisfaction with the settlement and its settings.

The community dynamic of these towns is fast, especially as revealed through population turnover; though this appears to be decreasing. They have an urban paced system without the large population mass or high densities and large geographical extent of much larger population concentrations. High rates of ecological mobility, most of which have their cause in the high rates of population turnover may be a *vital* feature. an essential component of the remote mining settlement's ecology, especially in the early stages of establishment. Perhaps we are dealing with a new settlement form, a new human ecological phenomenon in which *geographical and ecological movement* are dominants (Parkes 1983c). They have to be incorporated into the design phase of the physical environment, and into the ecological environment so that spatial and temporal flexibility of day to day behaviour is maximised.

Brealey and Newton (1981) have proposed commuter mining from a large central place to dispersed mine sites, as an alternative way to develop the mining industry's settlements in the future, acknowledging the role of movement in the ecological structure of settlement systems. However their examples, taken from the USA where miners travelled up to 210 kms per day from cities such as Tucson and Phoenix, are rather meaningless in the northern Australia context. The total population of these two cities alone is more than twice the size of the total population of northern Australia. Less than 80,000 people live in northern Western Australia, a region five times the size of Texas. Remoteness is likely to be an ecological fact of life for many years to come and while central places should be considered as one possible way to develop mining regions in future, it remains necessary to continue to evolve purpose built habitats which serve their function of contributing to the production efficiency of the mine operation. Perhaps landbased rigs, akin to those which operate on the oil and gas marine platforms can be adapted to form semi-permanent settlements with a physical environment quality far removed from that of the mining camp. Indeed a built environment complex along the lines of *time-shared resort developments* have probably got some potential. The general concept of time-sharing and periodically leased space by members of a single enterprise is worth further study as a procedure for remote settlement development, in which the residents share directly in the profits of the enterprise. The integration of mining and resort functions for periodic visitors including the families and friends of the workers,

into a totally *new form of mining settlement* should be
considered as a *reality* (sic.) by joint venture between mining
industry and leisure resort industry consortia.   (The Wittenoom region of the Pilbara provides one example in which this
concept could be adapted, more effectively than by the development of Wittenoom as a single function resort township.)

III.   ASPECTS OF BUILT ENVIRONMENT AND POPULATION

   1. Paraburdoo.   The basic features of the town plan are
shown in Figure 3.   Situated at an elevation of 365 metres, the
township is about 400 km from Dampier the port town for iron
ore mined in Tom Price and Paraburdoo.   The rail link to
Dampier was constructed by Hamersley.   A fleet of about 50
locomotives pull three trains per day, each over half a kilometre in length and weighing upwards of 27,000 tonnes, using
six locomotives for the uphill climb between Paraburdoo and
Tom Price.   A private, unsealed, company service road along
the length of the railway provides the best route for private
vehicle transport from Paraburdoo to the coast, but permits
are required to use the road.   Most of the public route is
unsealed and the journey takes about five hours.   All routes

   *FIGURE 3.   Town Plan of Paraburdoo and Distribution of
Staff Housing and Cottage Industries*[α]

[α]*Zone marked 21 is single men's quarters;   11 is the mess;
7 is the hospital;   23 is supermarket, adjacent are bank,
post office and recreational hall;   17 is the school.*

to and from the town are flood prone during the wet season
when cyclones crossing the coast between Karratha and Port
Hedland pose a threat to property and communications.

Construction of Paraburdoo commenced in 1970. The township was planned and serviced by Hamersley Iron Pty. Ltd. The
600 dwellings in the town, excluding the single men's and the
single women's quarters, include fourteen different house-
types, twelve are of brick construction and two are system
built. All dwellings, three and four bedroomed are fully air-
conditioned and rents are 'nominal', including subsidised
power and all maintenance. During the recent past, along with
normalisation, a house purchase scheme at guaranteed re-
purchase prices by the company has been introduced. Garden
areas are substantial by standards elsewhere in Australia and
all roads are sealed, curbed and guttered. Street lighting
however is only provided in the central area of the town.   (I
have scars as evidence of the dangers of cycling along the
unlit streets on a bicycle without lights!)

In 1980 a detailed survey of attitudes to housing design
undertaken by the author in conjunction with the Architectural
Research Development Unit at the University of Newcastle,
on behalf of Hamersley as part of a long term programme for
improved design, especially in relation to thermal efficiency
of dwellings, revealed almost complete satisfaction with all
aspects of the residential environment and of the built
environment of the town in general. Spacial areas of lawn,
high density tree cover (Plate 2) and extensive and well
maintained sports ovals, swimming pool, tennis courts, squash,
bowling and basket ball facilities, all contribute to the
creation of an environment for outdoor and indoor active and
passive recreation which it would be hard indeed to find in
any metropolitan suburb or even in most leisure resort
developments. Also of great importance was the conventional,
typically Australian, sub-urban flavour of houses and townscape. The residential environment (Plate 2) is akin to that
of middle class suburban Melbourne, Adelaide or Perth: the
sort of environment which the majority of residents in
Paraburdoo might aspire to in those cities.

An indication of the climatic regime is given by the
first two graphs in Figure 4. The third graph indicates the
water consumption regime of the townsite and that of the
townsite and the mine combined. Domestic water is supplied
from local bores pumping into two 9 million litre tanks.
Between winter and summer water consumption increases by a
factor of 3. The greatest increase being at the mine site,
about 7 kilometres south and west of the township. Electric
power is generated by Mirlees Blackstone Diesel fuel engines,
each with a maximum output of 4.5 megawatts. Total station

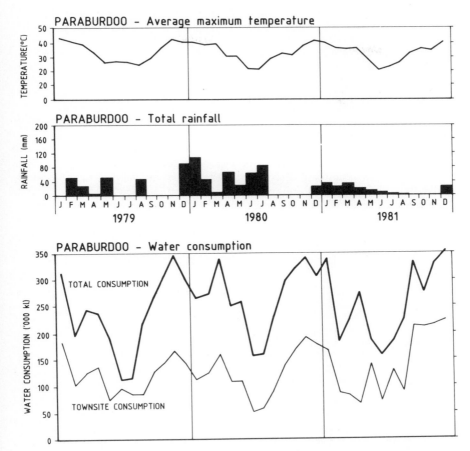

*FIGURE 4.   Paraburdoo Average Maximum Temperatures,
Rainfall and Water Consumption 1979-81[a]*

[a]*Based on graphs and figures provided by Town Administration,
Hamersley Iron, Paraburdoo.*

capacity is about 30 megawatts generating between 11kv. and
33kv. to the mine and townsite.   Fuel is supplied from
Dampier, by the world's largest rail tankers and stored in
500,000 litre tanks at the Shell Fuel Farm.   Each consignment
of fuel oil to drive the electric power supply of the township is 600,000 litres.   Aviation fuel for the site planes and
commercial jet aircraft is hauled by road from Port Hedland.
    In 1981 census counts recorded a population of 2,357
people in Paraburdoo.   In 1976 the population was 2,402;   a
population decline of 1.9%.   However, by January 1983 the town
population had fallen further to 2,102 (Hamersley monthly

survey of town population), but by July 1983 had risen again
to 2,189.   Population size in Paraburdoo is determined by the
vitality of the international iron ore market.   All ores are
exported.   A reduction in the number of company employees
leads to a lagged decline in the number of teachers, contract
workers and a range of public services, including nurses for
the 17-bed hospital, built by the mining company but not
maintained by the State Medical Department.

Table 1 summarises selected population characteristics
from the 1976 and 1981 census count data.   To ease comparison
with Tennant Creek, which I discuss in the next section,
values for both towns have been presented together here.   The
Tennant Creek figures appear in brackets in Table 1.   Because
there are only three census district subdivisions for
Paraburdoo, one of which includes the single men's quarters
and the single women's quarters, housing about 400 people, of
whom about 370 are men, when all accommodation is occupied,

TABLE 1. *Population and Housing Comparisons Paraburdoo and Tennant Creek*

	1976 %		1981 %			
OVERSEAS BORN	18.5	(17.6)	31.4	(15.8)	+	(−)
ABORIGINES AND ISLANDERS	0.5	(18.6)	0.6	(15.4)	+	(+)
EMPLOYED AS % TOTAL POPULATION	52.7	(40.3)	52.1	(45.7)	−	(+)
UNEMPLOYED AS % TOTAL POPULATION	0.9	(1.7)	0.8	(2.7)	−	(+)
NOT IN LABOUR FORCE (AGED 15 YEARS AND OVER)	26.7	(21.1)	15.1	(21.8)	−	(+)
OCCUPIED PRIVATE DWELLINGS	88.7	(82.2)	93.0	(90.7)	+	(+)
UNOCCUPIED PRIVATE DWELLINGS	10.8	(15.6)	7.0	(9.3)	−	(−)
NEVER MARRIED (AGED 15 YEARS AND OVER	19.2	(14.3)	21.1	(20.5)	+	(+)
SEPARATED AND DIVORCED	2.3	(5.9)	3.9	(6.5)	+	(+)
WIDOWED	0.3	(2.5)	0.6	(3.1)	+	(+)
SAME DWELLING OR SAME LOCAL AREA 5 YEARS AGO (1971 FOR 1976, 1976 FOR 1981)	14.7	(40.2)	36.8	(40.6)	+	(+)
AGE GROUP 0-4	15.0	(14.0)	10.7	(10.4)	−	(−)
AGE GROUP 5-9	13.5	(14.1)	11.8	(11.1)	−	(−)
AGE GROUP 10-14	5.3	(8.8)	9.5	(8.2)	+	(−)
AGE GROUP 15-19	3.8	(5.1)	5.6	(6.8)	+	(+)
ADULT POPULATION, AGE GROUP 15 AND OVER	66.2	(63.1)	68.0	(70.3)	+	(+)
AGE GROUP OVER 60	0.7	(5.2)	0.7	(6.6)	=	(+)
TOTAL MALES/100 FEMALES	148.7	(111.7)	145.1	(112.5)	−	(−)
MALES/100 FEMALES 18 YEARS AND OVER	174.6	(113.5)	194.2	(116.1)	+	(+)
ADULT MALES 15 YEARS AND OVER/100 ADULT FEMALES, NEVER MARRIED	670.0	(209.7)	352.7	(168.9)	−	(−)
SELF-EMPLOYED AND EMPLOYER	0.3	(4.2)	1.2	(4.3)	+	(+)
PRIVATE SECTOR EMPLOYEES	94.2	(68.2)	88.5	(57.2)	−	(−)
IMPROVISED DWELLINGS[a]	0.0	(6.0)	0.0	(2.9)	=	(−)

[a]*Improvised dwellings 1976, as % of dwellings, 1981, as % of households*

it is not really very helpful to give attention to internal differences within the town, based on census data. However field sample surveys in 1980 and further field study in 1983 reveal no demographic, socio-economic or built environment differences, from one part of the town to another. From Figure 3 it will have been noticed perhaps that the distribution of staff employees is regular across the settlement's 600 detached dwellings. The distribution of cottage industries, discussed earlier is an important ecological factor. Their location is determined by the residence which is allocated to employees and therefore the pattern of cottage industry location will be variable, and flexible in accord with the ecological dynamics of the settlement. If a resident wishes to establish a cottage industry, it will locate at the residence, except in one or two instances, such as the town's 'milk-bar'.

In August 1983 there were 573 people on the full-time award and apprenticeship roll of Hamersley in Paraburdoo. These are the so-called *wages* employees. A further 171 people are *staff* employees; a ratio of 1 staff to 3.4 wages personnel. The ratio of mining company employees to total population 16 years and over is approximately 1:2. To total population it is approximately 1:3.

Field surveys in 1980 indicated an average period of residence of 3 years and 4 months. The mode was exactly 4 years. At that time the shortest period of residence reported in the sample was 1 month and the longest was 9 years and six months.

Further reference to the population characteristics will be made later after brief discussion of Tennant Creek.

2. *Tennant Creek.* The town plan of Tennant Creek is shown as an aerial photograph in Plate 3. Unlike Paraburdoo there is no overall design plan and nor is there any observable thematic continuity in the townscape. The Stuart Highway is the ecologically dominant physical feature, elongating the built up area along the highway, without regard to dwelling and general building orientation in relation to shelter from the sun as in Paraburdoo, with its east west alignment (Figure 3). The Stuart Highway is the only north-south route between Alice Springs, 507 km to the south and Darwin, 1,035 km to the north. Katherine the nearest larger settlement is 679 km to the north and had a population of 3,737 people in 1981. In 1976 its population was 3,127, a 19.5% increase. The Stuart Highway is a sealed road, but over most of the 1,542 km between Alice Springs and Darwin it is narrow, single land and prone to severe flooding. Floods can be sufficiently serious to require that air-borne supplies

have to be dropped into the townships along the route.
    Tennant Creek has been associated with gold mining since
the 1930's and more recently with copper through the Peko
Wallsend Pty. Limited. In terms of mining activity at present
Tennant Creek is most closely linked with copper mining at
Warrego, which is a small mining township about 50 km north
and west of Tennant. Many miners now commute into Warrego
from Tennant Creek along a narrow sealed road built by the
mining company. A copper smelter between Warrego and Tennant
Creek no longer operates and though there is a care and maintenance unit at the site it seems unlikely that copper will be
smelted again unless world prices rise substantially. The
copper mined at Warrego is road hauled to Mt. Morgan in the
south east of northern Queensland. Gold is still mined at
Noble's Nob, now a large open-cut mine following a collapse of
the original underground mine. A number of smaller gold mines
have intermittent burst of activity. Employment is small.
    During the past decade and especially since 1978 there
has been a marked growth in public service functions. They
have succeeded the mining industry as the ecological dominant
of the settlement. Recently introduced statutory planning
controls over land use, built environment form and materials
are beginning to influence the scale and the quality of the
town's ecological environment. New Town Council buildings of
dramatic pyramid shape, new N.T. Government offices, a new
high school under construction in September 1983 (see Plate
3) and a growing awareness of the potential held by the
tourist industry, all demonstrate the changing economic base
of the settlement. From Figure 5 it will be seen that
Tennant Creek lies more or less in the centre of the Northern
Territory, but this geographical location alone is not likely
to ensure the future of the town as a service centre unless
roads are upgraded and a rail link north and south is
constructed.
    A substantial and growing aboriginal population is resi-
dent in the town, the built environment reflects their
presence in two observable ways; the established and officially designated town camps and the informal fringe camps.
Many Aborigines are also housed in normal residential accommodation in the settlement and there is an aboriginal housing
council, the Warramunga Pabulu Council which advises on
appropriate housing designs for aboriginal people. Drakakis-
Smith (1982) has discussed the town camps in Alice Springs in
some detail. The general situation in Tennant Creek appears
to be rather similar to that in the much larger settlement of
Alice Springs. The locations of Tennant Creek town camps are
shown in Plate 3 as white disks.

PLATE 3.   *Aerial View of Tennant Creek*[α]

[α]*Composite compiled by author from frames supplied by the Northern Territory Department of Lands, September 1983.*

In 1981 the population of Tennant Creek was 3,118;   an
increase of 39.4% over the 1976 figure. About 15% of the
population are Aborigine or Torres Islanders. Between 1976
and 1981 the census counts show an increase of Aborigines and
Torres Islanders of 58.2%, from 304 to 481. By September
1983, based on estimates from various official and private
sources, the aboriginal population resident in the town varies
periodically between a low of 400 and high of 1,000 people.
There is a very high rate of inward and outward flow of
Aborigines from cattle stations, reserves (e.g. Ali Curung
about 130 km south of Tennant Creek, Figure 5) and from out-
stations on *homelands*. Apart from the increasingly large
proportion of the population which is aboriginal, as well as
the growing absolute size of that population there are
language, dialect and other complex cultural differences among
the aboriginal settlers. Some people include Alyawarra,
Kaititj, Warlpari, Warlmanpa and Warramunga;   five or more
linguistic groups may be involved.

Two years before the granting of self government to the
Northern Territory, the Aboriginal Land Rights (Northern
Territory) Act 1976, passed by the Federal Government in
Canberra, provided "for Aboriginals in the Northern Territory
to gain title to existing reserve land and to make claim to
other land - aboriginal-owned pastoral leases and vacant
Crown Land - on the basis of traditional ownership" (Department of Aboriginal Affairs, 1982). Figure 5 shows the
distribution of aboriginal freehold land and land under claim
in the Northern Territory at July 7th 1982. If all the land
under claim is passed to the traditional owners, then a
substantial proportion of the Northern Territory will then
be held by the aboriginal peoples. The future impact of these
land claims on the ecological structure of towns like Tennant
Creek will depend to a great extent on the manner in which the
Aborigines change their present patterns of residence and
movements. At present it is the older aboriginal people who
have the strong, affective ties with their traditional lands.

The Population Projections Group of the Northern
Territory Government has prepared projections for Tennant
Creek, up to 1990. They have been projected from the 1981
census counts and in 1982 were calculated as follows:

1981	1982	1983	1984	1985	1986	1987	1988	1989	1990
3,118	2,806	2,890	2,991	3,063	3,134	3,206	3,280	3,346	3,419
	-10.0	3.0	3.5	2.4	2.3	2.3	2.3	2.0	2.2

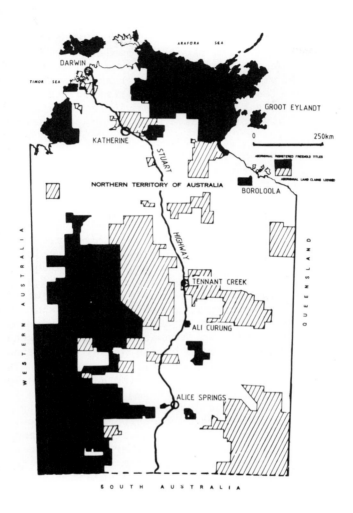

FIGURE 5.   Aboriginal Freehold Land and Land under
Claim, July 1982[a]

[a]Generalised by author from a map prepared by the Department
of Lands and Housing Survey, Northern Territory Government,
Darwin.

Percentage change is shown in the third row. These projections seem likely to be underestimates. A population count
undertaken in Tennant Creek on July 21st 1982 by the Australian Bureau of Statistics showed a population of 3,170, an
increase of 52 people over the 1981 June census count. These
figures include visitors and exclude residents who were out
of town at the time of the count. Estimated *resident*
population in Tennant Creek, i.e. excluding visitors, was
2,401 in 1976, by 1981 it was 3,087.

   3. Comparisons between Paraburdoo and Tennant Creek.
From Table 1 some general observations can be drawn which may
also indicate differences between the newer purpose-built
mining towns and the older remote mining towns in northern
Australia.

   The proportion of overseas born is higher in the new
settlement, but aboriginal people are hardly represented and
the townscape shows no evidence of their presence (Newman
[0.2%] and Tom Price [0.6%] show similar low levels).
Employment levels are higher, but this is a consequence of a
single employer and the tied housing policies which prevail.
Proportions of never married adults are similar but of
interest here is the increase in this category since 1976 in
Tennant Creek. Proportions of separated and divorced people
are almost twice as high in Tennant Creek and the generally
older population is indicated by the higher proportion of
widowed people. The growing maturity of the newer mining
towns is evident in the decreasing size of differences, on
many characteristics, between the two settlements. In the
ten years between 1971 and 1981, population mobility in
Tennant Creek, as indicated by the proportion of people who
were also resident there five years earlier, has remained
unchanged. In paraburdoo, on the other hand, there has been
a marked change, with an increase from 14% in 1976 to 36.8% in
1981. Neil *et al* discuss aspects of population transience in
the next chapter, as well as reasons which people have for
moving to, and from, remote mining towns.

   Age profiles in the two settlements are also converging,
with a general ageing of the population indicated. However,
both settlements have substantial proportions of children
under the age of 15;   32% in Paraburdoo and 28.7% in Tennant
Creek. A signpost, at the entry to Paraburdoo, exhorts
drivers to drive carefully, and the litter of bicycles at the
school indicates why! (Plate 4).

   New mining towns are dominated by high male to female
ratios. In Paraburdoo in 1981 there were 145 males to every
100 females of all ages, but there were 194 males to every
100 females 18 years and over and this was an increase over

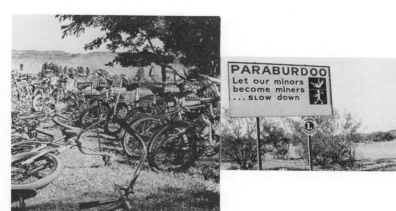

(Photos: Author)

PLATE 4.   *Paraburdoo Minors*

the 1976 ratio. The greatest difference occurs in the age groups 20-24, 25-29 and 35-39 with ratios of 167 : 100, 158 : 100 and 192 : 100 respectively. Tennant Creek, however, had a much more balanced profile of 116 males over 18 to 100 females over 18. In the same three age groups the ratios were 101 : 100, 106 : 100 and 126 : 100. A fall in the ratio of never married males to never married females has also occurred in both towns between 1976 and 1981, most noticeable in Paraburdoo from 670 : 100 to 352 : 100.

In the five years between 1976 and 1981 there has been an increase in the number of employees and self-employed people in Paraburdoo but the proportion is still only 25% of that found in Tennant Creek.

IV.   IMPLICATIONS OF UNDERMANNED SETTINGS

"1. A behaviour setting is a standing (observable) pattern of behaviour that occurs over and over again in a given place and at a given time. You can go to the *place* where it occurs at the *time* it occurs and see the behaviour repeated each time the setting happens.
2. Yet behaviour settings, even though they are defined as separate entities, are a part of the flow of behaviour in a community. People move in and out of settings but the settings do not disappear when different people arrive; they have a life of their own. Yet when the community changes, settings also change." (Brechtel 1977).

In periodically stressed climates there are marked differences in molar behaviour according to season.   In northern Australia the 'dry' and the 'wet' become *markers* and *pacemakers*[1] for the establishment of differing behaviour settings.   In Paraburdoo and in Tennant Creek, during the 'wet' between October and April daily temperatures rise above 40°C, people engage in less public, group-related outdoor activity.   This is also known as 'the silly' season.   The matron of the Paraburdoo hospital, Margo Latham explained that it is during the hot summer season that people take refuge in their homes;   see much less of each other;   have more difficulty coping with young children.   If they are going to feel lonely, this is when it is most likely to happen.   It follows perhaps that the time to move to a remote settlement such as Paraburdoo is during the 'dry';   not simply because of transport related factors but because behaviour settings are most abundant at this time, access to them is good and the chances of *settling* into the new ecological environment with fewer emotional problems are also higher. This is the time when the activity density of the town is at its peak.   The seeds of future behaviours can be sown in an ecologically fertile environment.   The cooler months are also known as the 'granny season'.   Grandparents and relatives visit during this season and perhaps as many as one in ten households have resident visitors at this time.   There are consequent and significant changes to behaviour routines which are multiplied through the community.

In Tennant Creek the impact is perhaps rather less significant because it is open to through traffic up and down 'the track' between Alice Springs and Darwin and there is usually a fairly substantial number of tourists in the town, either on camp sites or on short stop-overs from coaches and private vehicles.   Once again, however, the dry, cool season has higher levels of collective, public activity, a richer range of settings, paced by climatic factors driving the tourist industry, among others, which in turn initiate movement to and within the township.

In remote communities where there is limited opportunity to participate in a behaviour setting located in another town, there is a tendency for a rich and varied range of settings to occur.   The very fact of remoteness seems to lead to a relatively large number of settings, prompting the frequently made comment that "there is plenty to do here if you want to become involved". However behaviour settings in remote communities are likely to be characterised by *undermanning*.   Undermanning means that there are less people to carry out the necessary program in a setting, including its maintenance;   less people that is than the optimum.

However undermanned settings can and do exist and if they persist they do so because self-regulatory mechanisms have been developed to cope with the condition. As time passes the setting will be threatened either by a further decline in manning levels or because it is unable to satisfy the *regulations* which govern similar settings elsewhere: as for instance according to some regional or national Association, e.g. Scouts, Guides, football and others.

It is a frequently reported feature of mining and other remote settlements that *social relations* have a very favourable standing (Brealey, 1972; Brealey and Newton, 1978; Parkes and Burnley, 1981; Parkes, 1981 and 1983a). The wide range of behaviour settings and generally easy access to them in terms of spatial, temporal and regulatory conditions provide early opportunity for leadership roles. This is partly because of the undermanning which tends to occur. It is also partly because of the relatively high turnover rates among the population. The implication is that population turnover, coupled to undermanning has certain positive roles in the day to day life of remote settlements.

It is possible that one of the reasons why a sense of isolation develops from time to time is that a particular behaviour setting has collapsed because of a lack of people who are able or willing to take over leadership roles;   due for instance to shift work committments. The collapse of a setting means the removal of behavioural markers and pace-makers which may take some time to be replaced. The linkage between the timing system of the mine operation and the development and survival of behaviour settings is probably often very close. If high turnover of staff is a cost to the mining company then the timing of production related processes in the mine, as for instance with shift rosters, should be as flexible as possible to allow the maintenance of behaviour settings. Paradoxically perhaps, the relative abundance of independent behaviour settings is itself also a cause of the undermanning which occurs.

Wicker (1979) provides a useful summary of Barker's Theory of Manning as it relates to behaviour settings (Wicker 1979), a part of this summary is presented below. It is drawn from Barker (1968) and Barker and Schoggen (1973). The undermanned and the optimally manned behaviour setting are here contrasted in terms of the predicted consequences of undermanning. The reader will note that there are many positive consequences of undermanning. Pehaps there are implications for an improved understanding of northern Australia's remote settlements, especially as they might relate to the decisions which have to be made by mining

companies, 'planners' and 'managers' of remote habitats.
Three categories of consequences are proposed.

"1.   *Primary differences:*
Among occupants of undermanned settings,
(a)   actions to carry out setting programs are
more frequent, more vigorous, more varied;
(b)   actions to deal with threats to the
setting are more frequent, more vigorous, more
varied;
(c)   actions to correct or shape the inadequate
behaviour of other setting occupants are more
frequent;
(d)   actions to eliminate or eject from the
setting other occupants whose behaviour is
inadequate are less frequent;
(e)   actions to induce others to help deal with
threats to the setting are more frequent;
than among occupants of optimally manned settings.

2.   *Secondary differences:*
Occupants of undermanned settings more frequently
(a)   serve in responsible positions;
(b)   engage in actions that are difficult for
them;
(c)   engage in actions that are important to
the setting;
(d)   engage in a wide range of different activities;
(e)   act in response to important actions of others;
than occupants of optimally manned settings.

3.   *Psychological differences:*
Persons who occupy undermanned behaviour settings
tend to
(a)   see themselves as being more important to
the setting;
(b)   feel greater responsibility for the setting;
(c)   feel more insecure about the fate of the
setting;
(d)   work harder to support the setting;
(e)   feel more versatile;
(f)   be less sensitive to, and less evaluative
of, individual differences among people;
(g)   see themselves and others more often in
terms of the jobs they do, and less often in terms
of personality characteristics;
than persons who occupy optimally manned behaviour
settings."

(Wicker 1979, pp.72-73)

Although most of the items in the three categories above appear to have positive qualities, undermanning does also impose strains on individuals and settings. I am not able to expand on these features here. The purpose in discussing undermanning has been to suggest that the small remote community has certain special ecological characteristics which are a function of small population, remote location, high turnover of population, periodically stressed climate and that these well known factors are not necessarily the cause of the social pathologies which are so frequently reported upon in published research.

Referring especially to the *new* mining towns in Australia, Sharma (1983) writes that "the new mining towns are noticeable for their tranquility and order" (p.178). The order is observable in the thematic structure of the built environment. It is also observable in the ecological environment through the *persistence* of a wide range of behaviour settings, often undermanned, in spite of high population turnover.

NOTES

[1]The notion of *markers* and *pacemakers* as mechanisms in the ecological environment was first proposed by Parkes and Thirft (1975). These mechanisms operate to facilitate the scheduling of human behaviour and personal interaction with other people and necessary objects in the ecological environment, *including* the best achievable spatial and temporal locations. They operate by assigning points or zones of time to use in a particular space or in some ordered set of alternative spaces. For example, an appointment at the dentist has an absolutely fixed spatial location, a slightly elastic time location (one might be five minutes late but hardly an hour), which marks a part of the town and a part of the day, *entraining,* possibly for some days in advance the scheduling of personal activity. A higher density marker and a more powerful pacemaker is the starting time 'at work'. Annual periodic markers also occur, they are for instance festivals or fairs. These markers assume considerable importance in remote locations and *pace* the behaviours of many people, often for days, weeks and months before the marker arrives. Seasonal markers also exist and in remote areas of northern Australia the 'wet' and the 'dry' operate as ecological mechanisms. More detailed discussion of these and other ecological mechanisms will be found in Parkes and Thrift (1979) and Parkes and Thrift (1980).

360   Don Parkes

ACKNOWLEDGEMENTS

I am grateful to the *Internal Research Allocations Committee*, University of Newcastle, N.S.W. for a grant to support part of my field work. Also the Department of Geography, University of Newcastle, N.S.W. for a grant from C.T.E.C. funds towards field work expenses. I also wish to acknowledge with thanks the Council of the University of Newcastle for granting me leave on an Outside Studies Programme, between July and December 1983.

REFERENCES

Australian Bureau of Statistics 1981. *Census of Population and Housing, June 30th 1981*, Canberra.

Barker, R.G. 1968. *Ecological Psychology: Concepts and Methods for Studying the Environment of Human Behaviour*, Stanford University Press, Stanford.

Barker, R.G. 1979. Influence of Frontier Environments on Behaviour, *The American West: New Perspectives, New Dimensions*, (ed.) Jerome O. Steffen, University of Oklahoma Press, pp.61-93.

Barker, R.G. and Schoggen, P. 1973. *Qualities of Community Life*, Jossey-Bass, San Francisco.

Barrows, H.H. 1923. Geography as Human Ecology, *Ann. Ass. Am. Geogr.* 13, pp.1-14.

Brealey, T.B. 1972. *Living in Remote Communities in Tropical Australia*, CSIRO Division of Building Research Report TB27.1.

Brealey, T.B. and Newton, P.W. 1978. *Living in Remote Communities in Tropical Australia: The Hedland Study*, CSIRO Division of Building Research, Special Report.

Brechtel, T.B. 1977. *Enclosing Behaviour*, Dowden, Hutchinson and Ross Inc, Stroudsburg.

Carlstein, T. 1982. *Time Resources, Society and Ecology. On the Capacity for Human Interaction in Space and Time*, Volume 1, George Allen and Unwin, London, p. 444.

Carlstein, T., Parkes, D.N. and Thrift, N.J. (eds.) 1978. *Human Activity and Time Geography*, Edward Arnold, London, p.286. This is Volume 2 of a three volume series by the same editors, *Timing Space and Spacing Time*, Edward Arnold, London.

Department of Aboriginal Affairs 1982. *Aboriginals in Australia Today*, Australian Government Publishing Service, Canberra, p.7.

Drakakis-Smith, D.   1982.   The Alice Springs Town Camps, *Town Populations*, (eds.) E.A. Young and E.K. Fisk, Development Studies Centre, The Australian National University, Canberra, Australia and Miami, Florida, pp.107-132.

Laut, P., Firth, D. and Paine, T.A.   1980.   *Provisional Environmental Regions of Australia*, CSIRO Australia in collaboration with the Department of Science and the Environment, Melbourne

Newton, P.W.   1981.   New Towns in Isolated Settings in Australia, *Settlement Systems in Sparsley Populated Regions: The United States and Australia*, (eds.) R.E. Lonsdale and J.H. Holmes, Pergamon Press, New York, pp.169-188.

Parkes, D.N. and Thrift, N.J.   1975.   Timing Space and Spacing Time, *Environmental and Planning A*, 7, pp.651-670

Parkes, D.N. and Thrift, N.J.   1979.   Time Spacemakers and Entainment, *Transactions of the Institute of British Geographers*, N.S.4, pp.

Parkes, D.N. and Thrift, N.J.   1980.   *Times, Spaces and Places: A Chronogeographic Perspective*, John Wiley and Sons, Chichester, p.527.

Parkes, D.N.   1981.   Living in the Southern Centre of the North: Alice Springs, *Transactions of the Menzies Foundation* 2, pp.95-111.

Parkes, D.N.   1983a.   "Hello Alice!" Ecological Aspects of Settings and Life in Alice Springs, Australia, *Research Papers in Geography, No. 26 University of Newcastle, N.S.W.*, p.53.

Parkes, D.N.   1983b.   The Future Development of Human Settlement, *The Northern Territory of Australia: Present Indicative and Future Conditional. A Selection of Papers on Northern Development*, (ed.) D. Giese, Northern Territory University Planning Authority, Darwin, pp.23-34

Parkes, D.N.   1983c.   The Place of Movement and the Movement of Place in the Australian Arid Lands, *Proceedings of the 6th World Congress of Engineers and Architects in Israel: Development of the Desert and Sparsley Populated Areas Policies, Planning, Architecture and Industry*, Tel Aviv, December 18-23.

Parkes, D.N.   (forthcoming) *Human Ecology of Urban Places in Australia's Arid Zone*, Longman Cheshire, Sydney.

Parkes, D.N. and Burnley, I.H.   1981.   *Urbanisation and the Australian Arid Zone. A Focus on Alice Springs*, Report to the United Nations University, Tokyo (unpublished).

Sharma, P.C.   1982.   The New Mining Towns, *Mining and Australia*, (eds.) W.H. Richmond and P.C. Sharma, University of Queensland Press, St. Lucia, pp.150-180.

Tuxworth, H. (M.B.E.)   1980.   *Tennant Creek Yesterday and Today*, Tennant Creek (no Publisher named).   Further details from Hilda Tuxworth Collection, Fryer Memorial Library for Australian Literature, University of Queensland, St. Lucia.

Wicker, A.W.   1979.   *An Introduction to Ecological Psychology*, Brooks/Cole Publishing Company, Monterey, a Division of Wadsworth, Inc., especially Chapter 5, pp.70-82, The Theory of Manning.

19

POPULATION STABILITY IN NORTHERN AUSTRALIAN RESOURCE TOWNS: ENDOGENOUS VERSUS EXOGENOUS INFLUENCES?

C. C. NEIL
J. A. JONES
T. B. BREALEY
P. W. NEWTON

CSIRO Division of Building Research
Melbourne, Victoria

I. INTRODUCTION

This chapter looks at recent trends in population stability in a number of northern Australian mining towns, and discusses the relative impact on these trends of exogenous as opposed to endogenous influences. It is based on data drawn from a number of recent surveys of different towns in Western Australia, Northern Territory and Queensland, variously described elsewhere in Neil *et al,* (1982, 1983), Neil (1983) and Brealey *et al,* (1983). Although the surveys were designed for a variety of purposes, they all contained a number of overlapping questions, and were compatible in terms of methodology (personally delivered self-administered questionnaires) and sampling procedures (households in which either husband or wife was employed by the mining company, selected randomly either from town maps or from accommodation lists provided by the mining companies). Table 1a gives some basic details of the towns' locations, age, population and economic base – whilst Table 1b provides data on mean ages and family size. The representativeness of the towns surveyed is discussed in more detail below.

NORTHERN AUSTRALIA
ISBN 0 12 545080 X

TABLE 1.   Description of Survey Towns
a.   *Town Characteristics.*

| Town | Town characteristics | | | | |
	Date commenced as mining community	Road distance from coast km	Road distance from nearest large population centre km #	Economic base of town	Level of facilities +
Dampier	1966	0	1565	Iron-ore loading port	Moderate level-easy access to Karratha
Karratha	1971	4	1547	Government administrative centre, housing for port workers and salt processors	High level
Paraburdoo	1970	397	1439	Iron ore extraction	Moderate level
Tom Price	1966	308	1564	Iron ore extraction	Moderate level
Pannawonica	1972	221	1439	Iron ore extraction	Basic level - no easy access to other towns
Wickham	1972	12	1581	Iron ore processing and loading port	Moderate level
Port Hedland* South Hedland*	1970	0	1770	Government administrative centre and iron ore loading port	High level
Nhulunbuy	1971	0	1204	Govt. admin. centre. Bauxite extraction and loading port.	High level
Emerald		311	276	Rural, regional centre housing coal mine workers	High level
Moura		190	216	Coal extraction	Moderate level
Capella	1977	364	329	Rural town housing some some coal mine workers	Basic level- one hours access to Emerald

* Port Hedland and South Hedland combined.

Defined as an urban locality in excess of 25,000 persons at 1981 census.

+ Classification based on total number of following being available: Doctor, Medical Clinic, Hospital, Dentist, Primary and High School, Accommodation for outsiders, Restaurant, Diversity of shops, Recreational facilities and child care.

TABLE 1 (cont'd)   Description of Survey Towns

b.   Population Characteristics

	Total pop. (1981 census)	Male w'force (1981 census)	Males in w'force (Mining Co. figs)	Now married as % of male workforce (1981 census)	Male staff personnel# amongst sample(%)	Mean age# of males	Mean age# of females	Mean No.# children per household	Households# with no children (%)	Households# with children under 15 yrs (%)	Sample size: Households
Dampier	2471	1163	857	50.0	52	39.0	35.7	1.5	25	67	53
Karratha	8341	3224	575	62.6	37	36.0	33.3	1.9	15	77	108
Paraburdoo	2357	931	846+	54.0	28	34.8	32.2	1.7	19	75	82
Tom Price	3450	1216	1336+	59.6	29	35.7	32.3	1.5	22	71	159
Pannawonica	1170	599	453+	43.1	29	35.9	32.2	1.4	25	67	106
Wickham	2387	852	667+	57.9	32	34.8	32.4	1.5	30	68	66
Pt Hedland	12948*	4490*	1523*	57.0*	42	37.8	36.2	1.2	34	58	89
Sth Hedland					60	35.4	32.7	1.6	33	68	91
Nhulumbuy	3879	1381	796	62.2	52	34.1	34.2	1.7	26	69	140
Emerald	4628	1626	272	56.6	28	31.3	29.5	1.9	10	86	91
Moura	2871	1015	1080	60.8	17	36.1	32.8	2.2	7	89	105
Capella	660	232	55	63.8	17	30.8	28.4	1.9	10	86	31

*Port Hedland and South Hedland combined

+Does not include railways employees

#Survey data

A.   *Background to the Problem*

One of the most significant changes in northern Australia
in recent times has been its rate of population increase.
Over the past decade its population has grown by 30 per cent
compared to 14 per cent for southern Australia. Much of this
growth has been due to the development of industries linked
to mineral extraction, refinement and shipment. The location
of ore bodies of world ranking in remote, sparsely settled
regions of northern Australia has led to the construction
of more than 20 new towns. With a virtually non-existent
local labour market, mining companies were faced with the
prospect of encouraging workers from southern parts of
Australia, and from overseas, to take up residence where
relatively few Australians had previously lived. As the
survey data show (Table 2), the majority of new residents are
still recruited from the south although the proliferation of
new mining towns in northern Australia over the past twenty
years has provided increased opportunity for migration between
remote mining communities. The data in Table 3, based on
surveys discussed below, suggest a migration circuit amongst
new mining towns may have grown to some extent since the early
seventies, particularly in the Pilbara region (see also
Brealey and Newton, 1980). However, as discussed later it
appears that the absolute numbers of current residents who
intend to move to another mining town within their present
region are very low, either because respondents see little
opportunity for changing jobs within the mining community in
their region and so are staying in their particular town,
and hence particular job, because of a lack of opportunity
elsewhere, including within their region, or because they
moved to the region with the intention of staying in one
place until their target (money, length of time etc.) is
reached and then leaving the region entirely. The observed
evidence of intra-regional migration may therefore be
accounted for by employees moving between towns whilst in
the employ of the same company, and maintenance of the
population of those towns may still depend largely on migration from the south. The migration streams that have formed
have had, and to some extent still do have, a number of
distinctive characteristics. They are particularly youthful,
as the net migration data in Table 4 reveal. They also have
a high proportion of males. A number of studies have highlighted other distinctive characteristics (in relation to
family structure, birthplace, education etc.) amongst those
responding to recruitment initiatives of mining companies
(Sharma, 1983;   Brealey and Newton, 1980, Newton 1983, Neil
et al, 1982).

TABLE 2. *Previous Place Lived In by Survey Respondents*

Previous place lives in

Town	Mining town in northern Australia Males %	Females %	Non-mining towns in N. Australia Males %	Females %	Mining town elsewhere in Australia Males %	Females %	Non-mining town elsewhere in Australia Males %	Females %	Metropolitan area in Australia Males %	Females %	Outside Australia Males %	Females %	Missing Males N	Females N	Samples size: Households
Dampier	27*	22*	2	2	0	0	31*	27*	27*	37*	14	12	4	4	53
Karratha	37	33	8	7	1	2	14	16	35	39	7	6	8	9	105
Paraburdoo	9	6	1	4	5	5	36	42	40*	33	8	11	5	1	82
Tom Price	11	6	5	3	0	0	31	30	43	49	10	12	6	9	159
Pannawonica	15	9	8	7	4	4	15	16	45	50	14	16	10	3	106
Wickham	15	14	11	16	0	0	13	8	49*	50*	11	13	5	2	66
Port Hedland	9	7	2	1	4	3	8	8	59	57*	17	25	6	13	89
South Hedland	11	8	2	1	2	0	11	13	56	59	17	18	3	7	91
Nhulunbuy	8	7	5	7	0	0	21	18	55	52	10	16	12	5	140
Emerald	38	39	33	32	1	1	24	22	5	4	0	1	11	1	91
Moura	10	11	50	51	0	0	27	25	6	9	6	4	9	4	105
Capella	8	10*	60*	72*	0	0	28*	17*	4	0	0	0	6	2	31

* Standard error >5.0%, Otherwise standard error <5.0%

TABLE 3.   *Percentage of Respondents Who Had Ever Previously Lived in a New Mining Town in Northern Australia (Married couples only)*

Town	Previous experience		No previous experience		Missing Cases		Sample Size: Household
	Males %	Females %	Males %	Females %	Males N	Females N	
Dampier	29*	20*	71*	80*	4	4	53
Karratha	43	40	57	60	6	8	108
Paraburdoo	25	17	75	83	5	1	82
Tom Price	29	15	71	85	4	9	159
Pannawonica	44	24	56	76	10	2	106
Wickham	32*	15	68*	85	4	1	66
Port Hedland	18	6	82	94	5	11	89
South Hedland	24	13	76	87	3	6	91
Nhulunbuy	10	4	90	96	12	5	140
Emerald	30	31	70	69	10	0	91
Moura	9	4	91	91	9	0	105
Capella	12*	16*	88*	84*	6	0	31

* Standard error >5.0%, Otherwise Standard Error ≤5.0%

Early surveys of the evolving townships also suggested that population growth through recruitment of labour for the mining industry initially brought with it similar problems to those observed in new resource towns in Canada and USA: towns with high labour turnover, impermanence, an omnipresence of the company and lack of local autonomy (Neil, *et al*, 1982; Sharma, 1983).

However, a number of the northern Australian resource towns built in the sixties that have been recently surveyed (see below for details) now show reduced turnover, and increased length of residence and greater stability (Table 5). They also show a more balanced demographic structure, with an increase in the numbers of residents aged over 40, lower rates of masculinity, and reduced percentages of residents born overseas (Newton and Sharpe, 1980; Neil, *et al*, 1982; Newton, 1983). The increased stability of the population as reflected in the retrospective index (see Table 5a) is reinforced with survey data relating to prospective mobility intentions (Table 6). In most towns, and for both male and

TABLE 4.   *Net Migration Pattern for Northern Australia:*
1971-81

| | Net migration | | | |
| Age cohort | 1971–76 | | 1976–81 | |
	N	%	N	%
5–9	3731	9.9	3266	7.4
10–14	638	1.7	629	1.4
15–19	1152	3.1	2238	5.0
20–24	16573	43.9	16816	37.9
25–29	7734	20.5	7838	17.7
30–34	1939	5.1	3518	7.9
35–39	1945	5.2	2305	5.2
40–44	70	0.2	225	0.5
45–49	-117	-0.3	506	1.1
50–54	331	0.9	1420	3.2
55–59	195	0.5	865	1.9
60–64	1686	4.5	2113	4.8
65–69	1002	2.7	1008	2.3
70+	862	2.3	1653	3.7
Total	37741		44399	

Notes: (a) Northern Australian comprises the following Statistical
           Divisions as defined in 1971: North-West, Pilbara,
           Kimberley (WA): all NT; Rockhampton, Central Western,
           Far Western, Mackay, Townsville, Cairns, Peninsula,
           North-Western (Queensland); and comparable areas in
           successive censuses.

       (b) Analysis based on cohort survival model described in
           Newton (1982).

female groupings, fewer than 10 per cent of respondents
indicate a likely stay of less than 4 years.   This represents
a dramatic break with earlier trends (Brealey and Newton,
1978) where only 10 per cent of households (in Hedland) were
planning long term residence.   No consistent pattern is
apparent between townships nor is there much variation between
male and female respondents, excepting that surveyed females
expected to spend, in general, less time in the towns than
did their male counterparts.

   Nonetheless, consistent with earlier studies (Brealey
and Newton, 1980;   Brealey *et al*, 1981) the dominant reason
the majority of respondents in the survey had come to the
remote mining towns was work-related: a place to earn and
save, a place where work was available;   a place which
offered prospects for career development (Table 7) (although
because a majority of the females accompany their husbands
to such towns as wives rather than employees, a consistently
higher proportion of women indicated that they had little
choice in the decision to move).   For the most part the

TABLE 5a.   Labour Turnover Amongst Non-Professional Personnel (Males and Females)

Town	Year	
	1974 %	1981 %
Dampier + -------- - -	82	23 **
Karratha+		
Paraburdoo	61	34 **
Tom Price	61	31**
Pannawonica	112	22 **
Wickham	124	30 **
Port Hedland* -----------	83	17
South Hedland*		
Nhulunbuy	50 **	23 **
Emerald	na	7
Moura	na	8
Capella	na	7

*Port Hedland and South Hedland combined

+Dampier and Karratha combined

**Male figures only

attraction of the social or physical environment, or the desire for new experiences was a consideration for only a minority of immigrants to the towns.

Two alternative interpretations of these trends are however possible. Different schools of thought exist concerning the dynamics of new resource towns (Lea and Zehner, 1983). Lucas, for example, in a detailed study of Canadian single industry communities similar to the Australian resource towns suggested that despite the initial problems of these towns:

> 'The one-industry community grows from the initial construction stage, through the periods of recruitment, settling in and transition and finally reaches fairly stable maturity, (although) .... the maturity and stability are only comparative; the stability is precarious because of the forced migration of youth, and the ultimate uncertainty of the single economic base'. (Lucas, 1971:390).

TABLE 5b.   Current Length of Residence

Town	Two years or less Male %	Two years or less Female %	Three to five years Male %	Three to five years Female %	Six to eight years Male %	Six to eight years Female %	More than eight years Male %	More than eight years Female %	Missing Male N	Missing Female N	Sample size: Households
Dampier	10	14	18*	20*	24*	22*	47*	43*	4	4	53
Karratha	29	29	30	34	25	22	17	16	7	7	108
Paraburdoo	31	35	32	33	19	16	17	16	5	2	82
Tom Price	26	33	29	25	19	18	26	24	6	6	159
Pannawonica	25	37	41	38	22	17	12	8	9	4	106
Wickham	20	25	34*	39*	28*	17	18	19	5	2	66
Port Hedland	8	14	24	22	19	11	50*	53*	9	17	89
South Hedland	21	26	24	25	28	32	27	18	2	6	91
Nhulunbuy	22	24	28	28	22	24	28	24	11	5	140
Emerald	59*	59	31	31	3	0	8	10	11	0	91
Moura	12	12	16	16	19	21	54	51	10	2	105
Capella	52*	58*	28*	19*	8	3	12*	19*	6	0	31

*Standard error >5.0%, Otherwise standard error <5.0%

TABLE 6. Total Anticipated Length of Residence of Respondent

Town	One year		Two years		Three or four years		Five to eight years		Over eight years		No idea		Missing		Sample Size: Household
	Male %	Female %	Male %	Female %	Male %	Female %	Male %	Female %	Male %	Female %	Male %	Female %	Male N	Female N	
Dampier	0	0	0	0	4	4	4	8	81*	77*	10	10	5	5	53
Karratha	0	0	0	0	9	6	18	16	56	57	17	20	15	15	108
Paraburdoo	1	0	4	5	9	12	22	18	46*	43*	18	21	8	6	82
Tom Price	1	1	1	1	7	10	24	16	54	53	12	19	12	14	159
Pannawonica	0	0	2	4	9	19	32	26	42	34	15	16	10	7	106
Wickham	0	2	0	0	7	10	20	24	58*	53*	15	11	6	4	66
Port Hedland	0	0	1	1	4	3	15	14	68*	67	13	14	9	19	89
South Hedland	0	0	1	2	6	7	14	8	68	65	11	17	3	8	91
Nhulunbuy	1	1	0	0	7	11	23	22	52	48	17	18	17	10	140
Emerald	0	0	1	4	10	8	13	8	49*	49*	27	33	21	11	91
Moura	0	0	1	1	1	2	7	5	83	83	9	9	13	4	105
Capella	0	0	0	5	10*	0	0	0	70*	65*	20*	30*	11	11	31

*Standard error >5.0%, Otherwise standard error <5.0%

Youthfulness of the town is characterized by high turnover and high masculinity ratios, with a large proportion of young married couples in the population, the transition stage by the transfer of the responsibility for town administration from company to local government, and maturity by low workforce turnover, and, consequently, an exodus of youth seeking work (Lucas, 1971).

Bradbury's work in new resource towns in British Columbia, however, suggests that:

'It is apparent from the kinds of changes which have been instituted in Instant Towns in British Columbia that we should give support to theory which embraces a cyclical youth and transition process which would include termination as a distinct possibility, rather than a linear youth to maturity process' (Bradbury, 1974:9).

Bradbury further argues that:

'perhaps the major obstacle to any fundamental change is the single enterprise nature of most Instant Towns. While communities remain the progeny of one large company they continue to be subject to a degree of control and paternalism. As well they continue to suffer from a sense of impermanence and uncertainty derived from their reliance on one primary industry, a finite resource base, and a competitive and fluctuating international market' (1974:8-9).

Considered in terms of these alternative models of the dynamics of new resource towns, two alternative interpretations of recent Australian trends are possible. The first is that the recent changes in population stability are at least in part the outcome of autochthonous changes within the community which have had either a direct influence by encouraging residents to stay in the towns, regardless of their initial intended length of residence, or an indirect influence through their impact on the reasons people migrate to the town and hence their intended length of stay. If this is the case current trends are likely to continue, unless market forces result in decisions that directly affect the stability of the towns. Alternatively, if Bradbury's model is a more appropriate model for the Australian scene, and changes in exogenous factors such as national or international

TABLE 7.   Reasons for Coming to the Town

Town	Economic				Availability of work				Quality of work			
	Males		Females		Males		Females		Males		Females	
	% giving reason	Missing N	% giving reason	Missing N	% giving reason	Missing N	% giving reason	Missing N	% giving reason	Missing N	% giving reason	Missing N
Dampier	73*	5	75*	5	10	5	2	5	50*	5	38*	5
Karratha	67	8	60	12	4	8	2	16	46	8	55	12
Paraburdoo	78	5	72	4	10	5	7	6	44	5	47*	6
Tom Price	80	6	74	10	13	8	14	14	40	8	49	14
Pannawonica	90	11	88	6	16	13	85	8	37	12	53	7
Wickham	83	6	84	3	23	6	29*	3	50*	6	57*	3
Pt Hedland	73	10	69*	22	17	11	21	28	56*	11	52*	26
Sth Hedland	83	10	80	13	17	12	12	15	43*	11	39*	14
Nhulunbuy	69	13	69	6	5	16	7	6	45	16	44	6
Emerald	71	19	77	9	7	21	10	11	70	18	71	9
Moura	79	19	77	12	18	22	14	15	57	20	56	15
Capella	78*	8	80*	6	19*	10	17*	7	61*	8	72*	6

TABLE 7 (cont'd). Reasons for Coming to the Town

| Town | Attracted by social environment | | | | Pushed by social pressures | | | | Desire for new experience | | | | Like the location | | | | Sample size: households |
| | Males | | Females | | Males | | Females | | Males | | Females | | Males | | Females | | |
	% giving reason	Missing N	% giving reason	Missing N	% giving reason	Missing N	% giving reason	Missing N	% giving reason	Missing N	% giving reason	Missing N	% giving reason	Missing N	% giving reason	Missing N	
Dampier	4	5	25*	5	25*	5	15	5	40*	5	54*	5	21*	5	27*	5	53
Karratha	6	8	12	12	27	8	24	12	35	8	43	13	18	8	19	13	108
Paraburdoo	5	5	19	3	25	5	33	6	54*	5	41	5	16	5	16	7	82
Tom Price	2	8	20	13	25	6	24	12	45	7	40	13	22	7	22	13	159
Pannawonica	10	12	31	7	25	12	28	7	29	12	37	7	18	12	21	7	106
Wickham	13	6	31*	4	28*	6	27	3	48*	12	44*	3	28*	6	31*	4	66
Pt Hedland	7	12	21	26	20	12	20	25	38*	12	34*	25	12	12	14	26	89
Sth Hedland	20	12	18	15	34	12	22	15	45*	11	45*	15	11	11	11	16	91
Nhulunbuy	11	16	15	6	46	15	31	6	68	15	61	6	55	14	42	6	140
Emerald	3	19	10	11	24	19	24	12	33	18	34	11	11	19	12	11	91
Moura	18	30	20	14	22	20	20	14	27	20	24	15	1	20	2	14	105
Capella	10	10	16*	6	38*	10	37*	7	33*	19	32*	6	24*	10	17*	7	31

*standard error > 5.0%, otherwise standard error < 5.0%

trends in employment levels, demand for minerals and affordability of housing in the large cities (Brealey and Gribbin, 1981) then it is feasible to predict that reversal in the latter trends could bring a return to the earlier population characteristics in these towns.

Only longitudinal studies can attempt to answer the question of the extent to which the greater population stability and improved demographic balance observed more recently reflect a growing community 'maturity' or the extent to which it has been a response to external political and economic forces.

This chapter however provides some data on the relative impact of endogenous versus exogenous variables on the present motivation and future plans of residents employed by mining companies in northern Australian mining towns. It uses compatible household level data collected in various surveys conducted across eleven new resource towns in 1981-1982. The data have been used to explore how much variation in inhabitants' actual length of residence, and in their total anticipated length of residence, can be accounted for by:

(1)   the role residents perceive their attitudes to the town itself as having in their decision to continue living there;

(2)   their initial motivation for migration to the town; and

(3)   the role they see perceived lack of job and financial opportunities elsewhere as playing in their decision to stay in the town.

Before testing the above relationships residents' reasons for staying in these towns, and their plans for the future, were examined to see whether they provide support for either of the above interpretations of recent changes in population stability.

Reasons for staying.[1]   In contrast to the motivations for migrating to new mining towns, the reasons for staying in the towns were not strongly oriented to work or perceived change in external economic conditions inhibiting a planned move out of the town;   nor did they pertain to enhanced local conditions encouraging presence within the town, except for approximately one quarter of all households who indicated that it would be difficult for them to finance a lifestyle similar to that enjoyed in the mining towns were they to move elsewhere.

In view of the low volume of recruitment of mining personnel for northern Australian towns at the time of the survey, the respondents' assessments of the importance they

attached to the following reasons for staying in the town
were recorded.
(1) I/we couldn't maintain the same standard of living
elsewhere;
(2) My wife's/my type of work is only available in
towns like this;
(3) My wife/I couldn't get a job with as much
responsibility and/or chance of advancement elsewhere;
(4) The experience I/my wife gained here is only useful
in a town like this; and
(5) I/my wife can't get a job elsewhere.
(all of which implied that respondents were to some extent
staying in the town because of perceived lack of opportunity
elsewhere). These were considered in terms of respondents'
intentions regarding moving to another mining community in
the same region after leaving the town in which they lived,
to give some indication of whether external factors were
influencing their stay in their present town in particular
or northern Australia in general. In every case, of those who
had cited 'exogenous influences' as the major reason for
staying, the vast majority had no plans to move to another
mining community in the same region after leaving their
present town.

To obtain some measure of the possible extent to which
those attaching importance to liking aspects of the community
as reasons for staying may have been rationalizing as a
result of having been 'trapped' in the town by changing
'external' economic circumstances, two new variables were
computed, one a count of the number of the five 'exogenous'
reasons listed above that were given as important, and one a
count of the number of reasons related to liking the town.
These two variables were than crosstabulated separately for
those who did or did not come for economic reasons. Amongst
those who had not moved for economic reasons there was no
relationship between satisfaction with the town and a feeling
of entrapment. Amongst those who had moved for economic
reasons, however, those experiencing some feelings of being
trapped were significantly more likely to report liking some
features of the town, suggesting that there may in fact be
some rationalizing by that particular category of respondent.

Future plans.[1] There was considerable diversity in the
popularity of various plans for the future, across both towns
and regions and to a lesser extent between males and females.
Very few respondents indicated any intention to move to
another mining community either within or outside their
present region when they left their present town. Moving to
a city was contemplated by around thirty per cent of the
Pilbara males and forty per cent of the females, but only

TABLE 8.   *Perceived Entrapment Versus Satisfaction With Town Features by Motivation for Migration*[a]

| Entrapment | Satisfaction | | | |
| | No features liked | | Some features liked | |
	n	%	n	%
No feeling of entrapment	45	(30)	64	(42)
Some feeling of entrapment	14	(9)	29	(19)

$x^2 = .99$ n.s.

b. Moved for economic reasons

| Entrapment | Satisfaction | | | |
| | No features liked | | Some features liked | |
	n	%	n	%
No feeling of entrapment	145	(27)	159	(30)
Some feeling of entrapment	63	(12)	159	(30)

$x^2 = 20.03$ $p < .001$

[a]*Hedland and Nhulunbuy excluded*

about twenty per cent of the Queensland and Nhulunbuy samples. Few respondents had intentions of moving to a rural area or country town, with the exception of Nhulunbuy and South Hedland males, where thirty per cent or more had such plans, and Nhulunbuy females, thirty-three per cent of whom had such plans.

Plans to buy a business or a farm were given by a range of about ten to twenty per cent of the samples, with no marked differences between male and female. In addition, of those males not planning to buy a farm or business, from zero to fifteen per cent were planning to set up as self employed. Taking up employment with another employer was contemplated by about a quarter of both males and females in the Pilbara towns and Nhulunbuy females. It was a less popular choice in the Queensland towns for the Nhulunbuy males. There were no marked differences across the towns in the percentages planning to retire on leaving their town, the range being roughly ten to fifteen per cent.

Respondents' reasons for staying and their plans for the

future do not therefore suggest that the towns are reaching maturity in the sense in which Lucas used the term. However, when respondents were asked to indicate their satisfaction with various aspects of life in the town on a five point scale ranging from pleased through mostly satisfied, mixed, mostly dissatisfied and very unhappy, the majority of respondents, on nearly every item, were either pleased or mostly satisfied, suggesting a high degree of satisfaction generally with the aspects of town life in question. However people have little intention of staying, despite being satisfied with life in the town. This may indicate the operation of exogenous factors in their decision making, but it offers no clear cut support for such an interpretation.

II.   ANALYSIS OF RELATIONSHIPS

A.   *Representativeness of Towns in the Data Base*

As mentioned earlier, because of the different objectives underlying the surveys that provided the data used here, the towns surveyed were not selected on either a random or a systematic basis. In fact as Sharma (1983) points out, no systematic classification of new mining towns in Australia has yet evolved. However despite their shared involvement with the growth of the mining industry in northern Australia, there are obviously many points of difference between these towns, such as harshness of their physical environment, degree of geographical isolation of the towns, age and size of town, and planning principles underlying town design. Overseas studies however suggest that two overlapping dimensions may be of major significance in the development of resource towns:

(1)   the spatial and temporal contexts of the towns i.e. the regional setting of the towns and the era of their construction (Sharma, 1983:166);   and

(2)   the process whereby capital is extracted from the company by the state and reallocated for provision of infrastructure at the community level (Newby, 1982:14–15).

Although the consequences of these factors have not been systematically empirically explored in new Australian resource communities, they could be expected to have wide reaching implications for the social structure of the emergent communities. The towns included in the data base used in this chapter varied across both dimensions.

Spatial and Temporal Contexts.   Three of the towns
discussed in this paper are sited in the Bowen Basin coalfields, seven in the Pilbara iron-ore region and one in the
bauxite region of the Gove Peninsula in Arnham Land.   Sharma
(1983) points out that the coal-based development of the
Bowen Basin differs to a marked degree from the iron ore based
developments in the Pilbara and the bauxite complex at Gove:

'in that mining development has occurred within
the context of an established (predominantly
agricultural) regional economy.   Agricultural
development in the other ... complexes has
generally been minimal due to their climate being
marginal for anything other than extensive
stockgrazing ... (Further) while the Bowen Basin
falls within the tropics it has a relatively mild
climate when compared to the Pilbara ... or the
bauxite mining areas of northern Australia.   The
region is not as remote as the other ... areas,
being within easy driving distance of coastal
regional centres such as Mackay and Rockhampton.
As many of the inhabitants of the Bowen Basin
mining towns plan to retire in these coastal
centres, they are able to arrange the final move
over a period of time, and this flexibility is
a factor in producing a relatively stable labour
force in the Bowen Basin'(Sharma, 1983:178).

Further:

'Most of the Western Australian and northern
Australian developments have taken place in the
context of a (virtually) non-existent regional
labour market.   In contrast developments in the
Bowen Basin have taken place in the context of
a labour market facing structural adjustment of
the regional economy.   Consequently, there was
little need to cast the recruitment net too far,
as the labour force which was being displaced
in the rural sector ... was absorbed in the
mining projects' (Sharma, 1983:166).

Infrastructure Provision.   Three distinct types of
communities can also be observed amongst the surveyed towns,
in terms of the provision of infrastructure for the new
communities or, as Newby (1982) has put it: the process
through which capital accumulation by mining companies has
been expropriated and reallocated at the local community
level for the provision of infrastructure.   Although the
provision of infrastructure has, in part been a function of

state policies (Brown, 1982), the difference in community
type resulting from provision of infrastructure in part, but
not entirely, coincides with the spatial context of the towns
and their era of development. The first such type of
community is the 'additive' town, reminiscent of the United
States boom towns of the sixties (Filmore, 1976;   Davenport
and Davenport, 1979):   existing towns with a non-mining
economic base in which mining companies have added personnel
equivalent to some years of the projected population growth
had this influx not occurred. Of the surveyed towns, Emerald
and Capella are both such towns.   (In Port Hedland and Moura
the initial population was so small compared to the magnitude
of influx of mining personnel that it seems unlikely that
its existence had any real impact on the emerging communities
- thus the towns were excluded from this classification.)   As
is typical of this sort of town, in Emerald and Capella,
mining companies have provided housing for their personnel,
including all the subdivision costs.   For the most part,
however, the extension of the social infrastructure has
remained the responsibility of the State and the Local Council
- the additional strain on the Council being financed initially by a lump sum contribution from the mining company and
subsequently by residential and industrial rates.   The size of
the initial contribution has been a matter of negotiation
between company and Council.   Provision of State-provided
social infrastructure, such as schools has continued, dependent on the criteria normally used, although in Emerald for
example, the company joined the Council in lobbying the State
over the provision of major roads - part of the infrastructure
referred to by Brown (1982:238) as the grey area of infrastructure: i.e. the infrastructure that is both social and
industrial:- 'the provision of roads, harbour developments and
maintenance, power, and water supply for general usage (including industrial), housing for support communities and public
works associated with the expansion of existing towns'.

     The second type of community that has developed amongst
the northern Australian new towns is that which the state has
either created to service the mining industry, as in the case
of Karratha, or for which it has undertaken an extensive
amount of the work in its construction - as in Hedland (Brown,
1982:248) or, for which the state has taken responsibility
for the provision of funding for major facilities, as in
Nhulunbuy.   In the case of these three towns, the State has
been concerned with creating a regional centre which will
service, *inter alia*, the mining industry, although substantial
numbers of mining company personnel are housed in each.   In
such towns administration is the responsibility of
organizations other than the company, the company financing

its share of the cost of town maintenance through rates levied on company owned property. A variant of this type of community occurs in Queensland, in the form of towns such as Moura, which lack a regional centre function, but whose entire social infrastructure has been provided by the state or financed by the raising of 40 year loans raised by local government and in part leased to the mining company. These loans are serviced by special additional rates on the houses they occupy and on rates on the mining lease - yield from the latter being applied to road works.

The third type of community represented among the surveyed towns is the company town, including Pannawonica, Tom Price, Wickham and Dampier. These towns are reminiscent of the Instant Towns of British Columbia, extensively studied by Bradbury (1974, 1980), although the latter were designed with local government status from the outset, primarily to obviate company control. Although the Government again financed much of the 'grey' area of social and industrial infrastructure in some of these towns (Brown, 1982:244ff) the social infrastructure, including housing, has been provided predominantly by the mining company, who have subsequently been responsible for the administration of the towns, and run them as closed communities. In return for their provision and maintenance of infrastructure, the companies have been exempt from payment of rates on industrial land (Brown, 1982: 243). A recent variant of this type of town is the 'normalized' company town, such as Paraburdoo in which the administration of the town, including maintenance of relevant infrastructure, has been handed over to a local municipal council (Thompson, 1981). The companies in normalized towns, however, are still exempt from industrial rating, but have so far in each case of normalization agreed to make an annual contribution to the new Council over a period of years.

Although little research has been carried out in Australia on the implications that the processes of expropriation of accumulated capital and reallocation to the local community for provision of infrastructure have for the social structure of the emergent community, differences in a number of areas are predictable: in the degree of local autonomy, in the extent and uniformity of individual's control over local affairs (for example, whether as in company towns powerful union members can exert more pressure than, for example, woman; (Williams, 1981:124), in methods for exerting pressure on those controlling town affairs (for example, voting versus striking), on individuals' sense of power, on the overt goals aspired to by those planning and designing town infrastructure, and on the locus of accountability of planners (those responsible for company profitability, versus

State or Council bodies with particular ideologies), degree
of opportunity for social mix, opportunities for entrepren-
eurial activity and for social mobility.

Thus towns differ across both these dimensions - but have
not been selected to be representative of the different
categories of towns. Since it was felt that the data
collected in the six Pilbara towns could be safely assumed
to be representative of the population of that region, an
arbitrary decision was made that when the relative impact
of endogenous versus exogenous variables on residents' stay
was being tested, the model would first be examined in context
of the Pilbara towns. Subsequently the model which provided
the best fit for these towns would be applied in turn to
Nhulunbuy and then to the three Bowen Basin communities, and
finally to the three sets of towns classified according to
source of infrastructure provision, to see whether the model
that emerged with respect to the Pilbara area was also an
appropriate model under these different but not necessarily
representative circumstances.

III.   ANALYSIS OF DATA

To test the relative impact of residents' assessments
of the importance of exogenous and endogenous variables on
stability of residence, two sets of analyses were performed
on the male sample data:
   (1)   multiple regression analyses were performed in
relation to actual length of residence;   and
   (2)   discriminant analyses (Nie, *et al*, 1975:434ff) were
undertaken on anticipated length of residence.

As mentioned previously, the model was initially tested
for the Pilbara towns excluding Port and South Hedland as
those surveys did not include data for all the independent
variables. The independent variables incorporated in the
initial model were as follows:

A.   Household Structural Characteristics

Included here were some of the more common measures of
household structural characteristics used in studies of
population stability; variables which represented the major
dimensions of life cycle stage, family type and economic
status:
   (1)   total number of dependent children living at home;
   (2)   age of respondent;
   (3)   work status of husband (whether staff or wages); and
   (4)   whether or not the respondent had a tertiary
education.

B.   *Exogenous Variables*

The set of 5 variables represented measures of the relative importance in the decision to stay of:

(5)   non-availability of a job elsewhere;

(6)   type of work for which respondent was qualified being available only in similar towns;

(7)   inability to get a job with as much responsibility, and/or chance of advancement elsewhere;

(8)   experience gained being useful only in a similar town;   and

(9)   inability to maintain the same standard of living elsewhere.

C.   *Endogenous (local) Conditions*

(10)   extent to which goals were being achieved in the town;

(11)   respondent's feeling about quality of life in the town;

(12)   wife's feeling about quality of life in the town;

(13)   whether the respondent had come to the town partly or wholly because of the nature of the work available;

(14)   whether the respondent's decision to come to the town had been strongly influenced by the desire to escape social relations or lifestyle elsewhere;

(15)   whether liking the climate was a major consideration in the decision to stay (in some Pilbara towns this question included climate and/or countryside);

(16)   whether having good friends in the town was a major consideration in the respondent's decision to stay;

(17)   whether enjoying the lifestyle in the town was a major consideration in the decision to stay;

(18)   whether still finding new experiences in the town was a major consideration in the decision to stay;   and

(19)   respondent's job satisfaction as measured by the Worker Opinion Survey (Cross, 1973).

Previous work using cluster analysis (Gribbin and Brealey 1981) had suggested that the population of the Pilbara towns could be dichotomized according to whether or not households came to the town principally for economic or non-economic reasons (see also Brealey and Newton, 1978). On the basis of these earlier results, it was predicted that the relative impact of local versus external conditions might vary for these two groups of residents, and separate discriminant analyses were carried out for those who came for economic reasons and those who did not, using as the dependent variable whether the respondent *intended* to stay in the town a total of eight years or less, or more than eight years. This division point was chosen since 'over eight years' was the highest category used in the surveys, and alone accounted for from forty-two to eighty-three per cent of male respondents (see Table 6). Similarly, separate multiple regressions were undertaken using the same independent variables for the two groups, with actual length of residence to date as the dependent variable. The result of the application of the model to the Pilbara sample is discussed first, followed by a discussion of its applicability to the other regions and community types in the sample.

The stepwise discriminant analysis carried out on responses from those who had moved to their present town for economic reasons showed that five of the nineteen independent variables were significant predictors of whether respondents intended to stay for a total of eight years or less or more than eight years.

These were, in order of importance: the respondent's age, the role of the climate in the respondent's decision to stay, the perception that the respondent's or his wife's type of work is only available in such towns, whether or not the respondent had a tertiary education, and the belief that the respondent could not maintain his standard of living elsewhere. Those who planned to stay longer were older, regarded the climate as more important in their decision to stay, were more likely to see their type of work only available in such towns, were less likely to have had a tertiary education and were more likely to believe they could not maintain their standard of living elsewhere. These five variables in combination allowed 76 per cent of cases to be correctly classified. Details of the discriminant function and multiple regression analyses appear in Appendix 1.[2]

For those respondents who had not moved for economic reasons the most important predictor of their total anticipated length of residence was the importance they gave to having good friends in the town as a reason for staying, with those giving it as important intending to stay longer. Next, in order of importance, were: whether the respondent had a tertiary education, the importance given to the climate as a reason for staying, and the respondent's age. These measures were all related to the intended length of stay in the same way as they were for those who came for economic reasons. Taken together these four variables could correctly classify 83 per cent of cases on the basis of their total anticipated length of residence.

These two discriminant function analyses were rerun on the same data, using only those variables from the full model which had proven significant, and including some extra cases which had some missing values on the full but not the reduced

variable set. Some changes in order were found. For those
moving for economic reasons, the order of importance was age,
the climate, maintaining the standard of living, tertiary
education and availability of work only in such towns. For
those not moving for economic reasons both the order in which
the variables contributed, and the direction of the relationship in the reduced set was different to that in the original
run. Since the addition of approximately twelve per cent more
cases caused such major changes, any relationships described
by the model were obviously unstable when applied to those
moving for other than economic reasons. For this reason, the
application of the model for this sub-group to the other
regions and community types will not be discussed.

Multiple regression analysis carried out on responses
from those who had moved for economic reasons, revealed four
variables to be significantly related to respondent's current
length of residence in the town. These were, in order of
importance: the respondent's age, the importance he attached
to the climate as a reason for staying, whether he had a
tertiary education, and his belief that he could not maintain
the same standard of living elsewhere. In combination, these
variables accounted for 22 per cent of the total variance.
Those who had been resident in the town for longer were,
perhaps not surprisingly, older, they were more likely to
rate the climate as important in their reason for staying,
less likely to have had a tertiary education and more likely
to believe that they could not maintain their standard of
living elsewhere.

Amongst those who had not moved for economic reasons,
their age and perception of the local climate were the two
variables most significantly related to their length of
residence, the direction of the relationship being the same
as for those who did move for economic reasons. The third
variable to be so related was whether the respondent was a
wages or staff employee, with wages personnel being more
likely to have been resident for longer. Together, these
three variables accounted for 34 per cent of the total
variance.

Age has proven more important than any of the attitudinal
variables in predicting length of stay, and this superiority
of structural characteristics is consistent with findings
of other researchers in the area of population stability
(Durant and Echart 1973; Speare *et al*, 1975; Newton, 1977).

The question of the relative importance of exogenous and
endogenous variables in determining length of stay in the
Pilbara towns remains indeterminate since only one endogenous
and two exogenous variables emerged as significant determinants. The fact that the importance a respondent attached to

climate as a reason for staying was consistently more important as a determinant than were either of the exogenous
variables indicates the superiority of that variable, but not
necessarily of that category, in predicting length of stay.

Clearly there are differences between those who moved
to the towns for economic reasons and those who did not in the
determinants of their actual length of stay. The exogenous
variable, the ability to maintain their standard of living
elsewhere, was an important consideration for those who had
come for economic reasons, whereas it was not important to
those who had come for other reasons.

Although not necessarily representative of their regions,
survey samples from the towns of Nhulunbuy in the Northern
Territory and Emerald, Moura and Capella in the Bowen Basin
of Queensland were used to test whether the model developed
above could be applied outside the Pilbara.

For those in the Nhulunbuy sample who had moved there for
economic reasons the predictor variables derived above held
as predictors of the total anticipated length of residence,
allowing 79 per cent correct classification. The independent
variables were related to the dependent variable in the same
way as for the Pilbara, with one exception. For the Nhulunbuy
sample those who gave as an important reason for staying the
belief that their type of work was only available in such i
towns, intended to stay for less time than those who did not.
The variables which were found to be significantly related
to respondent's current length of residence in the Pilbara
also held for Nhulunbuy, accounting for 42 per cent of the
variance for those who had gone for economic reasons, and 32
per cent of the variance for those who had gone for other
reasons.

Only two of the set of variables predicting the total
anticipated length of residence in the Pilbara however were
good predictors of this behaviour for those who had gone
to the Queensland towns for economic reasons. These were the
social-demographic variables, tertiary education and age.
However, they were related to the dependent variable in the
opposite direction to that of the Pilbara samples, with the
older respondents and those without tertiary education
planning to stay a shorter total time than those younger and
with tertiary qualifications. The climate did not appear to
contribute to the move-stay decision and, although the
other two variables were not statistically significant, the
four in combination allowed 82 per cent correct classification.
Age and tertiary education showed the strongest relationship
to current length of residence for those moving for economic
reasons. The other two variables made small but significant

contributions and in combination the four variables accounted for 28 per cent of the variance.   The three independent variables used in an examination of the length of residence of those moving for other reasons were all significant, and accounted for 46 per cent of the variance.   The importance of climate as a reason for staying was related to the dependent variable in the opposite direction to that of the Pilbara samples, however, with those rating the climate as important having lived in the town for less time than those not so rating it.

As far as can be tested with the data available, then, the model developed for the Pilbara region does not appear to hold consistently for Nhulunbuy or the selected Bowen Basin towns. Whilst some of the independent variables remain significantly related to the dependent variables, the direction of that relationship is not consistent across the three regions. This suggests that there may in fact be regional variations in the determinants of length of residence.

Finally, the model developed for the Pilbara region is applied to communities differing on the basis of infrastructure provision. Once again, the towns used in these analyses are not necessarily representative of their infrastructure provision type.

The first of these types of communities to be examined is the community in which the state has provided the infrastructure: the representative towns being Karratha, Hedland and Nhulunbuy.

All five independent variables from the initial model contributed significantly to predicting whether a respondent intended to stay a total of more than eight years, or less time, for those who had moved for economic reasons.   The relationships were all in the same direction as for the initial model, and 78 per cent of cases were correctly classified.

Current length of residence was also found to be related to the four variables from the Pilbara model found to be significant for those moving for economic reasons.   These variables were related in the same direction as for the Pilbara model, and in combination accounted for 31 per cent of the total variance.   Similarly, for those who came for other reasons, length of residence proved to be dependent on the same three variables as for the Pilbara, and in the same direction, but only 15 per cent of the total variance could be accounted for.

The second type of community to be examined here is that in which infrastructure is provided by the mining company: the representative towns including Dampier, Paraburdoo, Tom

Price, Pannawonica and Wickham. These are all Pilbara towns
and hence were all included in the initial analysis of the
model. The model developed on the full set of six Pilbara
towns held in all respects for this sub-sample of five
company towns.

The third type of cummunity studied here is the additive
community, and the Queensland towns of Emerald and Capella
are analysed in this context. The model did not hold for
either of the categories in either the discriminant function
analysis or the multiple regression.

Of the three types of towns investigated then, those
towns with state provided infrastructure were similar to the
original Pilbara sample towns with respect to the ability
of the same variables to explain respondents' total anticipated and current length of residence, although the amount
of explanation varied. Those towns in which the company had
provided infrastructure (Dampier, Paraburdoo, Tom Price,
Pannawonica and Wickham) were a subset of the original
Pilbara sample and hence did not provide the basis for an
independent test of the model. The analysis was included
merely to allow comparison with the other categories of
infrastructure provision.

Current and total anticipated length of residence in
those towns where the mining community had been added on to
a pre-existing infrastructure could not be predicted from any
of the measures derived from the original Pilbara sample.
However, since the two towns in this category were also in the
Bowen Basin sample, for which the model held in part these
findings indicate the existence of an intra-regional difference, the basis of which can only be investigated with data
from a sample of towns representative of the Bowen Basin,
and from 'additive' towns from other regions. Such an
investigation would indicate whether, in fact, the length of
stay in 'additive' towns is affected by different
considerations to those operating in towns with either state
or company-provided infrastructure.

IV. CONCLUSION

The search for an explanation of changing population
stability in northern Australian mining towns, using the data
and analysis pertaining to the Pilbara and subsequent surveys
fails to indicate any clear ordering of the relative

importance of endogenous and exogenous variables in predicting length of stay in the new mining towns although age, a structural characteristic, consistently proved to be the most important of the variables in the model. Age, feelings towards the climate and the inability to maintain the same standard of living elsewhere, representing each of the three categories of variable, performed strongly in most of the analyses. Climate, the only endogenous variable to contribute significantly to explanation is, of all the endogenous variables postulated, least amenable to intervention strategies and least likely to account for change in population stability. Variables such as the respondents' feelings about the quality of life in the town, and the extent to which their goals are being achieved through living in the town, which would seem intuitively to be good predictors of length of stay, did not contribute to explanation in the Pilbara models.

The exogenous factors which did contribute to explanation for that category of household who came for economic reasons, the perception that the respondent's type of work was only available in such towns and the perception that standard of living could not be maintained elsewhere reflect the objective situation of contraction in the mining industry and high levels of unemployment in other sectors of the labour market, leading to decisions to stay in both the job and the town. This is reflected in both increased population stability and reduced labour turnover.

The importance of age, and, to a lesser extent, tertiary education in determining population stability may reflect the importance of personal strategies concerned with both career and life cycle. Migration between mining towns within a region was anticipated by very few respondents, but there is some evidence that it does occur, suggesting that it is largely the result of employee movements (transfers) within a company, such as promotion, requiring relocation from one town to another. Those with a tertiary education expected to stay in a town for less time than those without, again suggesting the possibility of company-initiated moves, a situation known to occur within the companies operating in the Pilbara. Age is also a good indicator of life cycle stage and hence the need for financial and job security. The relatively high proportion of adults with young families in the mining towns reflects the opportunity that a shift (possibly of moderate duration) to such communities can offer young households. The family life cycle also imposes constraints on the length of residence in the town as maturing children require educational facilities and job opportunities not always available locally. There is evidence reported elsewhere (Brealey *et al*, 1983) that lack of job prospects for children is an important factor in deciding when to leave northern mining towns.

The comparisons of towns differing in their spatial location, and in terms of infrastructure provision, show that in fact they also differ in terms of other determinants of length of stay of their residents. However these differences deal mainly with the contribution of individual characteristics, rather than the endogenous and exogenous variables of interest.

It seems, therefore, that the results are inconclusive with respect to the relative impact of these two types of variables on the measures of population stability used here, highlighting instead the greater importance of individual characteristics and perhaps personal strategies.

NOTES

[1]*Editor's comment*. The authors presented detailed tabulation of survey results relating to *Reasons for Staying* as discussed in this sub-section and also for *Future Plans*, discussed in the next sub-section. It was an editorial decision not to include them owing to limitations of space. They are available on request from the authors. See address against list of contributors.

[2]*Editor's comment*. Appendix 1 has not been included. This was an editorial decision. As editor I apologise for any inconvenience caused to the reader. The authors will provide copy of the constants and coefficients referred to in the text. See address against list of contributors.

REFERENCES

Bradbury, J.H. 1974. *From Company Town to Instant Town in British Columbia, Canada,* paper presented to IGU Regional Conference, Palmerston North, New Zealand, 1974.

Bradbury, J.H. 1980. Instant Resource Towns Policy in British Columbia: 1965-1972, *Plan Canada,* 20, pp.19-38.

Brealey, T.B. and Newton, P.W. 1978. *Living in Remote Communities in Tropical Australia. The Hedland Study,* CSIRO Division of Building Research, Melbourne.

Brealey, T.B. and Newton, P.W. 1980. Migration and New Mining Towns, *Mobility and Community Change in Australia,* (eds) I. Burnley, R. Pryor and D. Rowland, University of Queensland Press, St. Lucia.

Brealey, T.B., Gribbin, C.C. and Jones, J.A. 1981. Reasons Given for Migration to Some Mining Towns in the Australian Tropics, *Transactions of the Menzies Foundation,* Volume 2: Living in the North.

Brealey, T.B., Neil, C.C. and Jones, J.A. 1983. An
    assessment of the effects of endogenous changes on
    population stability in two remote Australian mining
    towns, *Proceedings 6th World Congress of Engineers and
    Architects in Israel 1983 'Development of the Desert and
    Sparsely Populated Regions'*, Tel Aviv.
Brown, J. 1982. Infrastructure Policies in the Pilbara,
    *State, Capital and Resources in the North and West of
    Australia,* (eds) E.G. Harman and N.W. Head, University
    of Western Australia Press, Nedlands.
Cross, D. 1973. The Worker Opinion Survey: A Measure of
    Shopfloor Satisfaction, *Occupation Psychology,* 47,
    pp.193-208.
Davenport, J.A. and Davenport, J. 1979. *Boom Towns and Human
    Services,* University of Wyoming Publications Vol. XLIII,
    University of Wyoming, Laramie.
Durant, R. and Eckart, D. 1973. Social Rank, Residential
    Effects and Community Satisfaction, *Social Forces,*
    pp.52, 74-85.
Gilmore, J.S. 1976. Boom Towns May Hinder Energy Resource
    Development, *Science,* 191, pp.335-540.
Gribbin, C.C. and Brealey, T.B. 1981. *An analysis of reasons
    given for migration to some mining towns in the
    Australian tropics between 1971 and 1981,* presented
    to Conference 'State, Capital and Labour', Murdoch
    University, Perth, Western Australia.
Harman, E.J. and Head, B.W. (eds) 1982. *State, Capital and
    Resources in the north and west of Australia,* University
    of Western Australia Press, Nedlands.
Lea, J.P. and Zehner, R.B. 1983. *Democracy and Planning in
    a Small Mining Town: The Governance Transition in
    Jabiru, Northern Territory,* paper presented to Conference
    on 'Economy and People in the North', North Australian
    Research Unit, Darwin, 2 December 1983.
Lucas, R.A. 1971. *Minetown, Milltown, Railtown,* University
    of Toronto Press, Toronto.
Neil, C.C., Brealey, T.B. and Jones, J.A. 1982. *The
    development of single enterprise resource towns,*
    Occasional Paper 25, Centre for Human Settlement,
    University of British Columbia.
Neil, C.C., Brealey, T.B. and Jones, J.A. 1983. The
    Deligitimization of Mental Health Myths of New Remote
    Mining Communities in Australia, *Community Health
    Studies,* VII, 1, pp.42-53.

Neil, C.C.   1983.   *Attitudinal Constraints on Innovation in New Resource Boom Towns in Arid Australia*, paper prepared for Colloque <<Technologies Innovantes dans le Batiment>>, ENPC-DCCAI Paris, September, 1983.

Newby, H.   1982.   A Sociological Approach, *Energy Resource Communities*, (eds) G.F. Summers and A. Selvik, Institute of Industrial Economics, Bergen.

Newton, P.W.   1977.   Choice of Residential Location in an Urban Environment, *Australian Geographical Studies*, 15, pp.3-21.

Newton, P.W.   1982.   Rapid Growth from Energy Projects: Assessing Population and Housing Impacts in the Gippsland Energy Resource Region, *The Building Economist*, 21(3), pp.99-107.

Newton, P.W.   1983.   The Problems and Prospects of Remote mining Towns: National and Regional Issues, *Proceedings of Australian Institute of Urban Studies Annual Conference*, Brisbane.

Newton, P.W. and Sharpe, R.   1980.   Regional Impacts of the Mining Industry in Northern and Western Australia, *Papers of the 5th Australian and New Zealand Regional Science Conference*, Tanunda, South Australia.

Nie, N.H. *et al*,   1975.   *Statistical Package for the Social Sciences (2nd ed.)*, McGraw-Hill.

Sharma, P.C.   1983.   The New Mining Towns - 'Outback Suburbias', *Mining and Australia*, (eds) W.H. Richmond and P.C. Sharma, University of Queensland Press, St. Lucia.

Speare, A., Goldstein, S. and Frey, W.   1975.   *Residential Mobility, Migration and Metropolitan Change*, Ballinger Publishing Company, Cambridge.

Thompson, H.M.   1981.   "Normalization" Industrial Relations and Community Control in the Pilbara, *Australian Quarterly*, 253, 3, pp.301-324.

Williams, C.   1981.   *Open Cut. The Working class in an Australian mining town*, George Allen and Unwin, Sydney.

20

INFECTIOUS DISEASE: HUMAN ECOSYSTEMS AND HEALTH IN NORTHERN AUSTRALIA

NEVILLE STANLEY

Department of Microbiology
University of Western Australia
Nedlands, Western Australia

I. INTRODUCTION

The ecological approach to the understanding and control of infectious disease requires a close examination of the behaviour and environment of a number of non-human vertebrates as well as *Homo sapiens*, and of many 'non-infectious' factors that relate to health such as nutrition, urbanization, stress, alcoholism, life styles and the patterns of health care - in short, an examination of the many ecosystems in which human beings participate. Indeed, it should be understood that 'health' is a complex state of physical, mental and social well-being irrespective of the constant presence of infective agents. I propose to examine some aspects of the changing health scene associated with the development of northern Australia, using infectious disease only as an obvious example of the need for an ecological approach to understanding, control and preventive medicine. In particular, it is essential to consider the special health problems of our aboriginal people since these original Australians are in relatively large numbers in the north. Also, our north is ecologically part of South-East Asia and in many ways our health and disease problems relate more to that area with its millions of people than to the highly urbanized South-East Australia. Irrespective of the data presented, I agree with Dr. Keith Fleming (1982) when he states "... the best way to improve health in the Northern Territory is actively to encourage people to assume a more responsible attitude to their personal and community

NORTHERN AUSTRALIA
ISBN 0 12 545080 X

well being, and so to become integrated with preventive
effort." Although this approach should apply to all who
comprise the multiracial human society of the northern half
of this continent, it would be misleading to suggest that
the medical system has immediate solutions to every health
problem or that it has a supply of health stored away that
can be given to any ethnic group.

II. CHANGING PATTERNS OF ABORIGINAL HEALTH AS A RESULT OF CAUCASIAN INTRUSION

Continuous 'white' contact with our Aboriginal people
has been established for almost 200 years. The health of
Aborigines prior to this is a matter of debate and only a
little useful information has been derived from paleopathological study and from the accounts of early explorers such
as Dampier (Kamien, 1980). A critical question to ask is
which diseases, if any, were endemic prior to effective
contact with Eurasians. Two possible examples are trachoma
and yaws. According to Mann (1957), the eye disease,
trachoma, could have been derived from Macassan trepang
fishermen. On the contrary, Abbie (1969) and White (1977)
suggest that the disease may have been endemic for many
thousands of years. Yaws, caused by a treponeme similar to
the one causing syphilis, was prevalent until penicillin
therapy helped to control it. Hackett (1936) described
yaws-like lesions in old aboriginal bones and further
paleopathological investigations are now being undertaken.
Infectious diseases could have come and gone and then been
reintroduced. For example, malaria and leprosy may have
been introduced about 400 years ago by Macassan fishermen,
and definitely more recently when Chinese and Pacific
Islanders were brought to Australia as labourers (Cook,
1927). [The current import of people with malaria is
discussed later.] Whatever the situation was, the general
consensus of students is that the Aborigines had reached a
satisfactory equilibrium with their environment. This
ecosystem provided them with the framework for a satisfying
human existence and a state of health probably better than
that existing in England at the time.
The aboriginal population fell from an estimated 300,000
in 1788 to 67,000 in 1933 (Kamien, 1980). This decline was
associated with 'new' diseases, dispossession of land, and a
breakdown of their cultural support system. The diseases
were numerous and fearful - smallpox, syphilis, tuberculosis, influenza, measles, to name a few. Add to this

alcoholism and malnutrition and we begin to realize the shame of our initial 'health' impact. Some diseases have disappeared, but others remain, the main problems being associated with diarrhoeal diseases, sexually-transmitted infections and respiratory tract infections - all too frequently accompanied as I have already said by malnutrition. How then has our health care system met this challenge, complicated as it has been by social disintegration, the cultural chasm, the confusion set up by an alien bureaucracy and an unfeeling institutionalization?

In 1981, the Northern Territory Department of Health established a Division of Aboriginal Health with the immediate need to improve the life-style of aboriginal people for survival and growth. The objectives are:

"To promote amongst aboriginal people an awareness for the need for and active commitment to health and fitness as a means of survival and growth.

To ensure aboriginal people have access to primary and secondary health care.

To ensure active participation by aboriginal people in all aspects of the health care system." (Fleming, 1982)

Malnutrition and infections resulting in enteric disorders and pneumonia comprise an important part of aboriginal ill health, particularly in infants. Alcohol problems contribute to this medical problem and sexually-transmitted diseases also play a significant role. These aspects have been recognized by health authorities in Queensland, the Northern Territory and Western Australia. Understanding and overcoming the problems of aboriginal ill health need more than the standard and formal approaches currently used in urban Western society. Effective activity appears to come from a greater knowledge of aboriginal life-style and the recruitment of specially trained staff and nursing sisters, including a selection of aboriginal people for training as health assistants. This approach has been adopted in all three States and the morbidity and mortality data from each shows clearly that it works (see Fig. 1).

Many people believe that, for Caucasian medical knowledge and health services to become more effective, a much greater involvement of aboriginal people themselves is essential. This is clearly the direction in the immediate future.

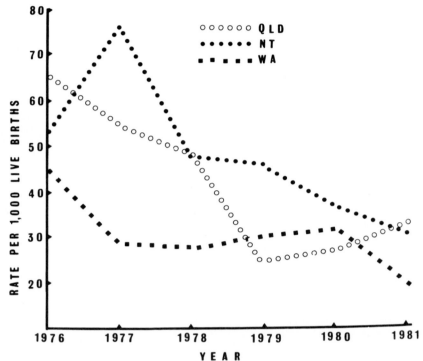

FIGURE 1. *Infant mortality rate of Aborigines for three States.*

III.   RELEVANT ENVIRONMENTAL CHANGES ASSOCIATED WITH NORTHERN DEVELOPMENT

Due to effective quarantine, Australia is a continent that is free from many important human, animal and plant diseases. Our quarantine procedures have recently been considerably modified due to the enormous increase in air-traffic and the speed with which humans, animals, insects and plants (with their microbes) can now be moved between continents (Stanley, 1980, 1981). The planet is now one great microbial system in which equatorial and tropical pathogens of man, animals, plants, birds and arthropods can be transmitted in a few hours to highly susceptible populations.

When the white man came to Australia he brought with him animals of economic and domestic importance - in particular, cattle, horses, sheep, pigs, goats, donkeys, dogs, cats, rabbits, buffaloes, camels and poultry. Almost all of these now constitute a feral animal population of considerable magnitude; they have clearly changed the existing ecology and are establishing new situations. Northern Australia's proximity to South-East Asia and the need to develop the north places this part of the continent probably at greater risk for some diseases. In ecological terms we can think of northern Australia as southern South-East Asia and it is this area that will be required to accept larger numbers of people for such developments as mining, tropical irrigated agriculture and tourism. All these require the use and control of water, the establishment of town sites, the disposal of sewage and the creation of man-made lakes. It would appear to be a vulnerable and changing situation and has merited much recent discussion (ANZAAS Symposium 1972, "Population growth, immigration and the quality of the environment - or man's disturbance of natural ecosystems; ANZAAS Jubilee Congress Symposium 1980, "Passports for pathogens: the immigration of infectious diseases"; "Transactions of the Menzies Foundation, 1981, Vol. 2, Living in the North"; "Transactions of the Menzies Foundation, 1982, Vol. 4, Towards a School of Health Research in the Northern Territory").

Man's use and control of water in tropical areas is critical for the control of tropical infectious disease (see Stanley and Alpers, 1975). This not only involves obvious water-borne diseases such as cholera, schistosomiasis, amoebiasis and hepatitis A (faecal-oral dissemination) but also the more subtle and complex one of arthropod-borne diseases (malaria, dengue, Japanese B encephalitis, Australian encephalitis, Australian polyarthritis and various animal diseases). Perhaps the most interesting diseases that illustrate the complexities of the ecology are three virus infections currently being spread by mosquitoes in Australia - dengue, Australian encephalitis and Australian polyarthritis. Strangely enough these are by no means the most important from a public health point of view.

Movement of people across the Timor Sea area could be hazardous if it continues to be illegal; but this is now lessened and to some extent controlled.

IV. SPECIFIC HEALTH PROBLEMS

Here we will briefly examine three separate categories - (A) diarrhoeal and respiratory tract disease, (B) sexually-transmitted diseases, and (C) the mosquito-transmitted agents just mentioned above, and see how both Caucasians and aboriginal people are involved with research, surveillance

and health services in an endeavour to maintain and improve
specific areas in environmental health, occupational health
and medical entomology.

A.   *Diarrhoeal and Respiratory Tract Disease*

Both types of infection are associated with
malnutrition; but of all childhood diseases including
measles, diarrhoeal diseases exert the most profound
influence on the nutritional status because of associated
poor appetite and the common practice of withholding food
from children with diarrhoea (Gracey, 1980).   Several
microbes are clearly concerned and some that have been
implicated are toxigenic *E. coli, Aeromonas* sp. and
Rotavirus.   There is generally a microbial overgrowth in the
upper intestine in childhood malnutrition and this must be a
significant factor.   Whatever the agents may be, the key to
prevention is to improve water supply, hygiene and
sanitation.   This requires increased public awareness and
education at all levels.   Our aboriginal children are
primary targets as they are also for respiratory tract
infections.   In the latter regard, one particular
microorganism - *Chlamydia trachomatis* - is specifically
involved in otitis media as well as trachoma and there is
now data showing that it is sexually-transmitted and is a
separate entity from the Chlamydia causing lymphogranuloma
venereum.   It is becoming clear that different serotypes of
C. trachomatis may be involved in these different disease
manifestations.   Infections usually respond to adequate
tetracycline therapy.   On the other hand, virus infections
cannot be effectively treated with any antibiotic and
prevention of disease by a variety of different procedures
is still the only effective answer.

B.   *Sexually-Transmitted Diseases*

More than twenty different microbes comprising bacteria,
mycoplasma, yeasts, protozoa, chlamydia and viruses can
cause venereal disease.   With the exception of the viruses,
the common diseases of gonorrhoea and syphilis have
effective and safe therapy - yet sexually-transmitted
diseases (STD) have recently increased in most Western
communities.   Epidemiological data confirm that eras of
sexual freedom are inseparably associated with high and
rising venereal disease rates (Morton, 1980).   The increase
in Australia is part of a world-wide trend to which we are

connected by rapid air travel. Approximately 70% of
infections occur in the 15-30 year age group.   STD should
really not be a greater problem in the north than in the
south, except in two regards - aboriginal people numbers and
population dispersion over a large geographic area. This is
particularly relevant to the Kimberley region of Western
Australia, where effective methods for the containment of
venereal diseases have been developed (Gollow, 1981). If
northern development demands increasing populations of young
people, then one could expect an increase in morbidity of
STD. The role of the prostitute as a vector of STD is
claimed to account for up to 80% of the gonorrhoea occurring
in males in South-East Asia. Figures are not available for
northern Australia.

C. Mosquito-Transmitted Agents

Three diseases of importance in Australia are malaria,
dengue and Australian encephalitis. Malaria is a protozoal
infection and the others are virus diseases.

1. Malaria. There have been no indigenous cases of
malaria in Australia since 1973 in spite of the presence of
Anopheline mosquito vectors. It is important to know that
the receptive area for transmission is north of 19 S
latitude. Between 1969 and 1981, 3,885 people with malaria
entered Australia but only 50% of these remained in the
receptive area (see Fig. 2). The increase in the number of
cases exceeds the rate of immigration from malaria-endemic
areas such as Papua New Guinea (Black, 1982). Most of the
parasites were Plasmodium vivax or Plasmodium falciparum.
The majority of the latter were acquired in Papua New
Guinea. Many of the strains of P. falciparum are
resistant to chloroquin. Outbreaks in the past were related
to mining, which is rapidly increasing in northern
Australia. It is therefore essential that if the malaria-
receptive area of Australia is to remain free from
indigenous malaria, there must be continued screening of
people moving from endemic malaria regions to Australia, as
well as effective mosquito control. Even well carried-out
malaria eradication programmes have been disrupted by a
great influx of people who are carriers of the parasite and
in other countries malaria epidemics repeatedly occur in
spite of our knowledge and its application. A malaria
vaccine may be the ultimate answer, and although Australia
is at the forefront in this area it is likely to be many
years before the results of clinical trials which may
result from current research become available.

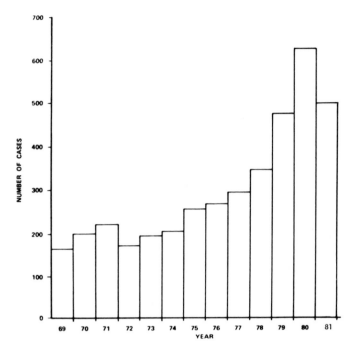

*FIGURE 2.   Cases of malaria imported to Australia
(1969-1981).   (After Black, 1982.)*

   2. Dengue.   Although the dengue fever virus may persist
in the jungles of South-East Asia in tree-top monkeys and
mosquitoes, it may persist, as does yellow fever, in the
human population by a very common dwelling-type mosquito -
Aedes aegypti.   At the end of the last century and well into
this one, dengue fever occurred in virtually all the
northern coastal areas of Australia - from Broome to Cairns.
As there is no therapy or vaccine, the only known effective
control is to stop the breeding of *Aedes aegypti* (Stanley,
1982).   Probably the best documented history of dengue in
Australia comes from Queensland, where the first recorded
outbreak was in 1879 in Townsville.   As in the Northern
Territory and Kimberley regions, haemorrhage with death was
recorded in these early outbreaks.   Between 70% and 80% of
the local populations were infected in a series of
outbreaks, with 15,000 cases occurring in Townsville in the
1953-55 outbreak.
   Dengue is once again endemic in North-East Queensland,
with the current outbreak starting early in 1981.   This was
the first evidence of the spread of dengue in Australia for

a quarter of a century. There are four antigenic types of
dengue and the current Queensland epidemic is due to type 1.
There were almost 400 cases in 1981-82, but there was also
some evidence of type 2 dengue in one patient. Cairns,
Townsville, Atherton and Thursday Island were particularly
affected - and it is in these areas almost down to the
New South Wales border that *Aedes aegypti* still persists.
Many of the early cases were probably missed or clinically
confused with diagnoses of rubella or influenza. Western
Australia and the Northern Territory recorded no indigenous
dengue and in neither of these States has *Aedes aegypti* been
reported in recent years. Great success was achieved in the
Northern Territory by the medical entomologists eliminating
the mosquito vector through a continuous surveillance
programme. This has undoubtedly contributed to the absence
of dengue in the Northern Territory, while the virus and its
mosquito vector persist in Northern Queensland. There is
also little doubt that the rapid movement of large numbers
of humans may move dengue viruses from one part of the world
to another. This is probably how it has spread throughout
the Western Pacific as well as in the areas of the southern
U.S.A., the Caribbean and Mexico, where all four antigenic
types have been detected.

  3. *Australian encephalitis*. Ecologically this disease
and the virus (Murray Valley encephalitis - MVE) causing it
represent a more complex picture than either dengue or
yellow fever. Its closest overseas relatives are Japanese B
encephalitis, existing over most of South-East Asia,
including India and China, and St. Louis encephalitis,
occurring in North America. Our studies have shown that MVE
virus persists in North-West Australia, where it is enzootic
(Stanley, 1975, 1981, 1982). For each clinical case there
may be up to 3,000 subclinical infections. For many years
it has been assumed that the virus can maintain itself in a
water-bird *Culex annulirostris* cycle, with feral animals
acting as amplifier hosts (see Fig. 3). In the dry season,
persistence may well be by transovarial transmission - but
that is still hypothetical and has not yet been established
in the field. With one exception, all virus isolates have
come from the mosquito - *Culex annulirostris* - which is
distributed widely over the entire continent and also in
Papua New Guinea. Aboriginal people in the North-West
develop a natural immunity by becoming infected at an early
age. The most severe epidemics have occurred in South-East
Australia; but no case has been recorded there since the
last outbreak in 1974, although the virus has been active in
the north. The current picture is probably more clearly

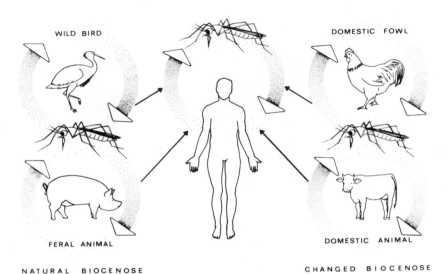

WILD BIRD

DOMESTIC FOWL

FERAL ANIMAL

DOMESTIC ANIMAL

NATURAL BIOCENOSE

CHANGED BIOCENOSE

FIGURE 3.   Arbovirus ecology.

seen by examination of Table I and Figs. 4 and 5.   The
proposals for developing Australia's north must take into
account the use and control of water, for it appears that
the mosquito-bird cycle is a critical one in determining the
pulsations of virus activity and its epidemic spread south
(see also Fig. 6).

V.   FUTURE HEALTH 'ACROSS THE TOP'

Understanding and control of health problems of northern
Australia now and in the future demand renewed national
surveillance both prior to and along with major environmental changes induced by man.   Of particular significance
are those changes which directly or indirectly affect the
use and control of water, for example - tropical irrigated
agriculture, mining, tourism, urbanization and sewage
disposal.   The types of surveillance should not only
include clinical observation and ecology studies, but also
the very sensitive and specific laboratory tests of
molecular epidemiology.   In this complex and changing scene
the movement of humans is critical, and because northern
Australia so closely relates to South-East Asia,

TABLE 1.   Numbers of Cases of Australian Encephalitis
in all States of Australia since 1917[a]

Year	WA	NT	Qld	NSW	Vic.	SA	Totals
1917			44	70	13		114
1918			5	49			67
1922			75				75
1925			11	10			21
1951				10	34	1	45
1956					3		3
1971			1	1			2
1974	1	5	10	5	27	10	58
1978	8						8
1979	2						2
1981	8	1	2				11
Totals	19	6	148	145	77	11	406

[a]The numbers refer to clinically recognizable cases,
usually confirmed by serology after 1951.

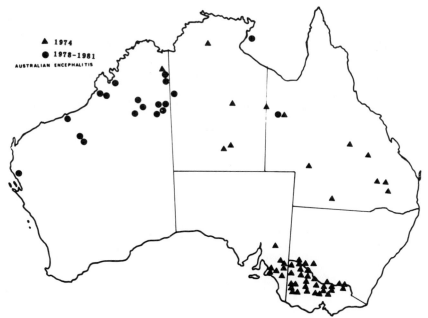

FIGURE 4.   Australian encephalitis cases since 1974.

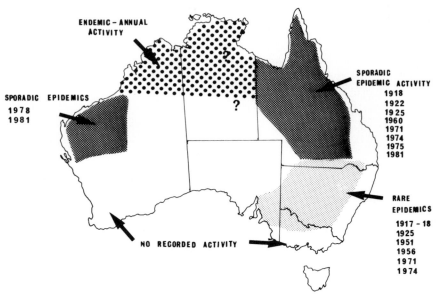

FIGURE 5.   Endemic-epidemic portrayal of Australian encephalitis.

surveillance should include examinations for the introduction of exotic diseases of man and animals such as
malaria, dengue, Japanese B encephalitis, rabies,
schistosomiasis, haemorrhagic fevers and other tropical
diseases.   In addition endemic disease patterns are
continuously changing and these changes demand monitoring
with particular emphasis on such aboriginal health problems
as diarrhoeal diseases, respiratory tract infections,
malnutrition and alcoholism.   State boundaries should not
impede the development of health services for northern
Australia and I would like to see more effective
collaboration and integration under stronger national
guidance.   The proposed Menzies School of Health Research
of the Northern Territory University, in collaboration with
southern Universities, could go some way in initiating and
developing the basic research required for this approach.

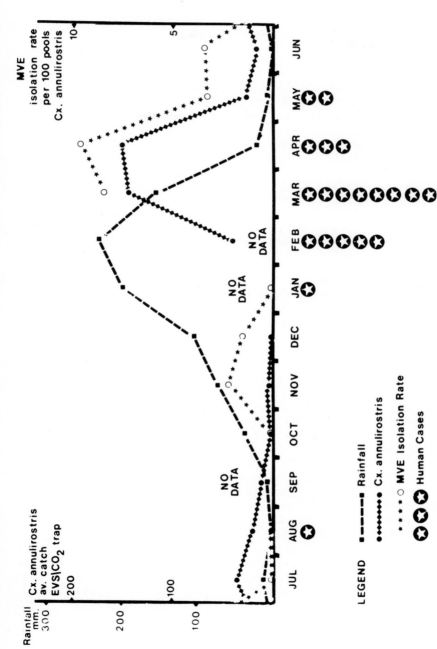

FIGURE 6. Arbovirus (MVE) ecology in Kimberley, Western Australia.

REFERENCES

Abbie, A.A.   1969.   *The Original Australians*, Reed, Sydney,
    p.86.
Black, R.H.   1982.   Malaria in Australia, 1981.   *Tropical
    Medical Technical Paper No. 8.   Commonwealth Institute of
    Health.*   Australian Government Publishing Services,
    Canberra.
Cook, C.   1927.   The Epidemiology of Leprosy in Australia.
    Commonwealth Department of Health.   Service Publication
    No. 38, Government Printer, Canberra.
Fleming, K.   1982.   In *Annual Report 1981/1982, Northern
    Territory Department of Health*, p.17.
Gollow, M.M.   1981.   In *Annual Report for the Year Ended
    December 31, 1981, Public Health Department*, W.A., p.99.
Gracey, M.   1980.   Malnutrition.   In *Changing Disease
    Patterns and Human Behaviour*, (eds.) N.F. Stanley and
    R.A. Joske, Academic Press, London, p.415.
Hackett, C.J.   1936.   A critical survey of some references
    to syphilis and yaws among the Australian Aborigines.
    Med. J. Aust. 1, 733.
Kamien, M.   1980.   The Aboriginal Australian experience.   In
    Changing Disease Patterns and Human Behaviour.
    (eds.) N.F. Stanley and R.A. Joske, Academic Press,
    London, p.253.
Mann, I.   1957.   Probable origins of trachoma in Australiasia.
    Bull. Wld Hlth Org. 16, 1165.
Morton, R.S.   1980.   Social determinants in venereal disease.
    In *Changing Disease Patterns and Human Behaviour.*
    (eds.) N.F. Stanley and R.A. Joske, Academic Press,
    London, p.205.
Stanley, N.F.   1975.   The Ord River Dam of tropical Australia.
    In *Man-made Lakes and Human Health.*   (eds.) N.F. Stanley
    and M.P. Alpers, Academic Press, London, p.103.
Stanley, N.F.   1980.   Man's role in changing patterns of
    arbovirus infections.   In *Changing Disease Patterns and
    Human Behaviour*, (eds.) N.F. Stanley and R.A. Joske,
    Academic Press, London, p.151.
Stanley, N.F.   1981.   Changing patterns of infectious disease
    and human behaviour - Australia's north.   In *Transactions
    of the Menzies Foundation, Vol. 2, Living in the North*,
    The Menzies Foundation, East Melbourne, p.147.
Stanley, N.F.   1982.   Virus diseases in Northern Australia.
    In *Viral Diseases in South-East Asia and the Western
    Pacific*, (ed.) J.S. MacKenzie, Academic Press, London,
    p.275.

Stanley, N.F. and Alpers, M.P. 1975. *Man-made Lakes and Human Health,* (eds.) N.F. Stanley and M.P. Alpers, Academic Press, London.

White, I.M. 1977. Pitfalls to avoid: the Australian experience. In *Health and Disease in Tribal Societies,* Ciba Foundation Sumposium 49 (new series), Elsevier/ Excerpta Medica/North-Holland, Amsterdam, p.269.

APPENDIX

NORTHERN AUSTRALIA
ISBN 0 12 545080 X

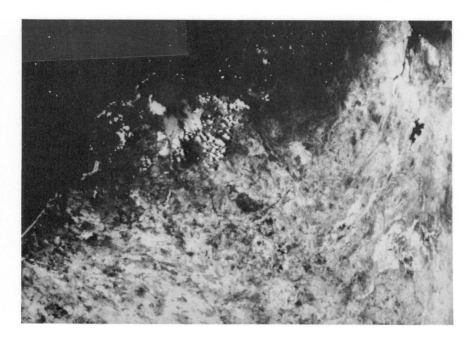

Plate 1.   NOAA-7 AVHRR band 1 (0.53 - 0.69 micrometres) image of northern Western Australia, acquired at 1511 hrs, 11 December 1983. The image, for the visible region of the spectrum, highlights the increasing vegetation density near the coast as a darker fringe. Small clouds appear white south of Derby, near the top centre of the image. Lake argyle appears as a black feature in the top right hand corner of the image. Eighty mile beach appears as a white strip along the coastline south-west of Broome. Fire scars are obvious in the spinifex in the Great Sandy Desert, south-east of the eighty mile beach. Large variations in the albedo of the image are primarily due to the different reflectances of the terrrain materials, and to the variation in vegetation density.

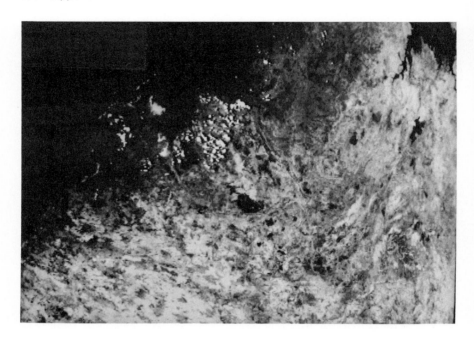

Plate 2.   N O A A - 7   A V H R R   b a n d   2   (0 . 7   -   1 . 1
micrometres) image for the area described in fig 1. With
the exception of areas of dense vegetation and water
areas, the visible (Fig 1) and infrared images are
strongly, positively correlated. Generally the infrared
images appear sharper than the visible bands due to
different atmospheric scattering. The dark boundary in
the top right hand corner of the image highlights the
considerable difference in soil colour and in vegetation
density on the soils associated with the volcanics
overlaying the sandstones on the western half of the
Kimberley Basin, compared to the soils derived from
sandstones on the Eastern half. Differences in
vegetation vigour are not apparent, primarily because
the darker soil, combined with the shadowing effect of
the open forest, result in a darker tone.

Plate 3.   N O A A - 7   A V H R R   b a n d   3   i m a g e .   The 3.5 - 3.9
micrometre band is difficult to interpret for daytime
imagery, as the radiance detected by the scanner is a
combination of reflected solar radiation, and radiation
emitted by the target materials. The image is
particularly "noisy", with high frequency variations in
brightness. The darker tone in the lower left hand
corner of the image indicates an apparent cooler, or
lower emissivity area associated with the Sandy Desert,
implying dominance of the detected signal by the emitted
radiation. Fire scar patterns are apparent in the desert
region.

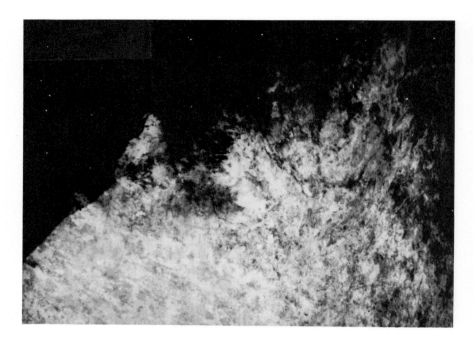

Plate 4.   N O A A - 7   A V H R R   b a n d   5   (1 1 - 5   -   1 2 . 5
micrometres) image. The thermal image presented here has
considerably less detail and contrast then the shorter
wavelength images of figs 1,2 and 3. Temperature
variations are generally more diffuse than changes in
albedo. The dark area east of Broome is most likely to
be an area which has had local rain. Cooler air near the
coast, combined with increasing atmosheric water vapour
content results in a gradual darkening of the image
towards the top. The small dark area at the western end
of the eighty mile beach is an area of water and
mangroves which represents the terminus of a paleoriver
system through the Sandy Desert. This feature is not
apparent on the shorter wavelength images.

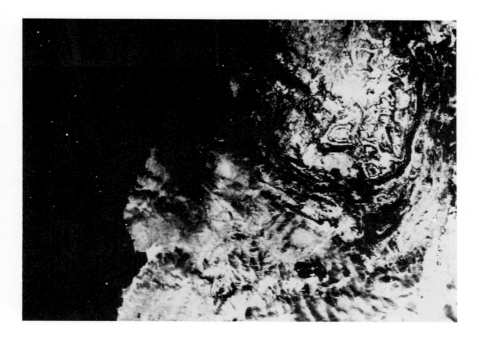

Plate 5. N O A A - 7   A V H R R   night-time   thermal   image,
acquired 0205 hrs. Little or no vegetation is apparent
on the night time image, presented here as a negative,
with warmer units appearing dark, and cold areas light
toned. The picture is dominated by the difference in the
temperatures of the various rock units. The darker
(hotter) volcanic units fringing the Hamersley Basin are
in stark contrast to the light (cool) sandstone units of
the Basin, and the sand areas of the north eastern Sandy
Desert. Combinations of day and night thermal images may
be used to estimate near surface soil moisture
differences.

Plate 6. NOAA-7 AVHRR vegetation index image.

(Band 2 - Band 1)

(Band 2 + Band 1)

"Vegetation Index" images are particularly useful
for delineating and monitoring subtle differences in
vegetation density and vigour. The tonal variations near
the coast in figs 1 and 2, due to combinations of darker
soils and/or increasing vegetation density and
subsequent shadow components, are cancelled by the
normalising effect of calculating the index. The areas
of darker volcanic soil and healthy, dense vegetation on
the western half of the Kimberleys are now particularly
bright, as are the areas of dense, healthy mitchell
grass on the black soil plains south east of Derby,
which appear as a bright, linear feature near the centre
of the image. The area of open water at Mandora Flats,
at the western end of Eighty Mile Beach, which was
apparent in the thermal image (fig 4) due to temperature
differences, but not in the reflected images (figs 1 and
2) due to low contrast of the target, is highlighted in
the vegetation index, as water has a zero value.

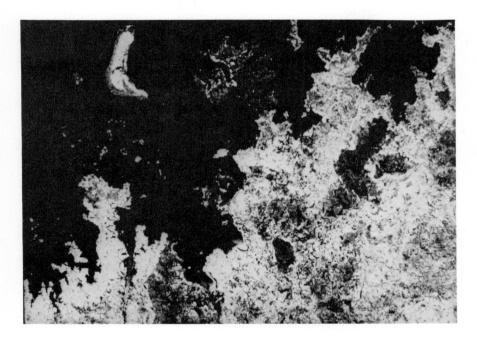

Plate 7. Landsat band 5 (0.6 - 0.7 micrometres) image. The most apparent vegetation / soil features in this image are the darker tones associated with the lateritic soils on the Mitchell Plateau, near the bottom centre of the image, and on Cape Bougainville, top centre. The dark area near the centre right of the image also appears due to forest/ dark soils. The shallow water on the shoals around Long Reef and Sandy Island, west of Cape Bougainville appears as a bright broad "L" shaped area.

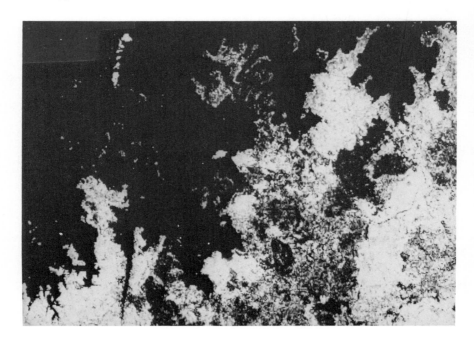

Plate 8.   Landsat band 7 (0.8 - 1.1 micrometres) image. Slight variations in tone compared to the red band (fig 6) are due to differences in vegetation density. As for the NOAA images, the total detected radiance for any area is frequently dominated by the soil contribution, resulting in a positive correlation of the two images. In the infrared channel, water areas are very dark, so that only exposed beach and reef appears in the area of Long Reef and Sandy Island.

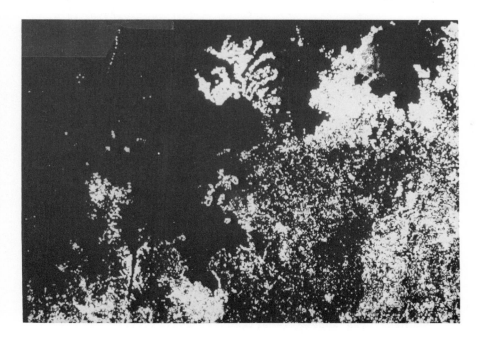

Plate 9. Vegetation Index image from Landsat data.

(Band 7 - Band 5)

(Band 7 + Band 5)

The vegetation index image highlights the dense, healthy vegetation on the lateritic soils of the Mitchell Plateau and Cape Bougainville. The dark area near the centre right of figs 6 and 7, which had a similar tone to the areas of healthy vegetation, now appears dark, indicating a lower vegetation density / vigour. The exposed sand / reef areas now only appear as a string of dots, indicating some patches of vegetation.

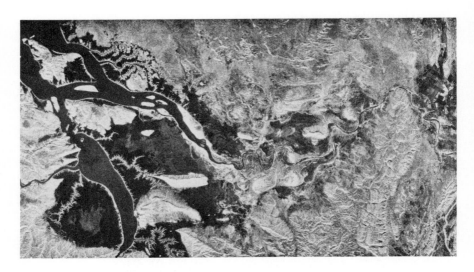

Plate 10. Shuttle SIR-A synthetic aperture radar image of Kununurra, Western Australia. Extensive areas of mangroves appear as bright fringes between the dark water areas, and the extremely dark salt flat areas inland of the mangroves. For the 25 cm L-band SIR-A radar, water areas and extensive areas of salt act as mirror reflectors, reflecting the microwave radiation away from the transmitter / receiver. Patterns of agriculture at the Kununurra research station are apparent. Drainage channels and small topographic variations are particularly highlighted by radar imagery. During the time of acquisition of these data (13 November 1982), the Kimberley region was completely cloud covered.

Index